Lecture Notes in Computer Science 14978

Founding Editors

Gerhard Goos
Juris Hartmanis

Editorial Board Members

Elisa Bertino, *Purdue University, West Lafayette, IN, USA*
Wen Gao, *Peking University, Beijing, China*
Bernhard Steffen , *TU Dortmund University, Dortmund, Germany*
Moti Yung , *Columbia University, New York, NY, USA*

The series Lecture Notes in Computer Science (LNCS), including its subseries Lecture Notes in Artificial Intelligence (LNAI) and Lecture Notes in Bioinformatics (LNBI), has established itself as a medium for the publication of new developments in computer science and information technology research, teaching, and education.

LNCS enjoys close cooperation with the computer science R & D community, the series counts many renowned academics among its volume editors and paper authors, and collaborates with prestigious societies. Its mission is to serve this international community by providing an invaluable service, mainly focused on the publication of conference and workshop proceedings and postproceedings. LNCS commenced publication in 1973.

Franco Bagnoli · Jan Baetens · Stefania Bandini · Tommaso Matteuzzi
Editors

Cellular Automata

16th International Conference
on Cellular Automata for Research and Industry, ACRI 2024
Florence, Italy, September 9–11, 2024
Proceedings

Editors
Franco Bagnoli
University of Florence
Florence, Italy

Stefania Bandini
University of Milano-Bicocca
Milan, Italy

Jan Baetens
Ghent University
Ghent, Belgium

Tommaso Matteuzzi
University of Florence
Florence, Italy

ISSN 0302-9743 ISSN 1611-3349 (electronic)
Lecture Notes in Computer Science
ISBN 978-3-031-71551-8 ISBN 978-3-031-71552-5 (eBook)
https://doi.org/10.1007/978-3-031-71552-5

© The Editor(s) (if applicable) and The Author(s), under exclusive license
to Springer Nature Switzerland AG 2024

This work is subject to copyright. All rights are solely and exclusively licensed by the Publisher, whether the whole or part of the material is concerned, specifically the rights of translation, reprinting, reuse of illustrations, recitation, broadcasting, reproduction on microfilms or in any other physical way, and transmission or information storage and retrieval, electronic adaptation, computer software, or by similar or dissimilar methodology now known or hereafter developed.
The use of general descriptive names, registered names, trademarks, service marks, etc. in this publication does not imply, even in the absence of a specific statement, that such names are exempt from the relevant protective laws and regulations and therefore free for general use.
The publisher, the authors and the editors are safe to assume that the advice and information in this book are believed to be true and accurate at the date of publication. Neither the publisher nor the authors or the editors give a warranty, expressed or implied, with respect to the material contained herein or for any errors or omissions that may have been made. The publisher remains neutral with regard to jurisdictional claims in published maps and institutional affiliations.

This Springer imprint is published by the registered company Springer Nature Switzerland AG
The registered company address is: Gewerbestrasse 11, 6330 Cham, Switzerland

If disposing of this product, please recycle the paper.

Preface

This volume contains a collection of original papers covering both applications and theoretical results on Cellular Automata, which were selected for presentation at the 16th International Conference on Cellular Automata for Research and Industry, ACRI 2024, held in Florence, Italy, from September 9–11, 2024. The event was organized by the Center for the Study of Complex Dynamics and the Department of Physics and Astronomy of the University of Florence.

The primary goal of the conference was to bring together researchers coming from many different scientific fields in order to favor and foster international collaborations on Cellular Automata and in general discrete dynamical and stochastic systems as well as to discuss their applications in several scientific areas: physics, biology, computer science, pure and applied mathematics, etc.

Cellular Automata are powerful computational models used for studying complex phenomena characterized by simple local interactions. They are discrete (in space and time as well as in state variables) models that constitute a paradigm of complex systems: simple models with unexpected collective behavior. Starting from their introduction in the middle of the 20th century, Cellular Automata have been studied both theoretically and for practical applications. This approach can be generalized in several ways, using continuous variables, i.e., coupled maps, geometries different from regular lattices, asynchronous updating, etc.

The ACRI conference series was first organized in Italy, namely, ACRI 1994 in Rende, ACRI 1996 in Milan, and ACRI 1998 in Trieste, which were followed by ACRI 2000 in Karlsruhe (Germany), ACRI 2002 in Geneva (Switzerland), ACRI 2004 in Amsterdam (The Netherlands), ACRI 2006 in Perpignan (France), ACRI 2008 in Yokohama (Japan), ACRI 2010 in Ascoli Piceno (Italy), ACRI 2012 in Santorini (Greece), ACRI 2014 in Kraków (Poland), ACRI 2016 in Fez (Morocco), ACRI 2018 in Como (Italy), ACRI 2020 in Łódź (Poland), and ACRI 2022 in Geneva (Switzerland).

This 16th edition of ACRI aimed at enlarging the traditional topics to include other areas related to or extending Cellular Automata. This allowed a larger community to have the opportunity to discuss their work in various related fields like, for example, complex networks, games, cryptography, lattice gas and lattice Boltzmann models, agent-based models, lattice maps, etc. We received 27 papers, and 20 of them were accepted. Each paper inside this volume was reviewed by at least two Program Committee members.

The papers are divided into three main sections, reflecting the scheduling of the conference: theory, physical and mathematical aspects; computer science and optimization; application to biological and social problems.

We would like to express our sincere thanks to the invited speakers who kindly accepted our invitation to give plenary lecture at ACRI 2024: Nazim Fatès, Giorgios Sirakoulis, and Alberto Dennunzio.

Moreover, we are grateful to the Program Committee and all the additional reviewers for their contribution in selecting the papers. We are also grateful for the financial and logistic support from the Department of Physics and Astronomy of the University of Florence, and in particular to Giovanna Pacini who helped a lot in the organization of the conference. Finally, we acknowledge the excellent cooperation from the Lecture Notes in Computer Science team of Springer for their help in producing this volume in time for the conference.

July 2024

Jan Baetens
Franco Bagnoli
Stefania Bandini
Tommaso Matteuzzi

Organization

Program Committee Chairs

Bagnoli, Franco	University of Florence, Italy
Bandini, Stefania	University of Milano-Bicocca, Italy
Matteuzzi, Tommaso	University of Florence, Italy
Pacini, Giovanna	University of Florence, Italy

Program Committee Members

Baetens, Jan	University of Ghent, Belgium
Bagnoli, Franco	University of Florence, Italy
Baia, Michele	University of Florence, Italy
Balbi, Pedro Paulo	Universidade Presbiteriana Mackenzie, Brazil
Bandini, Stefania	University of Milano-Bicocca, Italy
Bhattacharjee, Kamalika	National Institute of Technology Tiruchirappalli, India
Chopard, Bastien	University of Geneva, Switzerland
Das, Sukanta	Indian Institute of Engineering Science and Technology, India
Dennunzio, Alberto	University of Milano-Bicocca, Italy
Di Gregorio, Salvatore	University of Calabria, Italy
El Yacoubi, Samira	University of Perpignan Via Domitia, France
Fatès, Nazim	University of Lorraine, France
Hoffmann, Rolf	Technical University of Darmstadt, Germany
Lawniczak, Anna	University of Guelph, Canada
Martinez, Genaro	Instituto Politécnico Nacional, Mexico
Matteuzzi, Tommaso	University of Florence, Italy
Mauri, Giancarlo	University of Milano-Bicocca, Italy
Ninagawa, Shigeru	Kanazawa Institute of Technology, Japan
Pacini, Giovanna	University of Florence, Italy
Pikovsky, Arkady	University of Potsdam, Germany
Rechtman, Raul	Universidad Nacional Autónoma de México, Mexico
Vizzari, Giuseppe	University of Milano-Bicocca, Italy
Yanagisawa, Daichi	University of Tokyo, Japan
Leone, Pierre	University of Geneva, Switzerland

Reviewers

Baetens, Jan	University of Ghent, Belgium
Bagnoli, Franco	University of Florence, Italy
Baia, Michele	University of Florence, Italy
Balbi, Pedro Paulo	Universidade Presbiteriana Mackenzie, Brazil
Bandini, Stefania	University of Milano-Bicocca, Italy
Bhattacharjee, Kamalika	National Institute of Technology Tiruchirappalli, India
Chopard, Bastien	University of Geneva, Switzerland
Das, Sukanta	Indian Institute of Engineering Science and Technology, India
Dennunzio, Alberto	University of Milano-Bicocca, Italy
Di Gregorio, Salvatore	University of Calabria, Italy
Fatès, Nazim	University of Lorraine, France
Hoffmann, Rolf	Technical University of Darmstadt, Germany
Lawniczak, Anna	University of Guelph, Canada
Martinez, Genaro	Instituto Politécnico Nacional, Mexico
Matteuzzi, Tommaso	University of Florence, Italy
Mauri, Giancarlo	University of Milano-Bicocca, Italy
Ninagawa, Shigeru	Kanazawa Institute of Technology, Japan
Perrotin, Pacôme	Universidade Presbiteriana Mackenzie, Brazil
Pikovsky, Arkady	University of Potsdam, Germany
Rechtman, Raul	Universidad Nacional Autónoma de México, Mexico
Vizzari, Giuseppe	University of Milano-Bicocca, Italy
Yanagisawa, Daichi	University of Tokyo, Japan
Leone, Pierre	University of Geneva, Switzerland

Extended Abstracts

Self-stablisation in cellular systems

Nazim Fatès

Université de Lorraine, CNRS, Inria, LORIA, F-54000 Nancy, France
`nazim.fates@loria.fr`

Abstract. Remaining constant in a noisy environment is a difficult task, which is achieved "by design" – if we may say so – by living organisms. Cellular automata were invented to study self-reproduction; since then, this computing model have been the source of many problems regarding self-organisation, self-repair, self-diagnosis, and other "self-star" properties, which are found in living organisms. We propose to examine the problem of self-stabilisation as initially defined by Dijkstra [1, 2]: given a set of legal sets, how can a cellular automaton return to this set if it undergoes an external perturbation? [3] More generally we would like to ask why the reproduction of robust mechanisms found in Nature is so tedious. Would it be that there is no "Newton of the blade of grass"? (Kant). We do not aim to fully answer this question but simply illustrate it with some selected deterministic and probabilistic cellular automata.

Keywords: self-stabilisation · self-correction · probabilistic cellular automata

References

1. Dijkstra, E.W.: Self-stabilizing systems in spite of distributed control. Commun. ACM **17**(11), 643–644 (1974)
2. Dijkstra, E.W.: Self-stabilization in spite of distributed control. In: Selected Writings on Computing: A Personal Perspective. Texts and Monographs in Computer Science, pp. 41–46. Springer, New York (1982). https://doi.org/10.1007/978-1-4612-5695-3_7
3. Fatès, N., Marcovici, I., Taati, S.: Self-stabilisation of cellular automata on tilings. Fundamenta Informaticae **185-1** (2022), https://fi.episciences.org/9184

Patterns with Touching Loops

Rolf Hoffmann

Technical University Darmstadt, Germany
hoffmann@informatik.tu-darmstadt.de

Abstract. The objective is the design of a Cellular Automata rule that can form patterns with "touching" loops. A loop is defined as a closed path of 1-cells in a 2D grid on a zero background within a zero border. A path cell is connected with two of its adjacent neighbors. In *touching loops* a path cell is also allowed to touch another on a diagonal. A CA rule was designed that can evolve stable touching loop patterns. The rule tries to cover the 2D space by overlapping tiles. The rule uses so-called *templates*, 5×5 matching patterns which are systematically derived from the given set of 3×3 tiles. The rule checks the pattern being evolved against a list of templates. If the outer neighbors of a template match, then the cell's state is set to the template's center value. Noise is injected if there is no matching template, or the tiles are not properly assembled. Thereby the evolution is driven to the desired loop patterns.

Keywords: Cellular Automata · Pattern Formation · Loop Structure · Probabilistic Rule · Overlapping Tiles

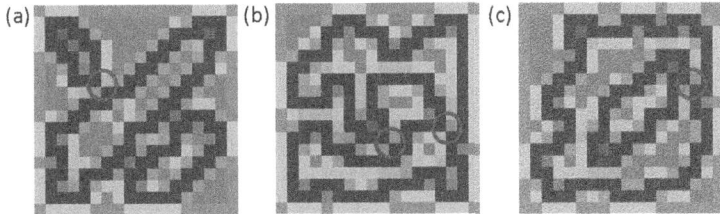

Fig. 1. (a) Three loops, externally touching. (b) One loop with self-touching points. (c) A loop enclosed by another, internally touching. Blue: loop path cells, red: uncovered, grey: border uncovered, yellow: cover level 1, green: cover level > 1.

Summary

Loop structures are of interest with regard to their emergence, construction principle, functionality, and various other properties. Therefore they are a topic of research in many disciplines. The challenge here was to generate globally "large" loops by local operations only, namely by a Cellular Automata (CA)rule.

In prior work [1] the pattern evolution was controlled by finite state machines of moving agents that were trained with a Genetic Algorithm. This technique yields good results but needs a lot of training effort and is not easy to manage for a variable grid size. Therefore we are now evolving patterns by CA rules which are directly constructed using problem specific overlapping tiles [2], and which are not sensitive against the grid size, the boundary or even obstacles. A first CA rule that can evolve loop pattern under cyclic boundary conditions was presented in [3]. The rule designed here is more simple, can evolve loops that touch each other, and is applied under fixed boundary conditions which is more difficult. The rule uses two types of overlapping tiles, tiles that are used to build straight path segments (0 111 0), and corner tiles (0 011 10). Templates (lcoal matching patterns) are systematically derived from the tiles. The CA rule tests the templates against the pattern being evolved. If there is a match in the outer neighborhood of a 5×5 window, the cell's state is adjusted to the templates center value, otherwise noise is injected. In addition the *path condition* has to be fulfilled, i.e. each path cell needs three tiles in sequence that overlap with cover level 3. Loops can touch each other internally (a loop within a loop) or externally (a loop alongside another). A loop can also be *self-touching* when path cells of the same loop touch each other. Some of the evolved 16×16 patterns are shown in Fig. 1. In future work the generating of Hamiltonian Cycles [4, 5] could be addressed.

References

1. Hoffmann, R.: How agents can form a specific pattern. In: Wąs, J., Sirakoulis, G.C., Bandini, S. (eds.) Cellular Automata. ACRI 2014. LNCS, vol. 8751, pp. 660–669.Springer, Cham (2014). https://doi.org/10.1007/978-3-319-11520-7_70
2. Hoffmann, R., Désérable, D., Seredyski, F. A cellular automata rule placing a maximal number of dominoes in the square and diamond. J. Supercomput.**77**, 9069–9087 (2021)
3. Hoffmann, R.: Generating loop patterns with a genetic algorithm and a probabilistic cellular automata rule. Algorithms **16**(7), 352 (2023)
4. Kwong, Y.H.: Enumeration of Hamiltonian cycles in P4 x Pn, and P5 x Pn.Ars Comb. **33** 87–96 (1992)
5. Umans, C., Lenhart, W.: Hamiltonian cycles in solid grid graphs. In: Proceedings 38th Annual Symposium on Foundations of Computer Science. IEEE (1997)

Exploring Damage Spreading in Cellular Automata and Boolean Networks Through Lyapunov Spectra Analyses

Milan Vispoel[iD], Aisling J. Daly[iD], and Jan M. Baetens[iD]

BionamiX, Coupure links 653, 9000 Gent, Belgium
jan.baetens@ugent.be

Abstract. This work explores the phenomenon of damage spreading in cellular automata (CA). Damage spreading involves investigating the spread and impact of defects or perturbations introduced into the initial state on the subsequent states of the CA. The study of damage spreading in CA is essential for understanding chaos and phase transitions in CA and complex systems in general. Here, we present a novel and comprehensive perspective on damage spreading in CA and Boolean networks. We introduce a methodology for computing the Lyapunov spectrum of both CA and Boolean networks which mirrors the well-established method employed in continuous-state dynamical system. We illustrate the versatility of the approach through the analysis of Lyapunov spectra for Elementary CA (ECA) and the Domany-Kinzel CA, showcasing its effectiveness in scenarios involving probabilistic update rules and network structures

Keywords: Damage spreading · Lyapunov exponents · Cellular automata

Contents

Theory, Mathematical and Physical Models

Theory of Cellular Automata: from the Past and Present to Some Path Towards the Future 3
   Alberto Dennunzio

Are Some Family Members Harmful? – A Study on Diploid Cellular Automata 10
   Souvik Roy, Harsh Modi, Rahil Patel, and Sumit Adak

Regional Controllability of Cellular Automata Through Preimages 22
   Sara Dridi, Franco Bagnoli, and Samira El Yacoubi

Pattern Formation by Collective Behavior of Competing Cellular Automata-Based Agents 34
   Miroslaw Szaban, Michal Seredyński, Rolf Hoffmann, Dominique Désérable, and Franciszek Seredyński

Effects of a Vanishing Noise on Elementary Cellular Automata Phase-Space Structure 45
   Franco Bagnoli, Michele Baia, and Tommaso Matteuzzi

A New Class of the Smallest 4-State Semi-symmetric FSSP Partial Solutions for 1D Arrays 58
   Hiroshi Umeo, Naoki Kamikawa, and Gen Fujita

Synchronization of Chains of Logistic Maps 72
   Franco Bagnoli, Michele Baia, Tommaso Matteuzzi, and Arkady Pikovsky

Fusing Different Cellular Automata Models for Surface Flows in SCURRI: Viscosity Extension Step 85
   Valeria Lupiano, Francesco Chidichimo, Paolo Catelan, Claudia R. Calidonna, and Salvatore Di Gregorio

Chaos in a Two-Dimensional Magneto-Hydrodynamic System 96
   Franco Bagnoli and Raúl Rechtman

Computational Aspects and Applications

Exploring Diverse Configurations of Cellular Automata Based S-Boxes
Using Reinforcement Learning 109
    A. Aravind, Anita John, and Jimmy Jose

Efficient Simulation of Non-uniform Cellular Automata
with a Convolutional Neural Network 121
    *Michiel Rollier, Aisling J. Daly, Odemir M. Bruno,
and Jan M. Baetens*

A Scheme for Symmetric Cryptosystem Using Large Cycle Reversible
Cellular Automata 132
    *Tarun Lywait, Kiran Srinivasan, Krishnadas Nair,
and Kamalika Bhattacharjee*

Reversible Decimal First Degree Cellular Automata for Data Classification 147
    C. J. Baby and Kamalika Bhattacharjee

Sentiment Analysis for Code-Mixed Data Using Cellular Automata
with Deep Learning Models 163
    M. J. Elizabeth, Avinash Krishna Kommineni, and Raju Hazari

Asynchronous Method of Generating Stream Ciphers in a Group of Robots
Based on Cellular Automata with Active Cells 177
    Volodymyr Mokhor, Stepan Bilan, and Volodymyr Samburskyi

Controlling Desertification Using Cellular Automata and Genetic
Algorithms 189
    Alassane Kone, Samira El Yacoubi, and Allyx Fontaine

Desertification Control Strategies: A Hybrid Approach Using Cellular
Automata and Reinforcement Learning 203
    *Amira Mouakher, Alassane Kone, Allyx Fontaine,
and Samira El Yacoubi*

Social and Biological Models

Global Analysis of a Lane Merging Strategy for Collaborative Autonomous
and Connected Vehicles 219
    Bastien Chopard, Pierre Leone, and Luka Lukic

Binary Hiking Optimization Algorithm 231
    Tahir Sağ

A Spatial Daisyworld Model 243
   Franco Bagnoli, Marco Bosi, and Tommaso Matteuzzi

A Reaction-Diffusion Cellular Automata Model for Mycelium-Based
Engineered Living Materials Evolution 253
   *Ioannis Tompris, Ioannis K. Chatzipaschalis,
Theodoros Panagiotis Chatzinikolaou, Iosif-Angelos Fyrigos,
Michail-Antisthenis Tsompanas, Andrew Adamatzky, Phil Ayres,
and Georgios Ch. Sirakoulis*

Mycelium-Based ELM Digital Twin Implemented in FPGA 265
   *Ioannis K. Chatzipaschalis, Ioannis Tompris, Konstantinos Rallis,
Theodoros Panagiotis Chatzinikolaou, Iosif-Angelos Fyrigos,
Michail-Antisthenis Tsompanas, Andrew Adamatzky, Phil Ayres,
Antonio Rubio, and Georgios Ch. Sirakoulis*

Author Index 277

Theory, Mathematical and Physical Models

Theory of Cellular Automata: from the Past and Present to Some Path Towards the Future

Alberto Dennunzio[✉]

Dipartimento di Informatica, Sistemistica e Comunicazione,
Università degli Studi di Milano-Bicocca,
Viale Sarca 336, 20126 Milan, MI, Italy
alberto.dennunzio@unimib.it

Abstract. This is an extended abstract about the research regarding theory of cellular automata: an overall overview of the past and current investigations along with an outlook on some promising research directions.

Keywords: Cellular Automata · Discrete Dynamical Systems · Theoretical Aspects · Complex Systems

Cellular Automata (CA) are discrete formal models introduced by J. von Neumann and S. Ulam in the late 1940s that at the same time define paradigmatic *Discrete Time Dynamical Systems (DTDS)* (due to the huge variety of distinct dynamical behaviors of general DTDS they exhibit) and describe *Complex Systems*, i.e., multitudes of elementary components which cooperate and produce emerging complex behaviors. For these reasons CA are used with success in many and different scientific fields for modelling phenomena and processes with complicated behaviors.

As a matter of fact, a CA is a pair $(S^{\mathcal{L}}, F)$, where S is a set called the set of states, or alphabet, of the CA, $\mathcal{L}$ is a d-dimensional regular lattice of elements (usually, vectors of $\mathbb{Z}^d$, and in particular $\mathcal{L} = \mathbb{Z}^d$), called cells, each of them associated with a state of S, and $F : S^{\mathcal{L}} \to S^{\mathcal{L}}$ is a function, called global transition map of the CA, or CA *global rule*, defined according to a *local rule* $f : S^{|\mathcal{N}|} \to S$ that updates the state of each cell on the basis of the states of a finite set $\mathcal{N}$, called neighborhood, of neighboring cells. The lattice $\mathcal{L}$ can be also viewed as the vertex set of a regular labelled graph. The states of all the cells are updated synchronously at each discrete time step. In this way, the global transition map F describes the overall updating of the states of all the cells, i.e., the updating of a certain element of $S^{\mathcal{L}}$, called configuration, and hence the sequence $\{F^t(c)\}_{t \in \mathbb{N}}$ is nothing but the *dynamical evolution*, or *orbit*, of the CA starting from the initial configuration $c \in S^{\mathcal{L}}$. Some CA variants were introduced, namely, as non-uniform CA [34–36,53,85] and asynchronous CA [10,21,29,43,67,75,84,86], obtained by relaxing the uniformity and synchrony in CA definition, i.e., in such a way that cells can update according to distinct local rules and in an asynchronous way, respectively. Probabilistic CA, i.e., CA defined by a probabilistic local rule also received a significant attention [42,45,63,82]. An infinite lattice $\mathcal{L}$ is often considered in order that CA and variants are able to capture some important features of the real-world.

A deep mathematical theory of CA was started in the late 1960s by the mathematician Gustav A. Hedlund who studied them in the context of symbolic dynamics [50]. Many researches followed about the *theoretic set* and *dynamical properties* of DTDS in the context of CA (over an infinite lattice) [2,8,9,11–13,19,24,39,40,60,61,71–74]: injectivity, surjectivity, openness, equicontinuity, almost equicontinuity, sensitivity to the initial conditions, positive expansivity, topological transitivity, topological mixing, topological chaos, topological entropy. The study of the CA dynamics allowed understanding their main features as DTDS: reachability, reversibility, stability, instability, chaos, periodic behaviors, *etc.*. Such features are nothing but typical behaviors seen in real-world phenomena and artificial processes used in applications. Therefore, when CA are used for modelling such phenomena and processes, they must exhibit the corresponding dynamical properties. Unfortunately, almost all the dynamical properties of DTDS raised by CA turned out to be *undecidable* [3,7,20,41,51,54–58,62,78]. This is an obstacle when one wants to model a certain phenomenon or an artificial process by a CA being sure that the latter is appropriate in the considered case of study. So, researches also focused and are still focusing on the identification and the investigation of classes of CA which are expressive enough (i.e., able to exhibit the complex behaviours of general CA) and, at the same time, allow one to *establish a given dynamical behavior* (i.e., in those classes the dynamical properties become decidable). *Linear CA*, *Additive CA*, and *Group CA* are just paradigmatic examples of such classes [14,15,26,38,52,59,65,66]. The current investigations concern the decidability of the dynamical properties and the efficiency issue of the decision algorithms [6,25,27,28,30,32,33], the latter being largely unexplored probably due to the fact that, as far as infinite lattices are concerned, the focus in CA research has been in (un)decidability.

As to CA and variants with a finite lattice $\mathcal{L}$, they are used too in applications for modelling several phenomena, including the diffusive ones [4,5,16–18,64,81]. From a theoretical point of view, due to the finiteness of $\mathcal{L}$, a crucial focus is on problems regarding the finite sets of fixed/periodic points, attractors, transients, the reachability and synchronization issues of the finite DTDS defined by such models [68–70,83]. Clearly, when the set of state is finite too, the finiteness implies that all the problems involving these aspects are decidable. However, the computational complexity landscape of solving such problems is currently largely unexplored (for some existing results, see [31,44,46–49,79], for instance). There is then a lack of results obstacles a more proper and most effective use of finite CA and variants in applications. Indeed, as an example, to face real scenarios one has to master every aspect around the global steady or cyclic global state of that phenomenon. This is reflected on the model used for describing the phenomenon in the need of being able to answer questions about the dynamics of the induced DTDS in an efficient way. So, researchers are still considering relevant problems over the dynamics of the DTDS defined by finite CA and variants along with finding out efficient algorithms that solve them: computing the number of fixed/periodic points and attractors, computing the lengths of cycles and transients (maximum, minimum length), establishing some property on all these, stating some reachability forms and deciding it, *etc.*.

Some meaningful future research directions in CA theory are:

- *Higher Dimensional Scenario.* Driven by applications that inherently also involve two and higher dimensional spaces (for instance, think of diffusive phenomena in real two/three-dimensional spaces or cryptographic protocols regarding two-dimensional data), DTDS raised by CA defined over a two and higher dimensional lattice should be considered in depth (for the existing results, see [1,22,23,37,80], for instance). Therefore, further new results are required especially for multidimensional CA. We stress that, due to its high complexity, the CA multidimensional scenario potentially offers security advantages over the one-dimensional setup, as for instance in the case of cryptographic protocols, where it deserves to be investigated also for dealing with one-dimensional data.
- *Control Theory in CA.* The complexity of real systems requires that aspects from control theory are considered too when studying formal models for complex systems. The problems of observability and controllability are among the main issues in the context of control theory. First important works dealing with observability and controllability in CA settings have been recently presented [76,77,87]. A comprehensive CA control theory integrating that of the dynamical properties is expected so that CA can be used to model real complex systems even more accurately.
- *Data-driven Synthesis of CA.* There is no general method that, given the description of a real complex system (which reasonably lends itself to be modelled by CA) in terms of data regarding its real dynamics, automatically provides the CA local rules, or a set of local rules, able to suitably model it. Theoretical results regarding parametrized families of CA to be learned are needed to develop one or a collection of such methods that would have a significant impact on the general modelling problem.

The current investigations on CA theory are carried out worldwide and through international collaborations. However, at our best knowledge, there are very few projects funded on an international scale. The themes illustrated in the above mentioned research directions could be the subjects of CA theory project proposals on an international scale. I believe that, due to their relevance, these subject would increase themselves the chance of proposals being selected for funding.

References

1. Acerbi, L., Dennunzio, A., Formenti, E.: Surjective multidimensional cellular automata are non-wandering: a combinatorial proof. Inf. Process. Lett. **113**(5–6), 156–159 (2013)

2. Acerbi, L., Dennunzio, A., Formenti, E.: Conservation of some dynamical properties for operations on cellular automata. Theoret. Comput. Sci. **410**(38–40), 3685–3693 (2009)
3. Amoroso, S., Patt, Y.: Decision procedures for surjectivity and injectivity of parallel maps for tesselation structures. J. Comput. Syst. Sci. **6**, 448–464 (1972)
4. Bandini, S., Mauri, G.: Multilayered cellular automata. Theor. Comput. Sci. **217**(1), 99–113 (1999)
5. Bandini, S., Mauri, G., Pavesi, G., Simone, C.: Parallel simulation of reaction-diffusion phenomena in percolation processes: a model based on cellular automata. Future Gener. Comput. Syst. **17**(6), 679–688 (2001)
6. Béaur, P., Kari, J.: Effective projections on group shifts to decide properties of group cellular automata. Int. J. Found. Comput. Sci. **35**(1&2), 77–100 (2024)
7. Bernardi, V., Durand, B., Formenti, E., Kari, J.: A new dimension sensitive property for cellular automata. Theor. Comput. Sci. **345**, 235–247 (2005)
8. Blanchard, F., Kůrka, P., Maass, A.: Topological and measure-theoretic properties of one-dimensional cellular automata. Physica D **103**, 86–99 (1997)
9. Blanchard, F., Tisseur, P.: Some properties of cellular automata with equicontinuity points. Ann. Inst. Henri Poincaré, Probabilité et Statistiques **36**, 569–582 (2000)
10. Bouré, O., Fatès, N., Chevrier, V.: Probing robustness of cellular automata through variations of asynchronous updating. Nat. Comput. **11**(4), 553–564 (2012)
11. Boyle, M., Kitchens, B.: Periodic points for cellular automata. Indag. Math. **10**, 483–493 (1999)
12. Cattaneo, G., Dennunzio, A., Margara, L.: Chaotic subshifts and related languages applications to one-dimensional cellular automata. Fund. Inform. **52**, 39–80 (2002)
13. Cattaneo, G., Finelli, M., Margara, L.: Investigating topological chaos by elementary cellular automata dynamics. Theor. Comput. Sci. **244**, 219–241 (2000)
14. Cattaneo, G., Dennunzio, A., Margara, L.: Solution of some conjectures about topological properties of linear cellular automata. Theor. Comput. Sci. **325**(2), 249–271 (2004)
15. Cattaneo, G., Formenti, E., Manzini, G., Margara, L.: Ergodicity, transitivity, and regularity for linear cellular automata over $\mathbb{Z}_m$. Theor. Comput. Sci. **233**(1–2), 147–164 (2000)
16. Chopard, B.: Cellular automata and lattice boltzmann modeling of physical systems. In: Rozenberg, G., Bäck, T., Kok, J.N. (eds.) Handbook of Natural Computing, pp. 287–331. Springer, Heidelberg (2012). https://doi.org/10.1007/978-3-540-92910-9_9
17. Chopard, B., Luthi, P.O.: Lattice boltzmann computations and applications to physics. Theor. Comput. Sci. **217**(1), 115–130 (1999)
18. Chopard, B., Masselot, A.: Cellular automata and lattice boltzmann methods: a new approach to computational fluid dynamics and particle transport. Future Gener. Comput. Syst. **16**(2–3), 249–257 (1999)
19. Codenotti, B., Margara, L.: Transitive cellular automata are sensitive. Am. Math. Mon. **103**(1), 58–62 (1996)
20. Culik, K., Yu, S.: Undecidability of cellular automata classification schemes. Complex Syst. **2**, 177–190 (1988)
21. Dennunzio, A., Formenti, E., Manzoni, L., Mauri, G.: m-Asynchronous cellular automata: from fairness to quasi-fairness. Natural Comput. **12**, 561–572 (2013)
22. Dennunzio, A., Guillon, P., Masson, B.: Sand automata as cellular automata. Theor. Comput. Sci. **410**, 3962–3974 (2009)

23. Dennunzio, A.: From one-dimensional to two-dimensional cellular automata. Fund. Inform. **115**(1), 87–105 (2012)
24. Dennunzio, A., Di Lena, P., Formenti, E., Margara, L.: Periodic orbits and dynamical complexity in cellular automata. Fund. Inform. **126**(2–3), 183–199 (2013)
25. Dennunzio, A., Formenti, E., Grinberg, D., Margara, L.: Chaos and ergodicity are decidable for linear cellular automata over $(\mathbb{Z}/m\mathbb{Z})^n$. Inf. Sci. **539**, 136–144 (2020)
26. Dennunzio, A., Formenti, E., Grinberg, D., Margara, L.: Dynamical behavior of additive cellular automata over finite abelian groups. Theor. Comput. Sci. **843**, 45–56 (2020)
27. Dennunzio, A., Formenti, E., Grinberg, D., Margara, L.: Decidable characterizations of dynamical properties for additive cellular automata over a finite abelian group with applications to data encryption. Inf. Sci. **563**, 183–195 (2021)
28. Dennunzio, A., Formenti, E., Grinberg, D., Margara, L.: An efficiently computable characterization of stability and instability for linear cellular automata. J. Comput. Syst. Sci. **122**, 63–71 (2021)
29. Dennunzio, A., Formenti, E., Manzoni, L.: Computing issues of asynchronous CA. Fund. Inform. **120**(2), 165–180 (2012)
30. Dennunzio, A., Formenti, E., Manzoni, L., Margara, L., Porreca, A.E.: On the dynamical behaviour of linear higher-order cellular automata and its decidability. Inf. Sci. **486**, 73–87 (2019)
31. Dennunzio, A., Formenti, E., Manzoni, L., Mauri, G., Porreca, A.E.: Computational complexity of finite asynchronous cellular automata. Theor. Comput. Sci. **664**, 131–143 (2017)
32. Dennunzio, A., Formenti, E., Margara, L.: An easy to check characterization of positive expansivity for additive cellular automata over a finite abelian group. IEEE Access **11**, 121246–121255 (2023)
33. Dennunzio, A., Formenti, E., Margara, L.: An efficient algorithm deciding chaos for linear cellular automata over $(\mathbb{Z}/m\mathbb{Z})^n$ with applications to data encryption. Inf. Sci. **657**, 119942 (2024)
34. Dennunzio, A., Formenti, E., Provillard, J.: Non-uniform cellular automata: classes, dynamics, and decidability. Inf. Comput. **215**, 32–46 (2012)
35. Dennunzio, A., Formenti, E., Provillard, J.: Local rule distributions, language complexity and non-uniform cellular automata. Theor. Comput. Sci. **504**, 38–51 (2013)
36. Dennunzio, A., Formenti, E., Provillard, J.: Three research directions in non-uniform cellular automata. Theor. Comput. Sci. **559**, 73–90 (2014)
37. Dennunzio, A., Formenti, E., Weiss, M.: Multidimensional cellular automata: closing property, quasi-expansivity, and (un)decidability issues. Theor. Comput. Sci. **516**, 40–59 (2014)
38. Dennunzio, A., di Lena, P., Formenti, E., Margara, L.: On the directional dynamics of additive cellular automata. Theor. Comput. Sci. **410**(47–49), 4823–4833 (2009)
39. Durand, B.: The surjectivity problem for 2D cellular automata. J. Comput. Syst. Sci. **49**(3), 718–725 (1994)
40. Durand, B.: Global properties of cellular automata. In: Goles, E., Martinez, S. (eds.) Cellular Automata and Complex Systems. Kluwer (1998)
41. Durand, B., Formenti, E., Varouchas, G.: On undecidability of equicontinuity classification for cellular automata. Disc. Math. Theor. Comput. Sci. **AB**, 117–128 (2003)
42. Fatès, N.: Stochastic cellular automata solutions to the density classification problem - when randomness helps computing. Theory Comput. Syst. **53**(2), 223–242 (2013)

43. Fatès, N.: A guided tour of asynchronous cellular automata. J. Cell. Autom. **9**(5–6), 387–416 (2014)
44. Fatès, N., Thierry, E., Morvan, M., Schabanel, N.: Fully asynchronous behavior of double-quiescent elementary cellular automata. Theor. Comput. Sci. **362**(1–3), 1–16 (2006)
45. Fuks, H.: Solving two-dimensional density classification problem with two probabilistic cellular automata. J. Cell. Autom. **10**(1–2), 149–160 (2015)
46. Goles, E., Lobos, F., Montealegre, P., Ruivo, E.L.P., de Oliveira, P.P.B.: Computational complexity of the stability problem for elementary cellular automata. J. Cell. Autom. **15**(4), 261–304 (2020)
47. Goles, E., Maldonado, D., Montealegre, P., Ollinger, N.: On the complexity of the stability problem of binary freezing totalistic cellular automata. Inf. Comput. **274**, 104535 (2020)
48. Goles, E., Montalva-Medel, M., Montealegre, P., Ríos-Wilson, M.: On the complexity of generalized Q2R automaton. Adv. Appl. Math. **138**, 102355 (2022)
49. Goles, E., Montealegre, P.: The complexity of the asynchronous prediction of the majority automata. Inf. Comput. **274**, 104537 (2020)
50. Hedlund, G.A.: Endomorphisms and automorphisms of the shift dynamical system. Math. Syst. Theory **3**, 320–375 (1969)
51. Hurd, L.P., Kari, J., Culik, K.: The topological entropy of cellular automata is uncomputable. Ergodic Theory Dyn. Syst. **12**, 255–265 (1992)
52. Ito, M., Osato, N., Nasu, M.: Linear cellular automata over $\mathbb{Z}_m$. J. Comput. Syst. Sci. **27**, 125–140 (1983)
53. Kamilya, S., Kari, J.: Nilpotency and periodic points in non-uniform cellular automata. Acta Informatica **58**(4), 319–333 (2021)
54. Kari, J.: The nilpotency problem of one dimensional cellular automata. SIAM J. Comput. **21**, 571–586 (1992)
55. Kari, J.: Reversibility and surjectivity problems of cellular automata. J. Comput. Syst. Sci. **48**, 149–182 (1994)
56. Kari, J.: Rice's theorem for the limit set of cellular automata. Theor. Comput. Sci. **127**(2), 229–254 (1994)
57. Kari, J.: Theory of cellular automata: a survey. Theor. Comput. Sci. **334**(1–3), 3–33 (2005)
58. Kari, J.: The nilpotency problem of one-dimensional cellular automata. SIAM J. Comput. **21**(3), 571–586 (1992)
59. Kari, J.: Linear cellular automata with multiple state variables. In: Reichel, H., Tison, S. (eds.) STACS 2000. LNCS, vol. 1770, pp. 110–121. Springer, Heidelberg (2000). https://doi.org/10.1007/3-540-46541-3_9
60. Kůrka, P.: Languages, equicontinuity and attractors in cellular automata. Ergodic Theory Dyn. Syst. **17**, 417–433 (1997)
61. Kůrka, P.: Topological and Symbolic Dynamics, vol. 11 of Cours Spécialisés, Société Mathématique de France (2004)
62. Lukkarila, V.: Sensitivity and topological mixing are undecidable for reversible one-dimensional cellular automata. J. Cell. Autom. **5**(3), 241–272 (2010)
63. Mairesse, J., Marcovici, I.: Around probabilistic cellular automata. Theor. Comput. Sci. **559**, 42–72 (2014)
64. Manenti, L., Manzoni, S., Bandini, S.: A stochastic cellular automata for modeling pedestrian groups distribution. J. Cell. Autom. **8**(5–6), 321–332 (2013)
65. Manzini, G., Margara, L.: Attractors of linear cellular automata. J. Comput. Syst. Sci. **58**(3), 597–610 (1999)

66. Manzini, G., Margara, L.: A complete and efficiently computable topological classification of d-dimensional linear cellular automata over $\mathbb{Z}_m$. Theor. Comput. Sci. **221**(1–2), 157–177 (1999)
67. Manzoni, L.: Asynchronous cellular automata and dynamical properties. Nat. Comput. **11**(2), 269–276 (2012)
68. Manzoni, L., Umeo, H.: The firing squad synchronization problem on CA with multiple updating cycles. Theor. Comput. Sci. **559**, 108–117 (2014)
69. Mariot, L.: Enumeration of maximal cycles generated by orthogonal cellular automata. Nat. Comput. **22**(3), 477–491 (2023)
70. Mariot, L., Manzoni, L.: A classification of s-boxes generated by orthogonal cellular automata. Nat. Comput. **23**(1), 5–16 (2024)
71. Maruoka, A., Kimura, M.: Conditions for injectivity of global maps for tessellation automata. Inf. Control **32**, 158–162 (1976)
72. Meyerovitch, T.: Finite entropy for multidimensional cellular automata. Ergodic Theory Dyn. Syst. **28**, 1243–1260 (2008)
73. Moore, E.F.: Machine models of self-reproduction. Proc. Symp. Appl. Math. **14**, 13–33 (1962)
74. Myhill, J.: The converse to Moore's garden-of-eden theorem. Proc. Am. Math. Soc. **14**, 685–686 (1963)
75. de Oliveira, P.P.B., Formenti, E., Perrot, K., Riva, S., Ruivo, E.L.P.: Non-maximal sensitivity to synchronism in elementary cellular automata: exact asymptotic measures. Theor. Comput. Sci. **926**, 21–50 (2022)
76. Plénet, T., Bagnoli, F., Yacoubi, S.E., Raïevsky, C., Lefèvre, L.: Synchronization of elementary cellular automata. Nat. Comput. **23**(1), 31–40 (2024)
77. Plénet, T., Yacoubi, S.E., Raïevsky, C., Lefèvre, L.: Observability and reconstructibility of bounded cellular automata. Int. J. Syst. Sci. **53**(14), 2901–2917 (2022)
78. Sutner, K.: De Bruijn graphs and linear cellular automata. Complex Syst. **5**, 19–30 (1991)
79. Sutner, K.: On the computational complexity of finite cellular automata. J. Comput. Syst. Sci. **50**(1), 87 (1995)
80. Theyssier, G., Sablik, M.: Topological dynamics of cellular automata: dimension matters. Theory Comput. Syst. **48**, 693–714 (2011)
81. Thorimbert, Y., Lätt, J., Chopard, B.: Coupling of lattice Boltzmann shallow water model with lattice Boltzmann free-surface model. J. Comput. Sci. **33**, 1–10 (2019)
82. Toupance, P., Chopard, B., Lefèvre, L.: System reduction: an approach based on probabilistic cellular automata. Nat. Comput. **23**(1), 17–29 (2024)
83. Umeo, H.: How to synchronize cellular automata - recent developments -. Fund. Informaticae **171**(1–4), 393–419 (2020)
84. Wacker, S., Worsch, T.: On completeness and decidability of phase space invertible asynchronous cellular automata. Fund. Informaticae **126**(2–3), 157–181 (2013)
85. Wolnik, B., Dziemianczuk, M., Baets, B.D.: Non-uniform number-conserving elementary cellular automata. Inf. Sci. **626**, 851–866 (2023)
86. Worsch, T.: Towards intrinsically universal asynchronous CA. Nat. Comput. **12**(4), 539–550 (2013)
87. Yacoubi, S.E., Plénet, T., Dridi, S., Bagnoli, F., Lefèvre, L., Raïevsky, C.: Some control and observation issues in cellular automata. Complex Syst. **30**(3), 391–413 (2021)

Are Some Family Members Harmful? – A Study on Diploid Cellular Automata

Souvik Roy[1], Harsh Modi[1], Rahil Patel[1], and Sumit Adak[2]( )

[1] School of Engineering and Applied Science, Ahmedabad University, Ahmedabad, India
{souvik.roy,harsh.m6,rahil.p}@ahduni.edu.in
[2] DTU Compute, Technical University of Denmark, Kgs. Lyngby, Denmark
suad@dtu.dk

Abstract. This paper explores diploid elementary cellular automata (ECA) systems where the rules of the cellular systems are acquired with a random mixing of two ECAs. However, here, we consider two ECAs from the same *family* following left to right, 0 to 1, and both transformations. Following the experimental approach, this study classifies the dynamics of (all possible) 300 diploid family couples following Wolfram's and Li and Packard's classification. We investigate the resistance of this diploid system against this family perturbation. As we will see, this study is interesting enough to provide the following rich phenomenon: (1) two-periodic family couple together show chaotic dynamics and vice-versa; (2) some diploid couple changes their class dynamics after a critical value of mixing rate, i.e. *class* transition; and (3) lastly, these diploid couples are also capable to show continuous or second-order *phase* transition dynamics.

Keywords: Diploid Cellular Automata · Chaos · Phase Transition · Class Transition · Classification

1 Introduction

According to the argument of Turing [16], the question of randomness is fundamental to understand the dynamics of living element. In this direction, cellular automata (CA) researchers have introduced the concept of stochastic cellular automata (or probabilistic cellular automata) [1,2,4,8,9,11], where we select the updating rules randomly from a set of rules. Moreover, the temporally stochastic model of cellular automata can be found in [13,15].

In this work, we explore the dynamics of stochastic systems considering elementary cellular automata (ECA) rules, i.e. one-dimensional three-neighbourhood two-state cellular automata. All cells update their state synchronously under periodic boundary conditions. Specifically, we consider *diploid* ECA as *Bernoulli* mixtures of two ECA rules. So far, only a few results have been obtained on diploid ECA [3,5,8]. In an early work, Fatès [8] has identified phase transition dynamics of the diploid system following numerical simulations after considering null, identity, and inversion ECAs as blind rules. In the same direction, the work of [5] has revisited this phase transition dynamics following mean field approximation and Dobrushin criterion after considering null ECA as blind rule. In a different direction, Bolt et al. [3] have introduced an algorithm for filling the missing cell states, i.e., identification problem, for diploid ECA.

Next, in the context of elementary cellular automata, Wolfram [17] has classified the ECA rules into four classes according to their dynamics and space-time diagrams. However, Wolfram classification is formally undecidable [6]. Hereafter, Li and Packard [10] have introduced the following more general classification of the ECA rules:

(A) Null rules: homogeneous fixed-point rules (denoted as N);
(B) Fixed point rules: inhomogeneous fixed-point rules (FP);
(C) Periodic rules (P);
(D) Locally chaotic: chaotic dynamics confined by the domain walls (LC);
(E) Chaotic rules (C).

There are 256 ECA rules. However, with the use of left-to-right, 0 to 1, and both of these transformations, it is possible to narrow down the 256 ECA rule space to 88 classes, each represented by the rule of smallest number, i.e. the minimal representative ECA rule [10]. In details, an elementary CA rule f can be equivalent to another rule f_1 (resp. f_2 and f_3) under the left to right transformation (resp. 0 to 1 transformation, joint operation of both)[1]. Therefore, the rules f, f_1, f_2, and f_3 belong to the *same family*, in other words, they are *family members*. Obviously, they belong to the same class following Wolfram and Li-Packard classification. Moreover, the space-time diagram of f_1, f_2, f_3 are exactly the same following the space-time diagram of f by a mirror reflection transformation, or a 0 to 1 transformation, or a combination of both. As an evidence, Fig. 1 shows the dynamics of ECAs where $f = 152$; $f_1 = 194$; $f_2 = 230$; and $f_3 = 188$.

With this background, we attempt to understand the dynamics of diploid ECA, where we use random mixing of two ECAs, say f and g. Importantly, f and g are the members of the same family, i.e. $g \in \{f_1, f_2, f_3\}$. Therefore, in the diploid ECA, we are mixing two ECAs with exactly the same space-time diagrams (or dynamics: convergence, reversibility, cycle structure properties) by a mirror reflection transformation, or a 0 to 1 transformation, or a combination of both. Here, we ask the following interesting question about the dynamics of diploid ECA: is it possible that during the evolution of the diploid system, we observe the behaviour of a different family (specifically, class)? More specifically, is it possible that two chaotic or locally chaotic (resp. periodic or fixed point) rules of the same family together depict a kind of closeness towards simplicity (resp. chaos) during the evolution of the diploid system? Another interesting aspect is whether two family members (rules) in this stochastic setting are able

---

[1] $f_1(x_{i-1}, x_i, x_{i+1}) = f(x_{i+1}, x_i, x_{i-1})$; $f_2(x_{i-1}, x_i, x_{i+1}) = \overline{f(\overline{x_{i-1}}, \overline{x_i}, \overline{x_{i+1}})}$; $f_3(x_{i-1}, x_i, x_{i+1}) = \overline{f(\overline{x_{i+1}}, \overline{x_i}, \overline{x_{i-1}})}$.

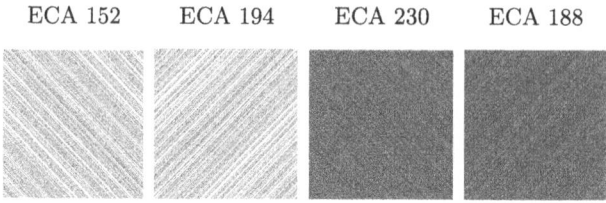

Fig. 1. Space-time digram of ECAs of same family where $f = 152$, $f_1 = 194$, $f_2 = 230$ and $f_3 = 188$.

to show the *phase transition* [8] dynamics or not. The same question is also applicable for *class transition* [15] dynamics. As we will observe, this study is interesting enough to provide this kind of many examples with potential applications in the study of natural (physical, biological, and chemical) systems.

2 Cellular Automata, Diploid Cellular Automata, and Experimental Protocol

In this work, we consider elementary cellular automata, i.e. one-dimensional three-neighbourhood (left, self, right) and two-state ($\{0,1\}$) cellular automata, following periodic boundary condition and synchronous updating scheme. In periodic boundary condition, cells are arranged as a ring, where the set of indices that represent each cell is denoted by $\mathcal{L} = \mathbb{Z}/n\mathbb{Z}$. The lattice size (the number of cells) is denoted by n. The collection of all states at a given time is called a configuration. Following this, $\{0,1\}^{\mathcal{L}}$ depicts the set of configurations. A cell updates its state according to its own state and the state of its neighbours (left and right) following the local transition function $f : \{0,1\}^3 \rightarrow \{0,1\}$. Next, one can define the global transition function $F : \{0,1\}^{\mathcal{L}} \rightarrow \{0,1\}^{\mathcal{L}}$, such that, the next configuration $y = F(x)$ of a configuration $x \in \{0,1\}^{\mathcal{L}}$ is given by $\forall i \in \mathcal{L}$, $y_i = f(x_{i-1}, x_i, x_{i+1})$. We express the local transition function in a look-up table format where we name each of the 8 combinations of x_{i-1}, x_i, and x_{i+1} as Rule Min Term (RMT), which is generally presented in its decimal equivalent. Finally, the decimal equivalent of the eight outputs is called "rule", i.e. $f(1,1,1) \cdot 2^7 + f(1,1,0) \cdot 2^6 + \cdots + f(0,0,0) \cdot 2^0$. We call an RMT active if it changes the state of a cell, i.e. $f(x_{i-1}, x_i, x_{i+1}) \neq x_i$; otherwise, it is passive. Recall that we have $2^8 = 256$ ECAs, out of which 88 are minimal representative ECAs and the rest are their equivalent.

If the local rule of a cellular system is stochastic, we use the notion of Markov chain on the configuration space [8]. Specifically, we directly introduce the special stochastic system, called diploid ECA, which is obtained as random mixtures of two different ECA rules. In detail, given the *mixing rate* $\lambda \in [0,1]$, and given two local rules f and g, the stochastic system defined by $\phi = (1-\lambda)f + \lambda g$ is called a diploid ECA. In other words, we can write the evolution of the system as follows: at time t for each cell $i \in \mathcal{L}$, we select either the rule f with probability $1-\lambda$ or the rule g with probability λ for an update following the state of the neighbours and self. We will write $\phi = (f, g)[\lambda]$ to represent this relation. Note that, the special cases $\lambda = 0, 1$ or $f = g$ represent the deterministic ECA. Moreover, if $\lambda' = 1 - \lambda$, then the diploid ECA $(f, g)[\lambda]$ is identical to $(g, f)[\lambda']$ which helps us to restrict to $\lambda \in [0, 0.5]$.

Following this, to understand the dynamics of diploid ECA, we follow the qualitative and quantitative experimental approaches from [14]: (a) Firstly, we need to observe the evolution of the system through space-time diagrams which can be able to provide an important qualitative visual comparison. We consider $n \in [90, 100]$ and evolve the system for 2000 time steps; and (b) Secondly, we need to calculate the density of a configuration c which can be written as $d_c = c_1/n$ (c_1 is the number of 1s in configuration c and n is the lattice size). We start with $n = 100$ considering initial density $d_{ini} = 0.5$; and evolve the system again for 2000 steps; and calculate the average density parameter value for (last) 100 time steps. It is also possible that the diploid system may show different dynamics for different runs. To resolve this issue, we observe the stochastic system's dynamics 20 times for each instance of $(f, g)[\lambda]$. The change in density during evolution of the system provides the formal quantitative results.

3 Main Results: Dynamics of Family Diploid ECA

This section introduces the dynamics of family diploid ECAs where we mainly follow Li and Packard's classification [10]. Let us denote $\mathbb{C}(f)$ as the class of ECA f following Li and Packard's classification. Therefore $\mathbb{C}(f) \in \{$N, FP, P, LC, C$\}$ (following the details of this classification in Sect. 1). Following this, $\mathbb{C}(f, g)$ depicts the dynamic (or class) of diploid ECA (f, g) considering the overall mixing rate $\lambda \in [0, 0.5]$. Obviously, here, $\mathbb{C}(f) = \mathbb{C}(g)$ because f and g comes from the same family. As a result, Table 1 shows the overall dynamics of all possible family diploid ECAs.[2] According to Table 1, following are the immediate two possibilities:

1. $\mathbb{C}(f) = \mathbb{C}(g) = \mathbb{C}(f, g)$. In other words, the family diploid ECA follows the dynamics of the participating ECAs. For evidence, $\mathbb{C}(8) = \mathbb{C}(32) = $ N, following this diploid $\mathbb{C}(8, 32)$ also converges to homogeneous point attractor; similarly, ECAs 90 and 165 follow chaotic dynamics, in the same direction, $\mathbb{C}(90, 165) = $ C. However, this possibility is quite natural. In Table 1, diploid ECAs in black show these possibilities; the rest are marked with blue.
2. In the opposite direction, $\mathbb{C}(f) = \mathbb{C}(g) \neq \mathbb{C}(f, g)$ where the diploid ECA follows different dynamics in comparison with the participating ECAs. Recall that, in this stochastic system, the original ECA (f) and the *noise* (perturbing) ECA (g) are part of the same family (or with exactly the same dynamics). Therefore, the above possibility depicts a interesting situation where noise of

---

[2] Note that, Table 1 excludes ECAs 23, 51, 77, 105, 150, 178, 204, 232 where $f = f_1 = f_2 = f_3$, i.e. these ECAs are not associated with any family members.

the same family or dynamics massively affects the system. We note the following two most peculiar dynamics following this situation: (a) The diploid couple of two locally chaotic rules can break the domain walls by the effect of noise. That is, it shows chaotic dynamics. For evidence, in Table 1, $\mathbb{C}(73)$ = $\mathbb{C}(109)$ = LC, however, $\mathbb{C}(73, 109)$ = C; (b) One step further, diploid couple of two simple (periodic) ECAs show kind of closeness towards chaos. For evidence, in Table 1, $\mathbb{C}(131)$ = $\mathbb{C}(62)$ = P, however, $\mathbb{C}(131, 62)$ = C.

Till now, we restrict the characterization of diploid ECAs following Li and Packard's classification [10]. That is, $\mathbb{C}(f, g) \in \{$N, FP, P, LC, C$\}$. However, these are cases where it is not possible to mark the system dynamics only following Li and Packard's classification. Here, we report the following interesting cases.

3. First, we introduce the simple class *Noise* (denoted by NO). Here, in the diploid ECAs, we have used a source of randomness to implement the random mixing rate λ (in our study, random number generator "rand"). For some diploid ECAs, this source of randomness dominates the system dynamics. As a trivial example, a diploid couple of ECA 0 (always moves to state 0) and ECA 255 (always moves to state 1) shows this *Noise* (NO) dynamics, which only reflects the dynamics of the changing the mixing rate λ using rand.
4. We also introduce a new class Pc (kind of "Periodic dynamics with more complexity"), which shows complex patterns in comparison with periodic ones. However, we are not able to identify them as complex/chaotic. Note that, we are still open about many issues of this new class (see Sect. 4).

Up to now, the characterization is independent of the mixing rate in the diploid system. That is, changing the value of λ has no effect on the dynamics. However, we can not avoid the importance of λ further; see the following cases.

5. Now, we introduce the *class transition* dynamics where there exists a critical value of mixing rate λ_c which distinguishes the system in two different classes – $\mathbb{C}(f, g)[0, \lambda_c] \neq \mathbb{C}(f, g)[\lambda_c, 0.5]$. We note the following most interesting situation where $\mathbb{C}(18) = \mathbb{C}(183) = $ C, however, their diploid couple shows locally chaotic dynamics after a critical value of mixing rate, i.e. $\mathbb{C}(18, 183)[\lambda_c, 0.5]$ = LC. In other words, a chaotic system with the effect of chaotic noise moves towards simplicity. In Table 1, CT denotes the class transition dynamics.
6. Lastly, let us introduce *phase transition* which is the most discussed phenomenon of different non-uniform cellular systems. In fact, the researchers have invested the most in identifying this property for diploid systems, see [5,8]. In Table 1, PT denotes the phase transition dynamics. For example, let us consider two fixed point ECAs ($\mathbb{C}(34) = \mathbb{C}(48) = $ FP), where $\mathbb{C}(34, 48)$ = PT.

Next, the following section revisits these different phenomena with examples and space-time diagrams after considering their computational ability.

Table 1. Classification of the family diploid ECAs. Here, ♠ (resp. ♣, ♡, ▼) denotes f (resp. f_1, f_2, f_3).

♠	♣	♡	▼	Class	(♠,♣)	(♠,♡)	(♠,▼)	(♣,♡)	(♣,▼)	(♡,▼)
0	255			P	P					
1			127	P						
2	16	191	247	FP	N	CT	CT	CT	CT	N
3			119	P						
4	223			FP	P_c					
5			95	P						
6	20	159	215	P	N	C	C	C	C	N
7	21	31	87	P						
8	64	239	253	N	N	NO	NO	NO	NO	N
9	65	111	125	P						
10	80	175	245	FP	N	CT	CT	C	C	N
11	47	81	117	P						
12	68	207	221	FP	FP	P	P	P	P	FP
13	69	79	93	FP						
14	84	143	213	P	P_c	P_c	P_c	P_c	P_c	P_c
15	85			P						
18	183			C	CT					
19	55			P						
22	151			C	C					
24	66	189	231	FP	N	CT	CT	CT	CT	N
25	61	67	103	P	P_c	P_c	P_c	P_c	P_c	P_c
26	82	167	181	LC	C	C	C	C	C	C
27	39	53	83	P						
28	70	157	199	P	P	P	P	P	P	P_c
29	71			P						
30	86	135	149	C	C	C	C	C	C	C
32	251			P	CT					
33	123			P						
34	48	187	243	FP	PT	P	P	P	P	P
35	49	59	115	P						
36	219			FP	CT					
37	91			P						
38	52	155	211	P	N	CT	CT	CT	CT	N
40	96	235	249	N	N					
41	97	107	121	C	LC	LC	P_c	P_c	P_c	LC
42	112	171	241	FP	PT	CT	CT	CT	CT	FP
43	113			P	CT					
44	100	203	217	FP	FP	CT	CT	CT	CT	FP
45	75	89	101	C	C	C	C	C	C	C
46	116	139	209	FP	C	P	P	P	P	PT
50	179			P						
54	147			C	CT					
56	98	185	227	FP	P	FP	FP	FP	P	FP
57	99			FP	CT					
58	114	163	177	FP	P_c	P_c	P	FP	P	P
60	102	153	195	C	PT	C	C	C	C	PT
72	237			FP	LC					
73	109			FP						
74	88	173	229	P	N	C	C	C	C	N
76	105			FP						
78	92	141	197	FP	FP					FP
90	165			C						
104	120	169	225	C	PT					
106	233			N						
108	201			P	PT					
128	254			N						
129	126			C	P					
130	144	190	246	FP	N	CT	CT	CT	CT	N
131	62	145	118	P	C	C	C	C	C	C
132	222			FP	PT					
133	94			P	P_c					
134	148	158	214	P	N	C	C	C	C	N
136	192	238	252	N	N	PT	PT	PT	PT	N
137	110	124	193	C	P	P	P	P	P	C
138	174	208	244	FP	N	N	N	N	C	FP
140	196	206	220	FP	FP	FP	FP	FP	FP	FP
142	212			C						
146	182			P						
152	188	194	230	FP	CT	N	PT	PT	LC	C
154	166	180	210	LC	PT	C	LC	LC	C	C
156	198			P						
160	250			N	PT					
161	122			C						
162	176	186	242	FP	P	P	P	P	P	P
164	218			C	CT					
168	224	234	248	N	N	PT	PT	P	PT	N
170	240			FP	P	PT	PT	PT	FP	PT
172	202	216	228	P	P	FP	FP	FP	FP	PT
184	226			FP	P					
200	236			FP						

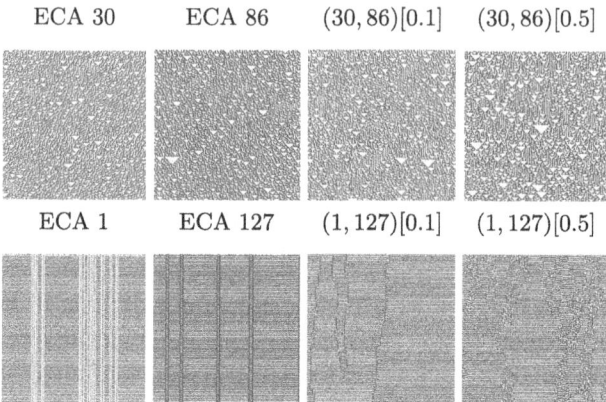

Fig. 2. The situation $\mathbb{C}(f) = \mathbb{C}(f,g)$ considering family diploid couples $(30, 86)$ and $(1, 127)$. In the space-time diagrams, when rule f and g are applied, the cell with state 1 is indicated by red and blue respectively. For both f and g, white shows state 0. This convention is kept in the rest of the text.

4 λ-Independent Dynamics of the System

For the λ-independent situation, if we progressively change the mixing rate λ, the system dynamics remain almost the same. Here, we discuss this possibility. In this situation, $\mathbb{C}(f,g) \in \{$N, FP, P, LC, C, NO, Pc$\}$ following Li and Packard's classification along with the notion of *Noise* (NO) and *Periodic dynamics with more complexity* (Pc) class. Under this umbrella, there are following two possibilities: $\mathbb{C}(f) = \mathbb{C}(f,g)$; and $\mathbb{C}(f) \neq \mathbb{C}(f,g)$. In this context, recall that, $\mathbb{C}(f) = \mathbb{C}(g)$ is always true for family diploid couples.

Let us first discuss $\mathbb{C}(f) = \mathbb{C}(f,g)$. As an overview, this instance captures the diploid couples which can resist this (family) perturbation. As a piece of evidence, Fig. 2 (top) shows the dynamics of chaotic ECAs 30 and 86; moreover, their diploid couple $(30, 86)$ also generates chaotic dynamics for any mixing rate. Next, Fig. 2 (bottom) depicts the dynamics of periodic ECAs 1 and 127 where some part of the system shows blocks of blinking state (blocks of 0 or consecutive state 0's move to blocks of 1 or consecutive state 1's after every iteration, and vice versa) and these synchronized blocks are separated by kind of walls. It again shows periodic dynamics for the diploid couple $(1, 127)$. However, the diploid system $(1, 127)$ evolves towards solving the global synchronization problem (i.e. every initial configuration moves towards a homogeneous blinking state) [7,12] where the walls are destroyed by the effects of perturbation. Note that, the above situation (solving global synchronization problem) is only true for small perturbation effects considering this limited experiment, for evidence see $(1, 127)[0.1]$ in Fig. 2 (bottom). Moreover, the same applies to ECAs 3, 7, 27 considering some of their family members. Indeed, this example displays evidence of the computational ability of this diploid system. To sum up, 99 diploid couples,

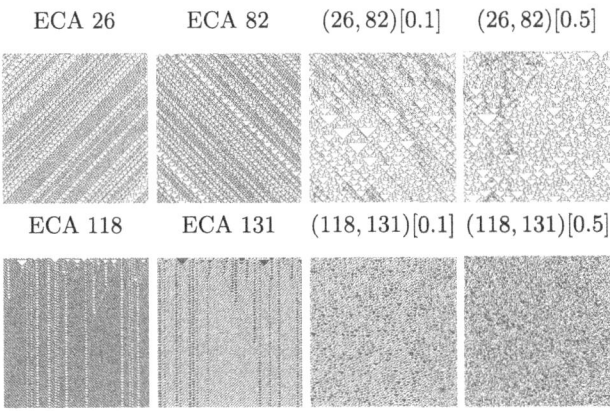

Fig. 3. $\mathbb{C}(f) \neq \mathbb{C}(f,g)$ considering family diploid couples $(26,82)$ and $(118,131)$.

out of 300 family couples, are able to resist this family perturbation, i.e. $\mathbb{C}(f) = \mathbb{C}(f,g)$, where 8, 20, 37, 2 and 32 belong to classes N, FP, P, LC, C respectively.

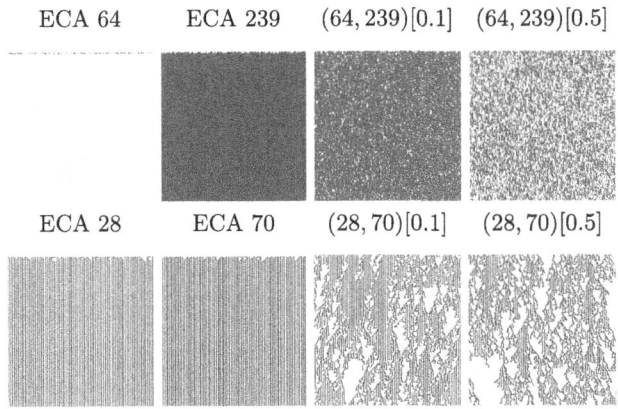

Fig. 4. (top): Noise, i.e. NO; (bottom): Periodic dynamics with more complexity, i.e. Pc behaviour of couples $(64,239)$ and $(28,70)$ respectively.

Next, we note the *"harmful"* family members, i.e. $\mathbb{C}(f) \neq \mathbb{C}(f,g)$, where noise ECA of the same family (g) massively affects the diploid system. For example, ECAs 26 and 82 show locally chaotic dynamics where the domain walls separate the chaotic dynamics. However, in Fig. 3 (top), these LC couple $(26,82)$ depict chaotic dynamics; in other words, locally chaotic noise destroys the domain walls of a locally chaotic system. Remark that, LC ECA 73 (along with family member) also reflects the same; however, LC couple $(154,210)$ shows resistance against this perturbation and remains in the LC class. Next, we discuss another extreme situation where ECAs 118 and 131 depict the P class (or simplicity). However, couple (118, 131) shows chaotic dynamics, see Fig. 3 (bottom) for example. To sum up, the above two evidences reflect the possibilities – $\mathbb{C}(f,g) = $ C where $\mathbb{C}(f) = $ LC; and $\mathbb{C}(f,g) = $ C where $\mathbb{C}(f) = $ P. Moreover, according to Table 1, following are the remaining interesting situations – $\mathbb{C}(f,g) = $ N where $\mathbb{C}(f) \in $ {FP,P}; and $\mathbb{C}(f,g) = $ P where $\mathbb{C}(f) = $ FP. Here, 20, 40, 2, 38 couples belong to classes N, P, LC, C respectively considering $\mathbb{C}(f) \neq \mathbb{C}(f,g)$ (Note that, FP = 0).

Following this, we discuss the situations where it is not possible to capture the dynamics of this stochastic system following Li and Packard's classification. Firstly, to implement the stochastic system (i.e. mixing rate λ), we use a random number generator **rand** as a source of randomness. In this context, some diploid couples show the dynamics of **rand** only, in other words, the dynamics of the system is independent of the ECA rules. We call *Noise* dynamics (NO). This dynamics is observed for trivial N (null) ECA couples. As an evidence, Fig. 4 (top) shows the NO dynamics for couple (64, 239). ECAs 0 and 8 (along with their family members) also show this trivial behaviour (the total number of couples is five). Next, we introduce *Periodic dynamics with more complexity* (Pc) class where the diploid couples show a closeness with complex phenomenon, however, we can not be able to identify them as complex/chaotic following the visual comparison. In other words, they show complex *tree* space-time with noisy *fractal* structure. As en evidence, in Fig. 4 (bottom), couple (28, 70) shows Pc class where $\mathbb{C}(28) = \mathbb{C}(70) = $ P. Out of 300 diploid family couples, 28 couples show Pc dynamics where $\mathbb{C}(f) \in $ {P,FP}. It indicates the desire of periodic and fixed point rules towards complex phenomenon with the effect of this perturbation. However, we are still open about this class's complexity (between periodic and complex) following the visual comparison.

Overall, the above situations include all possible λ-independent dynamics of the diploid system. Here, 236 couples, out of 300 family couples, are not dependent on the progressive change of the mixing rate. However, there are λ-dependent situations. The next Section depicts those possibilities.

5 λ-Dependent Dynamics: Class and Phase Transitions

There are diploid couples where a progressive change of mixing rate shows a massive impact on the system dynamics. Specifically, a situation where system shows class (say) X_1 dynamics for $\lambda \in [0, \lambda_c]$, on the other hand, for mixing rate $\lambda \in [\lambda_c, 0.5]$ the same system depicts class X_2 dynamics, where $X_1 \neq X_2$ and $X_1, X_2 \in $ {N, FP, P, LC, C, NO, Pc}, we call *class transition* dynamics. Here, the class dynamics of the diploid system changes for a critical value of mixing rate λ_c. For example, in Fig. 5 (top), (122, 161)[0.1] shows chaotic dynamics, however, the system introduces simplicity for increasing rate of perturbation, i.e. (122, 161)[0.5] depicts LC class. We denote $\mathbb{C}(122, 161) = $ {C $\rightsquigarrow$ LC}. Similarly, couple (33, 123)[0.1] shows periodic dynamics, however, perturbation adds complexity in the system where (33, 123)[0.5] reflects Pc class, $\mathbb{C}(33, 123) = $ {P $\rightsquigarrow$ Pc}. Here, 36 family couples show this class transition dynamics. Following the above two examples, the system depicts the following possibilities: $\mathbb{C}(f,g) = \{$P $\rightsquigarrow$ C$\}$; $\mathbb{C}(f,g) = \{$P $\rightsquigarrow$ LC$\}$; $\mathbb{C}(f,g) = \{$Pc $\rightsquigarrow$ C$\}$ and $\mathbb{C}(f,g) = \{$C $\rightsquigarrow$ LC$\}$.

(122, 161)[0.1] (122, 161)[0.3] (122, 161)[0.5] (33, 123)[0.1] (33, 123)[0.3] (33, 123)[0.5]

Fig. 5. Class transition dynamics of diploid couples (122, 161) and (33, 123).

Next, we introduce the *phase transition* dynamics where, after considering the changing λ parameter, it is possible to distinguish the system dynamics into two phases: firstly, the system oscillates around a non-zero/non-one density, i.e. active phase; and the system converges into all-0/all-1 configuration, i.e. passive phase. Here, the phase dynamics of the diploid system changes (active to passive, or vice-versa) for a critical value of mixing rate λ_c. Observe that, the phase transition is a special case of class transition where one class can be identified as null (N). It is one of the most interesting properties of different stochastic CA models [8,14,15] in the context of statistical physics. For example, in Fig. 6 (top), couple (132, 222)[0.1] converges to all-1 configuration (passive phase), however, the system shows non-convergent dynamics for (132, 222)[0.5] (active phase). Similarly, couple (60, 102) shows active phase for $\lambda = 0.1, 0.3$, on the other hand, the system depicts passive phase (all-0 configuration) for $\lambda = 0.5$, for evidence, see Fig. 6 (bottom). Figure 6 also depicts the profile of density d_c as a function of the mixing rate λ for couples (132, 222) and (60, 102). Here, 28 family diploid couples show this brutal phase change dynamics. Overall, if we consider both the class and phase transition dynamics, 64 couples (out of 300 family diploid couples) show λ dependent dynamics.

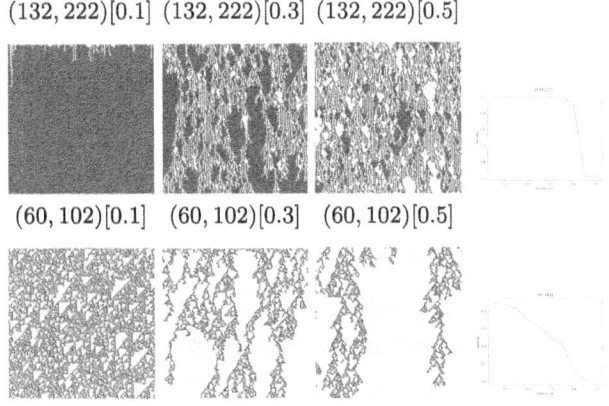

Fig. 6. Phase transition dynamics of diploid couples (132, 222) and (60, 102).

6 Conclusion

In this paper, we systematically explored the diploid ECA system, i.e. random mixing of two ECAs, where two ECAs are family members following left to right, 0 to 1, and both transformations. Following are the overall interesting observations of this study:

1. Some family diploid couples show solid resistance against this perturbation.
2. However, there are peculiar instances where noise of same family creates a massive impact on the system. Most importantly, $\mathbb{C}$(LC,LC) = C and $\mathbb{C}$(P,P) = C. In other words, two simple systems together show complex dynamics.
3. Here, we can not be able to classify few diploid couples following Wolfram's [17] and Li and Packard's [10] classification. In this direction, we have introduced new classes, Noise (LC) and Periodic dynamics with more complexity (Pc) for these systems.
4. We have identified class transition dynamics for many diploid family couples. Most importantly, $\mathbb{C}$(P,P) = {P $\leftrightsquigarrow$ C}, $\mathbb{C}$(P,P) = {Pc $\leftrightsquigarrow$ C} and $\mathbb{C}$(C,C) = {C $\leftrightsquigarrow$ LC}, i.e. class transition from simplicity to chaos, and vice versa.
5. We have also explored this simple system's (continuous or second-order) phase transition possibility.

However, we are still open to many questions following this first experimental study. The proper theoretical reasons behind brutal class and phase transition are still open to us. We have introduced a new class (Pc) to capture the dynamics of this system, however, the proper quantitative (i.e. entropy, Lyapunov exponent) understanding about this class is still open. Here, this study only includes 300 diploid couples, however, the proper understanding about the overall diploid ECA rule space with 8808 couples [8] is still open to the community. The above fascinating questions include the future direction of this study.

Acknowledgments. This work has been supported by the Danish Council for Independent Research through grant 10.46540/2032-00101B.

References

1. Arrighi, P., Schabanel, N., Theyssier, G.: Stochastic cellular automata: correlations, decidability and simulations. Fund. Inf. **126**(2–3), 121–156 (2013)
2. Bołt, W., Baetens, J.M., De Baets, B.: On the decomposition of stochastic cellular automata. J. Comput. Sci. **11**, 245–257 (2015)
3. Bołt, W., Bołt, A., Wolnik, B., Baetens, J.M., De Baets, B.: A statistical approach to the identification of diploid cellular automata based on incomplete observations. Biosystems **186**, 103976 (2019)
4. Bušić, A., Mairesse, J., Marcovici, I.: Probabilistic cellular automata, invariant measures, and perfect sampling. Adv. Appl. Probab. **45**(4), 960–980 (2013)
5. Cirillo, E.N., Nardi, F.R., Spitoni, C.: Phase transitions in random mixtures of elementary cellular automata. Phys. A **573**, 125942 (2021)
6. Culik II, K., Yu, S.: Yu, S.: Undecidability of ca classification schemes. Complex Syst. **2**, 177–190 (1988)

7. Fatès, N.: Remarks on the cellular automaton global synchronisation problem. In: Kari, J. (ed.) AUTOMATA 2015. LNCS, vol. 9099, pp. 113–126. Springer, Heidelberg (2015). https://doi.org/10.1007/978-3-662-47221-7_9
8. Fatès, N.: Diploid cellular automata: first experiments on the random mixtures of two elementary rules. In: Dennunzio, A., Formenti, E., Manzoni, L., Porreca, A.E. (eds.) AUTOMATA 2017. LNCS, vol. 10248, pp. 97–108. Springer, Cham (2017). https://doi.org/10.1007/978-3-319-58631-1_8
9. Fernández, R., Louis, P.Y., Nardi, F.R.: Overview: PCA models and issues. In: Probabilistic Cellular Automata: Theory, Applications and Future Perspectives, pp. 1–30 (2018)
10. Li, W., Packard, N.: The structure of the elementary cellular automata rule space. Complex Syst. **4**, 281–297 (1990)
11. Mairesse, J., Marcovici, I.: Around probabilistic cellular automata. Theor. Comput. Sci. **559**, 42–72 (2014)
12. Oliveira, G.M., Martins, L.G., de Carvalho, L.B., Fynn, E.: Some investigations about synchronization and density classification tasks in one-dimensional and two-dimensional cellular automata rule spaces. Electron. Notes Theor. Comput. Sci. **252**, 121–142 (2009)
13. Paul, S., Roy, S., Das, S.: Pattern classification with temporally stochastic cellular automata. In: Manzoni, L., Mariot, L., Roy Chowdhury, D. (eds.) AUTOMATA 20. LNCS, pp. 137–152. Springer, Cham (2023). https://doi.org/10.1007/978-3-031-42250-8_10
14. Roy, S.: A study on delay-sensitive cellular automata. Phys. A **515**, 600–616 (2019)
15. Roy, S., Paul, S., Das, S.: Temporally stochastic cellular automata: classes and dynamics. Int. J. Bifurcat. Chaos **32**(12), 2230029 (2022)
16. Turing, A.M.: The chemical basis of morphogenesis. Phil. Trans. Royal Soc. Lond. Ser. B Biol. Sci. **237**(641), 37–72 (1952)
17. Wolfram, S.: Cellular Automata and Complexity: Collected Papers, 1st edn. CRC Press, Boca Raton (1994)

Regional Controllability of Cellular Automata Through Preimages

Sara Dridi[1], Franco Bagnoli[2,3,4(✉)], and Samira El Yacoubi[4]

[1] University of Setif 1, Sétif, Algeria
sara.dridi@univ-setif.dz
[2] Department of Physics and Astronomy and CSDC, University of Florence, via G. Sansone 1, 50019 Sesto Fiorentino, Italy
franco.bagnoli@unifi.it
[3] INFN, Florence sect., Florence, Italy
[4] Team Project IMAGES - UMR ESPACE-Dev, University of Perpignan, Perpignan, France
yacoubi@univ-perp.fr

Abstract. We investigate a regional controllability problem applied to elementary Cellular Automata (CA). We first examine the conditions for boundary control, showing that, at least for small lattice sizes, only peripherally linear or affine CA can be fully controllable. Exploiting linearity, it is possible to develop an algorithm to construct the tree of preimages of a given configuration, therefore explicitly finding the optimal control for any given configuration. We apply then this method to non-linear CA.

Keywords: Boundary controls · regional controllability · non-linear elementary cellular automata · preimages

1 Introduction

The controllability problem is a fundamental concept in control theory, a branch of engineering and mathematics concerned with influencing the behavior of dynamical systems. In essence, it deals with the question of whether a given system can be driven from one state to another using a suitable control input within a specified time frame. A wide literature has been devoted to the control of dynamical systems with a continuous variables and continuous time evolution in both finite and infinite dimensional cases by means of differential or partial differential equations [1,2].

Cellular Automata (CA for short) are spatially extended systems that are widely used as effective computational tools for simulating complex systems [3].

Our general aim is to develop specific techniques for studying control problems on spatially extended systems for which the classical approaches cannot be used in general.

This paper addresses a classical problem related to the so-called regional controllability, introduced by Zerrik et al. [4], as a special case of output controllability [5]. The regional control problem consists in achieving an objective only on a subregion of the domain when some specific actions are exerted on the system, in its domain interior or on its boundaries.

Some regional control problems have been already studied using CA, see for instance Ref. [6–11]. In Refs. [11], a new approach has been proposed using Markov chains. Another approach based on graph theory notions was also investigated in Ref. [12]. The regional control problem has been studied on deterministic CA [9], using the concept of Boolean derivatives for deterministic one-dimensional CA. The problem of regional controllability and observability have been dealt using Kalman condition [6,13].

In this paper, we explore the problem of regional controllability of one-dimensional deterministic CA via boundary actions. More precisely, it consists of forcing the appearance of a given pattern in a region ω by applying the control on a part of its boundary. We concentrate on a special class of Boolean non-linear CA that may exhibit very rich and interesting behaviours despite their apparent simplicity.

A preimage algorithm to decide whether a desired configuration is reachable starting from an initial configuration is reported using a characterisation tool called the Controllability Tree. This paper extends the results of Ref. [13], introducing the algoithm for generating preimages.

The paper is organized as follows. In Sect. 2, we present definitions of classical elementary CA and give some of their important proprieties. Section 3 is devoted to Boolean derivatives of CA. The regional controllability problem is presented in Sect. 4, and the preimage method is presented in Sect. 5. The last section is devoted to conclusions and perspectives.

2 Elementary Cellular Automata

An elementary Cellular Automata (ECA) is formed by a one-dimensional lattice of L cells, each of which interacts with its two nearest neighbors. The state of each cell $i, \in \{1, \ldots, L\}$ at time t is given by a a Boolean variable $s_i^t \in \{0, 1\}$.

The evolution of cells is defined in terms of a local function f

$$x_i' \equiv x_i^{t+1} = f(\mathcal{N}_i^t),$$

where $\mathcal{N}_i^t = (x_{i-1}^t, \ldots, x_i^t, \ldots, x_{i+1}^t)$ is the state of the neighbourhood of cell i at time t.

We can also introduce the whole configuration $X = (x_1, \ldots, x_L)$ and introduce the function F such that

$$X' = F(lXr)$$

given the boundary conditions l and r. For instance $l = x_L$ and $r = x_1$ for periodic conditions.

Table 1. The look-up table of rules 150 and 22. $\mathcal{N}$ is the neighborhood in base-2 representation, $\mathcal{N}^{(10)}$ is the same in base-ten representation. The column $f^{(150)}$ is the output of rule 150 and $D_{\mathcal{N}}^{(150)}$ the derivative of the rule 150 in zero with respect to the ones in the neighborhood in base-two representation, same for $f^{(22)}$ and $D^{(22)}$ for rule 22.

$\mathcal{N} = (x_{-1}, x_0, x_1)$	$\mathcal{N}^{(10)} = 4x_{-1} + 2x_0 + x_1$	$f^{(150)}(\mathcal{N})$	$D_{\mathcal{N}}^{(150)}$	$f^{(22)}(\mathcal{N})$	$D_{\mathcal{N}}^{(22)}$
0, 0, 0	0	0	0	0	0
0, 0, 1	1	1	1	1	1
0, 1, 0	2	1	1	1	1
0, 1, 1	3	0	0	0	0
1, 0, 0	4	1	1	1	1
1, 0, 1	5	0	0	0	0
1, 1, 0	6	0	0	0	0
1, 1, 1	7	1	0	0	1

By relabeling cells, it is possible to express the neighborhood as $\mathcal{N}_i^t = (x_i^t, x_{i+1}^t, x_{i+2}^t)$, which is more practical for describing algorithms, as will be done in Sect. 5.

There are $256 = 2^{2^3}$ different ECA rules. Each of them can be defined by the array of values it takes for every configuration of the neighborhood, from $(0, 0, 0) = 0$ to $(1, 1, 1) = 7$. Reading the resulting array as a number in base-two representation, one gets that Wolfram code for the rule [14,15]. As an example, rule 150 corresponds to the array $1, 0, 0, 1, 0, 1, 1, 0$ which is equal to 150 in base-10, see Table 1.

One can apply some transformations on the rule that do not alter their behavior, such as left-right inversion and one-zero exchange. There are 88 unique rules after the application of these symmetries, called *minimal* CA. These are rules 0, 1, 2, 3, 4, 5, 6, 7, 8, 9, 10, 11, 12, 13, 14, 15, 18, 19, 22, 23, 24, 25, 26, 27, 28, 29, 30, 32, 33, 34, 35, 36, 37, 38, 40, 41, 42, 43, 44, 45, 46, 50, 51, 54, 56, 57, 58, 60, 62, 72, 73, 74, 76, 77, 78, 90, 94, 104, 105, 106, 108, 110, 122, 126, 128, 130, 132, 134, 136, 138, 140, 142, 146, 150, 152, 154, 156, 160, 162, 164, 168, 170, 172, 178, 184, 200, 204 and 232.

3 Boolean Derivatives and Linearity of ECA

It is possible to define the derivative of a Boolean function $f(x, y)$ [16] as

$$\frac{\partial f}{\partial x} = f(x \oplus 1, y) \oplus f(x, y)$$

which obeys many standard properties of derivatives line chain rule [17]. It is possible also to define higher-order derivatives.

By means of Boolean derivatives in zero one can obtain the Ring Sum Expansion [18] of a function

$$f(x_0, x_1, x_2) = D_0 \oplus D_1 x_0 \oplus D_2 x_1 \oplus D_4 x_2 \oplus D_3 x_0 x_1 \oplus \\ D_5 x_0 x_2 \oplus D_6 x_1 x_2 \oplus D_7 x_0 x_1 x_2, \quad (1)$$

where D_i is the derivative of f with respect to the bits that have value 1 in the binary representation of i and the symbol $\oplus$ stands for the sum modulo two or exclusive or.

For instance

$$D_5 = D_{1,0,1} = \frac{\partial^2 f}{\partial x_0 \partial x_2}(0) = f(1,0,1) \oplus f(1,0,0) \oplus f(0,0,1) \oplus f(0,0,0).$$

A rule which can be expressed only using constant ($f(0,0,0)$) and first-order derivatives is called affine, such as for instance rule 150, which can be expressed as

$$f^{(150)}(s_-, s, s_+) = s_- \oplus s \oplus s_+,$$

while rule 22, although quite similar to rule 150, is not affine

$$f^{(22)}(s_-, s, s_+) = s_- \oplus s \oplus s_+ \oplus s_- s s_+.$$

See also Table 1.

In other words, affine CA have higher-order derivatives D_3, D_5, D_6 and D_7 equal to zero

The Boolean derivatives in zero and the affinity of all minimal ECA are reported in Table 2.

Among ECA, there are 16 affine Boolean transition functions, e.g., rules 0, 60, 90, 102, 150, 170, 204, 240, 15, 51, 85, 105, 153, 165, 195, 255. All the remaining rule numbers represent nonlinear ECA with various degrees of deviation from linearity.

Another property which will be useful in the following is that of linearity with respect to the periphery of the neighborhood, which for ECA implies

$$s' = s_- \oplus g(s, s_+) \quad \text{or} \quad s' = g(s_-, s) \oplus s_+.$$

In term of Boolean derivatives, a rule is peripherally linear if $D_1 = 1$ and $D_3 = D_5 = D_7 = 0$ or $D_4 = 1$ and $D_5 = D_6 = D_7 = 0$. We shall denote these rules as peripherally linear CA, and for ECA these are rules 15, 30, 45, 60, 90, 105, 106, 150, 154, 170, as reported in Table 2.

Table 2. Minimal ECA properties for $L = 13$. R is the rule number, F is the look-up table and D are the derivatives in zero. Notice that the neighborhood configurations have been grouped according with the number of ones. A is a flag for affine rules, P is a flag for peripherally linear rules and h is the fraction of reachable configurations, for different times T.

R	f(N)								D_i								A	P	h(T)			
	0	1	2	4	3	5	6	7	0	1	2	4	3	5	6	7			52	53	54	55
0	0	0	0	0	0	0	0	0	0	0	0	0	0	0	0	0	0	0	0.000122	0.000122	0.000122	0.000122
1	1	0	0	0	0	0	0	0	1	1	1	1	1	1	1	0	0	0	0.000488	0.000231	0.000488	0.000231
2	0	1	0	0	0	0	0	0	0	1	0	0	1	1	0	1	0	0	0.023071	0.023071	0.023071	0.023071
3	1	1	0	0	0	0	0	0	1	0	1	1	0	0	1	0	0	0	0.231567	0.231567	0.231567	0.231567
4	0	0	1	0	0	0	0	0	0	0	1	0	1	0	1	1	0	0	0.000191	0.000191	0.000191	0.000191
5	1	0	1	0	0	0	0	0	1	1	0	1	0	1	0	0	0	0	0.000584	0.000348	0.000584	0.000348
6	0	1	1	0	0	0	0	0	0	1	1	0	0	1	1	0	0	0	0.171387	0.171387	0.171387	0.171387
7	1	1	1	0	0	0	0	0	1	0	0	1	1	0	0	1	0	0	0.069929	0.069913	0.069929	0.069913
8	0	0	0	0	1	0	0	0	0	0	0	0	1	0	0	1	0	0	0.000153	0.000153	0.000153	0.000153
9	1	0	0	0	1	0	0	0	1	1	1	1	0	1	1	0	0	0	0.134225	0.134232	0.134194	0.134242
10	0	1	0	0	1	0	0	0	0	1	0	0	0	1	0	0	0	0	0.087158	0.087158	0.087158	0.087158
11	1	1	0	0	1	0	0	0	1	0	1	1	1	0	1	1	0	0	0.144287	0.144287	0.144287	0.144287
12	0	0	1	0	1	0	0	0	0	1	0	0	0	1	0	0	0	0	0.000183	0.000183	0.000183	0.000183
13	1	0	1	0	1	0	0	0	1	1	0	1	1	1	0	1	0	0	0.000519	0.000514	0.000519	0.000514
14	0	1	1	0	1	0	0	0	0	1	1	0	1	1	1	1	0	0	0.195312	0.195312	0.195312	0.195312
15	1	1	1	0	1	0	0	0	1	0	0	1	0	0	0	0	1	1	1.000000	1.000000	1.000000	1.000000
18	0	1	0	1	0	0	0	0	0	1	0	1	1	0	1	0	0	0	0.237764	0.237764	0.237764	0.237764
19	1	1	0	1	0	0	0	0	1	0	1	0	0	1	0	1	0	0	0.000231	0.000231	0.000231	0.000231
22	0	1	1	1	0	0	0	0	0	1	1	1	0	0	0	1	0	0	0.701550	0.701550	0.701550	0.701550
23	1	1	1	1	0	0	0	0	1	0	0	0	1	1	1	0	0	0	0.000341	0.000341	0.000341	0.000341
24	0	0	0	1	1	0	0	0	0	0	0	1	1	1	1	0	0	0	0.026489	0.026489	0.026489	0.026489
25	1	0	0	1	1	0	0	0	1	1	1	0	0	0	0	1	0	0	0.407975	0.407975	0.407975	0.407975
26	0	1	0	1	1	0	0	0	0	1	0	1	0	0	1	1	0	0	0.608887	0.608887	0.608887	0.608887
27	1	1	0	1	1	0	0	0	1	0	1	0	1	1	0	0	0	0	0.357666	0.357666	0.357666	0.357666
28	0	0	1	1	1	0	0	0	0	1	1	0	1	0	1	0	0	0	0.000764	0.000764	0.000764	0.000764
29	1	0	1	1	1	0	0	0	1	1	0	0	1	0	1	0	0	0	0.000488	0.000458	0.000488	0.000458
30	0	1	1	1	1	0	0	0	0	1	1	1	0	0	0	0	0	1	1.000000	1.000000	1.000000	1.000000
32	0	0	0	0	0	1	0	0	0	0	0	0	0	1	0	1	0	0	0.000123	0.000123	0.000123	0.000123
33	1	0	0	0	0	1	0	0	1	1	1	1	1	0	1	0	0	0	0.001470	0.001318	0.001470	0.001318
34	0	1	0	0	0	1	0	0	0	1	0	0	1	0	0	0	0	0	0.074463	0.074463	0.074463	0.074463
35	1	1	0	0	0	1	0	0	1	0	1	1	0	1	1	1	0	0	0.351318	0.351318	0.351318	0.351318
36	0	0	1	0	0	1	0	0	0	0	1	0	1	1	1	0	0	0	0.000322	0.000322	0.000322	0.000322
37	1	0	1	0	0	1	0	0	1	1	0	1	0	0	0	1	0	0	0.756966	0.756966	0.756966	0.756966
38	0	1	1	0	0	1	0	0	0	1	1	0	0	0	1	1	0	0	0.202148	0.202148	0.202148	0.202148
40	0	0	0	0	1	1	0	0	0	0	0	0	1	1	0	0	0	0	0.000503	0.000503	0.000503	0.000503
41	1	0	0	0	1	1	0	0	1	1	1	1	0	0	1	1	0	0	0.608521	0.608521	0.608521	0.608521
42	0	1	0	0	1	1	0	0	0	0	0	0	0	1	0	0	0	0	0.382812	0.382812	0.382812	0.382812
43	1	1	0	0	1	1	0	0	1	0	1	1	1	1	1	0	0	0	0.376221	0.376221	0.376221	0.376221
44	0	0	1	0	1	1	0	0	0	1	0	0	1	1	1	0	0	0	0.000488	0.000488	0.000488	0.000488
45	1	0	1	0	1	1	0	0	1	1	0	1	1	0	0	0	0	1	1.000000	1.000000	1.000000	1.000000
46	0	1	1	0	1	1	0	0	0	1	1	0	1	0	1	0	0	0	0.038818	0.038818	0.038818	0.038818
50	0	1	0	1	0	1	0	0	0	1	0	1	1	1	1	0	0	0	0.000341	0.000341	0.000341	0.000341
51	1	1	0	1	0	1	0	0	1	0	1	0	0	0	0	1	1	0	0.000122	0.000122	0.000122	0.000122

(continued)

Table 2. (*continued*)

R	f(N)								D_i								A	P	h(T)			
	0	1	2	4	3	5	6	7	0	1	2	4	3	5	6	7			52	53	54	55
54	0	1	1	1	0	1	0	0	0	1	1	1	0	1	0	0	0	0	0.380138	0.380137	0.380138	0.380137
56	0	0	0	1	1	1	0	0	0	0	0	1	1	0	1	1	0	0	0.142944	0.142944	0.142944	0.142944
57	1	0	0	1	1	1	0	0	1	1	1	0	0	1	0	0	0	0	0.030770	0.030770	0.030770	0.030770
58	0	1	0	1	1	1	0	0	0	1	0	1	0	1	1	0	0	0	0.066904	0.066904	0.066904	0.066904
60	0	0	1	1	1	1	0	0	0	0	1	1	0	0	0	0	1	1	1.000000	1.000000	1.000000	1.000000
62	0	1	1	1	1	1	0	0	0	1	1	1	1	0	1	0	0	0	0.195884	0.197244	0.198145	0.198999
72	0	0	0	0	1	0	1	0	0	0	0	0	1	0	1	0	0	0	0.000322	0.000322	0.000322	0.000322
73	1	0	0	0	1	0	1	0	1	1	1	1	0	1	0	1	0	0	0.158539	0.158519	0.158534	0.158539
74	0	1	0	0	1	0	1	0	0	1	0	0	0	1	1	1	0	0	0.354492	0.354492	0.354492	0.354492
76	0	0	1	0	1	0	1	0	0	0	1	0	0	0	0	1	0	0	0.000191	0.000191	0.000191	0.000191
77	1	0	1	0	1	0	1	0	1	1	0	1	1	1	1	0	0	0	0.000341	0.000341	0.000341	0.000341
78	0	1	1	0	1	0	1	0	0	1	1	0	1	1	0	0	0	0	0.000769	0.000769	0.000769	0.000769
90	0	1	0	1	1	0	1	0	0	1	0	1	0	0	0	0	1	1	1.000000	1.000000	1.000000	1.000000
94	0	1	1	1	1	0	1	0	0	1	1	1	1	0	1	1	0	0	0.052288	0.052250	0.052194	0.052260
104	0	0	0	0	1	1	1	0	0	0	0	0	1	1	1	1	0	0	0.002035	0.002035	0.002035	0.002035
105	1	0	0	0	1	1	1	0	1	1	1	1	0	0	0	0	1	1	1.000000	1.000000	1.000000	1.000000
106	0	1	0	0	1	1	1	0	0	1	0	0	0	0	1	0	0	1	1.000000	1.000000	1.000000	1.000000
108	0	0	1	0	1	1	1	0	0	0	1	0	0	1	0	0	0	0	0.000519	0.000519	0.000519	0.000519
110	0	1	1	0	1	1	1	0	0	1	1	0	1	0	0	1	0	0	0.621095	0.621095	0.621095	0.621095
122	0	1	0	1	1	1	1	0	0	1	0	1	0	1	0	1	0	0	0.507102	0.507101	0.507102	0.507101
126	0	1	1	1	1	1	1	0	0	1	1	1	1	1	1	0	0	0	0.183428	0.183425	0.183428	0.183425
128	0	0	0	0	0	0	0	1	0	0	0	0	0	0	0	1	0	0	0.000123	0.000123	0.000123	0.000123
130	0	1	0	0	0	0	0	1	0	1	0	0	1	1	0	0	0	0	0.023082	0.023082	0.023082	0.023082
132	0	0	1	0	0	0	0	1	0	0	1	0	1	0	1	0	0	0	0.000340	0.000340	0.000340	0.000340
134	0	1	1	0	0	0	0	1	0	1	1	0	0	1	1	1	0	0	0.209128	0.209128	0.209128	0.209128
136	0	0	0	0	1	0	0	1	0	0	0	0	1	0	0	0	0	0	0.000244	0.000244	0.000244	0.000244
138	0	1	0	0	1	0	0	1	0	1	0	0	0	1	0	1	0	0	0.231567	0.231567	0.231567	0.231567
140	0	0	1	0	1	0	0	1	0	0	1	0	0	0	1	1	0	0	0.000274	0.000274	0.000274	0.000274
142	0	1	1	0	1	0	0	1	0	1	1	0	1	1	1	0	0	0	0.376221	0.376221	0.376221	0.376221
146	0	1	0	1	0	0	0	1	0	1	0	1	1	0	1	1	0	0	0.237766	0.237764	0.237766	0.237764
150	0	1	1	1	0	0	0	1	0	1	1	1	0	0	0	0	1	1	1.000000	1.000000	1.000000	1.000000
152	0	0	0	1	1	0	0	1	0	0	0	1	1	1	1	1	0	0	0.029779	0.029779	0.029779	0.029779
154	0	1	0	1	1	0	0	1	0	1	0	1	0	0	1	0	0	1	1.000000	1.000000	1.000000	1.000000
156	0	0	1	1	1	0	0	1	0	0	1	1	0	1	0	0	0	0	0.000714	0.000714	0.000714	0.000714
160	0	0	0	0	0	1	0	1	0	0	0	0	0	1	0	0	0	0	0.000198	0.000204	0.000198	0.000204
162	0	1	0	0	0	1	0	1	0	1	0	0	1	0	0	1	0	0	0.074477	0.074477	0.074477	0.074477
164	0	0	1	0	0	1	0	1	0	0	1	0	1	1	1	1	0	0	0.018845	0.018845	0.018845	0.018845
168	0	0	0	0	1	1	0	1	0	0	0	0	1	1	0	1	0	0	0.016104	0.016104	0.016104	0.016104
170	0	1	0	0	1	1	0	1	0	1	0	1	0	0	0	0	1	1	1.000000	1.000000	1.000000	1.000000
172	0	0	1	0	1	1	0	1	0	0	1	0	0	1	1	0	0	0	0.005635	0.005635	0.005635	0.005635
178	0	1	0	1	0	1	0	1	0	1	0	1	1	1	1	0	0	0	0.000341	0.000341	0.000341	0.000341
184	0	0	0	1	1	1	0	1	0	0	0	1	1	0	1	0	0	0	0.324951	0.324951	0.324951	0.324951
200	0	0	0	0	1	0	1	1	0	0	0	0	1	0	1	1	0	0	0.000191	0.000191	0.000191	0.000191
204	0	0	1	0	1	0	1	1	0	0	1	0	0	0	0	0	1	0	0.000122	0.000122	0.000122	0.000122
232	0	0	0	0	1	1	1	1	0	0	0	0	1	1	1	0	0	0	0.000341	0.000341	0.000341	0.000341

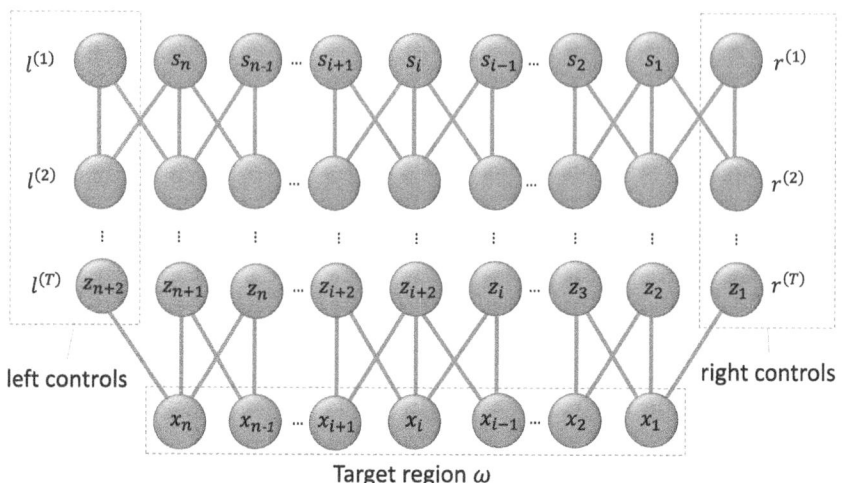

Fig. 1. Regional controllability of one-dimensional CA, via boundary actions.

4 Boundary Regional Control Problem

Let us consider a Boolean CA defined on a lattice $\mathcal{L} = \{l, x_1, \ldots, x_n, r\}$ that is assumed to be finite and composed of n interior cells and 2 boundary cells (l, r), which are the left and right control cells respectively.

Let $\omega = \{x_1, \ldots, x_n\}$ be the subset of the lattice denoted the target region. We are interested in finding the suitable sequences of controls acting on the boundary of the lattice, $(l^1, l^2, \ldots, l^T)$ and $(r^1, r^2, \ldots, r^T)$, in order to drive the system from a given initial state $S^0 = S$ to a desired configuration $S^T = X$ on the target region ω in T steps, see Fig. 1.

The problem of regional controllability was investigated in Ref. [8] and [10]. The degree of controllability of a rule f for a given size L of the target region can be computed in the following way. Let us indicate the state of a lattice of size n plus left and right controls as $Z = (z_{n+1}, z_n, \ldots, z_1, z_0) = (lYr)$, where $l = z_{n+1}$ and $r = z_0$ are the boundary controls and $Y = (z_n, z_{n-1} \ldots, z_2, z_1)$ is the target region. Let us apply the rule f to each site, generating the configuration $X = (x_n, \ldots, x_1)$ such that

$$x_i = f(z_{i+1}, z_i, z_{i-1}).$$

We shall denote compactly

$$X = F(Z) = F(lYr).$$

The configurations Z, Y and X can be read as base-two representation of numbers (this is the reason for the inverse-ordering of sites) and in the following we shall refer to them as numbers.

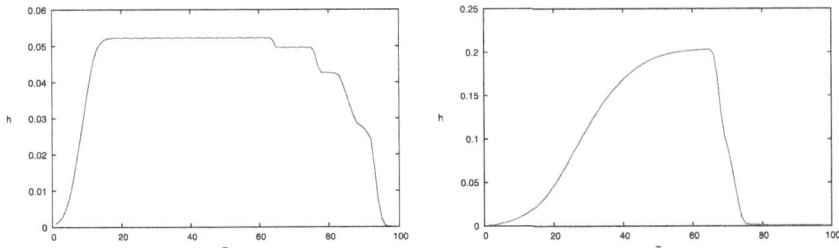

Fig. 2. Fraction of controllable configuration h with respect to time T for rule 94 (left) and 62 (right)

Let us now build the matrix $M_{XY} = M(X|Y)$ by summing over l and r, for each given X and Y,

$$M(X|Y) = \sum_{l=0,1} \sum_{r=0,1} [X = F(lYr)],$$

where $[\cdot] = 1$ if $\cdot$ is true and zero otherwise. $M(X|Y) = 0$ is zero if X cannot be reached in one step starting from Y for any value of l and r, while it is 4 if Y always gives X regardless of the control.

For instance, for rule 0, $M(0|Y) = 4$ since, for any values of l and r, $F(lYr) = 0$. For any other value of X, $M(X|Y) = 0$ since $F(lYr) = 0 \neq X$.

$MM(X|Y)$ can be seen as the number of ways of going from Y to X in one step given all possible boundary controls. The element of $(M^T)_{XY}$ therefore gives the number of ways of going from Y to X in T steps given all possible boundary controls at each of the T time steps.

We can define the fraction $h(T)$ of pairs X, Y for which there is at least one control sequence in T steps, i.e., such that $(M^T)_{XY} > 0$, as

$$h(T) = \frac{1}{2^{2L}} \sum_{XY} [(M^T)_{XY} > 0].$$

The values of $h(T)$ are reported in Table 2 for all minimal CA for $L = 13$ and some time T.

As noticed in Ref. [13] it can be seen that only peripherally linear cellular automata can always be boundary controlled, i.e., they have $h = 1$.

Notice that the controllability may change with L and T. We have chosen L prime so to avoid symmetries in the configuration, and reported large consecutive values of T so to put into evidence both the presence of an asymptotic state and the possibility that the control may be differently effective for different times T.

In particular, we see that most rules reach an asymptotic state except rule 1, 5, 13, 33, 54, 73, 122, 126, 146 and 160 which oscillates from odd to even times, and rules 62 and 94 which have not reached an asymptotic state for $T = 55$.

As shown in Fig. 2, Rule 94 shows a quick increase in h with T up to time $T = 20$, followed by a slow decrease. Rule 62 shows a sharp increase after an

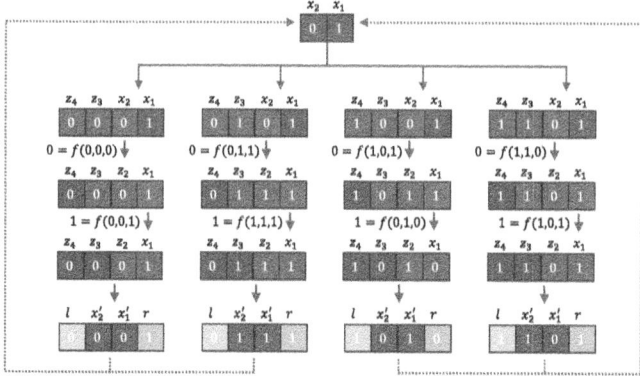

Fig. 3. One step of the generation of the preimages (control tree) of configuration 01 for rule 150. Since rule 150 is (doubly) peripherally linear, there is no failures nor additional forking.

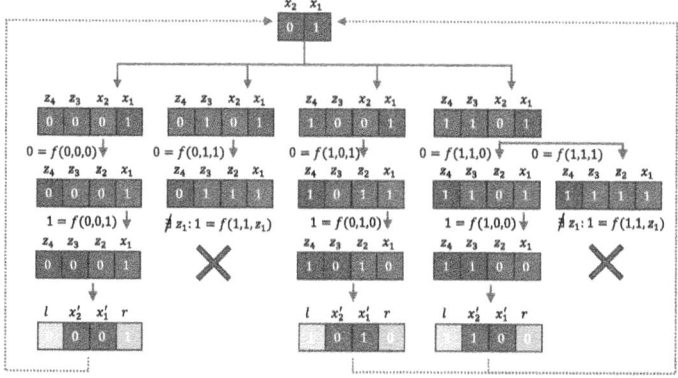

Fig. 4. One step of the generation of the preimages (control tree) of configuration 01 for rule 22. Since rule 22 is not peripherally linear, some configurations do not have preimages and others have more than one preimage.

initial slow growth. Both plots decrease quickly after $T = 64$ when presumably numerical errors in matrix multiplication start to appear.

5 Control by Preimages

As seen and as already reported in Ref. [13], only peripherally linear cellular automata can always be boundary controlled.

The idea is that for peripherally linear rules, a flip of a cell on the boundary always induces a flip of the future value of the computed cell, i.e. considering a left-linear rule

$$s'_i = s_{i-1} \oplus g(s_i, s_{i+1}) \tag{2}$$

replacing s_{i-1} with $s_{i-1} \oplus 1$ implies that s'_i is replaced by $s'_i \oplus 1$. The chain of flipping travels at the speed of light (i.e., a cell at a time), and therefore one can start from a given configuration of size n and let it evolve for a time $T = n + 1$. Then if the state of the rightmost cell is not the desired one, one has to flip the left control l^0. Then one can repeat it for the next-leftmost site, acting on l^1 and so on.

Therefore peripherally linear rules can be unconditionally driven acting on boundaries: for any given configuration one can find a sequence of boundary controls that can drive the system to a desired target configuration.

If the rule is doubly peripherally linear, one can drive the system within half time. For ECA, peripherally linear rule are affine rule, but the reverse is not true: rule 0, 51, 204 and 255 are affine but are not peripherally linear, see Table 2.

As we shall see, rules that are not peripherally linear cannot be unconditionally driven, for any given configuration there may be some target configuration that cannot be reached for any boundary control sequence.

However, even for peripherally linear CA, the task of finding the control sequence implies recalculating the evolution of the automata for each cell not in the desired state in the target configuration.

An alternative approach is that of (conceptually) computing the tree of all preimages of the target configuration, backtracking if the result (i.e., the starting configuration) is not the actual one.

This can be illustrated considering that Eq. (2) can be inverted

$$x_- = x' \oplus g(x, x_+)$$

which implies that one can start from the target configuration and build the preimages until the required start configuration is found.

For instance, for rule 150

$$x' = x_- \oplus x \oplus x_+ \quad \Rightarrow \quad x_- = x' \oplus \oplus x \oplus x_+,$$

while for rule 22

$$x' = x_- \oplus x \oplus x_+ \oplus x_- x x_+ \quad \Rightarrow \quad x_-(1 \oplus x x_+) = x' \oplus x \oplus x_+,$$

and only if $1 \oplus x x_+ = 1$ one can obtain x_-.

One has to explore the tree of all possible preimages, where the free options are the values of boundaries, and, for non-peripherally linear rules, the entries that give the same value for the same partial neighborhood, as we shall explain in the following. Each branch of the tree requires a "forking" of the algorithm.

The exploration of the tree can be done breadth-first, which is more convenient for checking the reachability of the desired starting configuration but requires keeping memory of intermediate configuration when forking, or depth-first, which is simpler to implement as a recursive function, but implies establishing a limit. These approaches can also be mixed, by repeating the depth-first exploration with increasing limit.

Let us denote by $x_n, x_{n-1}, \ldots, x_1$ the target configuration and set $t = 0$.

We start generating all possible values (and therefore forking) of two extra sites z_{n+2} and z_{n+1}, obtaining $z_{n+2}, z_{n+1}, x_n, \ldots, x_1$.

Then, for $i = n$ up to $i = 1$, find x_i such that $f(z_{i+2}, z_{i+1}, z_i) = x_i$. This can be done scanning the look-up table for non-peripherally linear rules, forking if more than one entries are found or returning is no entry is found, or using Eq. (2) for peripherally linear CA. The value z_i is inserted in place of x_i.

After n steps we have $z_{n+2}, z_{n+1}, z_n, \ldots, z_2, z_1$. We store the extreme values as controls, $l^{(t)} = z_{n+2}$ and $r^{(t)} = z_1$, rename $z_{n+1}, z_n, \ldots, z_2$ as $x_n, x_{n-1}, \ldots, x_1$, increase t and repeat the computation until either the start configuration is found or the time limit is reached.

Instead of scanning the whole look-up table (which can be costly for large neighborhood), one can take profit of the ring sum expansion, Eq. (1)

$$x_i = D_0 \oplus D_1 z_{i-1} \oplus D_2 z_i \oplus D_4 z_{i+1} \oplus D_3 z_{i-1} z_i \oplus$$
$$D_5 z_{i-1} z_{i+1} \oplus D_6 z_i z_{i+1} \oplus D_7 z_{i-1} z_i z_{i+1},$$

extracting

$$z_{i-1}(D_1 \oplus D_3 z_i \oplus D_5 z_{i+1} \oplus D_7 z_i z_{i+1}) =$$
$$x_i \oplus D_0 \oplus D_2 z_i \oplus D_4 z_{i+1} \oplus D_6 z_i z_{i+1},$$

as illustrated above for rule 22.

If $(D_1 \oplus D_3 z_i \oplus D_5 z_{i+1} \oplus D_7 z_i z_{i+1}) = 1$ then $z_{i-1} = x_i \oplus D_0 \oplus D_2 z_i \oplus D_4 z_{i+1} \oplus D_6 z_i z_{i+1}$, otherwise, if $x_i = D_0 \oplus D_2 z_i \oplus D_4 z_{i+1} \oplus D_6 z_i z_{i+1}$, one has to fork for $z_{i-1} = 0, 1$, else the recursion fails.

An example of this procedure is reported in Fig. 3 for a peripherally linear rule (rule 150) and in Fig. 4 for a nonlinear rule (rule 22).

6 Conclusions

We have investigated the application of regional controllability problem to elementary cellular automata (CA). We have shown that only peripherally linear or affine CA can be fully controllable, while other rules can only partially controlled, meaning that only a fraction of configurations can be driven to any desired target.

Peripherally linear CA can be driven by building the preimages of the target configuration, through a backtracking algorithm. We can apply this technique also to nonlinear CA, with additional forking phases and occasional failures of recursion.

The proposed algorithm can be applied to cellular automata with a larger range, which will be the subject of future investigations.

References

1. Lions, J.L.: Exact controllability for distributed systems: some trends and some problems. In: Applied and Industrial Mathematics, pp. 59–84. Springer, Heidelberg (1991). https://doi.org/10.1007/978-94-009-1908-2_7.

2. Curtain, R.F., Zwart, H.: An Introduction to Infinite-Dimensional Linear Systems Theory, vol. 21. Springer, Heidelberg (2012). https://doi.org/10.1007/978-1-4612-4224-6
3. Various editors. ed. Cellular Automata. https://link.springer.com/conference/acri. Lecture notes in Computer Science series (2002–2022)
4. Zerrik, E., Boutoulout, A., Jai, A.E.: Actuators and regional boundary controllability of parabolic systems. Int. J. Syst. Sci. **31**(1), 73–82 (2000). https://doi.org/10.1080/002077200291479
5. Russell, D.L.: Controllability and stabilizability theory for linear partial differential equations: recent progress and open questions. Siam Rev. **20**(4), 639–739 (1978). https://doi.org/10.1137/1020095
6. El Yacoubi, S., et al.: Some control and observation issues in cellular automata. Complex Syst. **30**(3), 391–413 (2021). https://doi.org/10.25088/ComplexSystems.30.3.391
7. Bagnoli, F., Rechtman, R., El Yacoubi, S.: Control of cellular automata. Phys. Rev. E **86**(6), 066201 (2012). https://doi.org/10.1103/PhysRevE.86.066201
8. Bagnoli, F., Dridi, S., El Yacoubi, S., Rechtman, R.: Regional control of probabilistic cellular automata. In: Mauri, G., El Yacoubi, S., Dennunzio, A., Nishinari, K., Manzoni, L. (eds.) ACRI 2018. LNCS, vol. 11115, pp. 243–254. Springer, Cham (2018). https://doi.org/10.1007/978-3-319-99813-8_22
9. Bagnoli, F., El Yacoubi, S., Rechtman, R.: Toward a boundary regional control problem for Boolean cellular automata. Nat. Comput. **17**, 479–486 (2018). https://doi.org/10.1007/s11047-017-9626-1
10. Bagnoli, F., et al.: Optimal and suboptimal regional control of probabilistic cellular automata. Nat. Comput. **18**, 845–853 (2019). https://doi.org/10.1007/s11047-019-09763-5
11. Dridi, S., Yacoubi, S.E., Bagnoli, F.: Boundary regional controllability of linear Boolean cellular automata using Markov chain. In: Zerrik, E.H., Melliani, S., Castillo, O. (eds.) Recent Advances in Modeling, Analysis and Systems Control: Theoretical Aspects and Applications. SSDC, vol. 243, pp. 37–48. Springer, Cham (2020). https://doi.org/10.1007/978-3-030-26149-8_4
12. Dridi, S., et al.: A graph theory approach for regional controllability of Boolean cellular automata. Int. J. Parallel Emerg. Distrib. Syst. **35**(5), 499–513 (2020). https://doi.org/10.1080/17445760.2019.1608442
13. Dridi, S., El Yacoubi, S., Bagnoli, F.: Kalman condition and new algorithm approach for regional controllability of peripherally-linear elementary cellular automata via boundary actions. J. Cell. Autom. **16**(3–4), 173–195 (2022)
14. Wolfram, S.: Statistical mechanics of cellular automata. Rev. Mod. Phys. **55**(3), 601 (1983). https://doi.org/10.1103/revmodphys.55.601
15. Wolfram, S., Gad-el-Hak, M.: A new kind of science. Appl. Mech. Rev. **56**(2), B18–B19 (2003)
16. Vichniac, G.Y.: Boolean derivatives on cellular automata. Physica D: Nonlinear Phenomena 45(13), 63–74 (1990). ISSN: 0167-2789. https://doi.org/10.1016/0167-2789(90)90174-n
17. Bagnoli, F.: Boolean derivatives and computation of cellular automata. Int. J. Mod. Phys. C **3**(02), 307–320 (1992). https://doi.org/10.1142/s0129183192000257
18. Wegener, I.: The complexity of Boolean functions. In: Applicable Theory in Computer Science. Vieweg & Teubner, Wiesbaden (1987). isbn: 3-519-02107-2

Pattern Formation by Collective Behavior of Competing Cellular Automata-Based Agents

Miroslaw Szaban[1(✉)], Michal Seredyński[1], Rolf Hoffmann[2], Dominique Désérable[3], and Franciszek Seredyński[1]

[1] Institute of Computer Science, University of Siedlce, Siedlce, Poland
{miroslaw.szaban,michal.seredynski,franciszek.seredynski}@uws.edu.pl
[2] Technische Universität Darmstadt, Darmstadt, Germany
hoffmann@informatik.tu-darmstadt.de
[3] Institut National des Sciences Appliquées, Rennes, France

Abstract. We propose a novel game-theoretic multi-agent system approach to create a desired 2D pattern. We interpret a pattern formation problem as a variant of the iterated Spatial Prisoner's Dilemma game, where evolutionary competing CA-based agents are used as learning machines. We design a payoff function reflecting a local goal of CA-based agent-players, and we show that the system of competing players is able to reach a Nash equilibrium, providing at the same time the maximization of a global criterion unknown for the agents that is related to the considered pattern formation problem. We provide experimental results showing a high performance of the pattern formation process.

Keywords: Collective behavior · Competing Cellular Automata · Nash equilibria · Pattern Formation · Spatial Prisoner's Dilemma game.

1 Introduction

Patterns (pictures) are one of the most important forms of communicating between human beings. They are rationally or emotionally interpreted, classified and used to take decisions. They reflect either a result of a creation by Nature (e.g. a mountain landscape, winter snowing) or are a result of a human activity (e.g. architecture of a town, painting or other objects of art, etc.). Today, more and more often they are products of an application of mathematical/computer science tools like Cellular Automata (CA), fractals or Artificial Intelligence. There is an essential difference between the way of creating, e.g. a winter snowing landscape by a painter and by Nature. While both approaches are very interesting we focus in this paper, to mimic a Nature approach, in a way as we guess it does it. We believe that phenomena like this is a result of emergent behavior of a huge number of elementary units, and in particular CA are a perfect tool to do it.

Pattern formation is an area of active research in various domains such as physics, chemistry, biology, computer science or natural and artificial life. CA are suitable and powerful tools for catching the influence of the microscopic scale onto the macroscopic behavior of such complex systems [2,3] The main issue while a CA-based approach is to be used to solve pattern formation problem and other similar problems is the translation of a global goal related to a problem into corresponding CA rules [4].

The arrangement of dominoes in a grid of cells is a special case of pattern formation, due to a number of possible applications such as loading boxes on pallets, arrangements of pallets in trucks, cargo stowage, etc. [1]. Different patterns were generated [5,6] by agents with embedded finite state control evolved by Genetic Algorithms (GA). In [7] domino patterns were formed by moving agents also controlled by a finite state machine, evolved by GA. In further works an afford was done to construct directly the required CA rule what resulted in a number of recent papers (see, e.g. [8]), where probabilistic rules working with templates derived from domino tiles were proposed.

This paper presents a novel approach based on game-theoretic model of self-organization and collective behavior of CA-based agents to solve a pattern formation problem. The approach applies a recently proposed [9] methodology which is based on three components: a) a multi-agent intepretation of a pattern formation problem, b) applying a variant of Spatial Prisoner's Dilemma (SPD) game as a model of interaction between agent-players, and using evolutionary competing CA-based agents as elementary reinforcement learning machines.

This approach was recently applied [10] to solve the Coverage Problem in Wireless Sensor Networks (WSN) in the form of a self-organizing system. One of the key issues to successively apply this methodology is designing a payoff function for CA-based agents representing local goals and reflecting the global criterion related to our problem. We will show that the system of competing agents with appropriate designed payoff function is able to reach collectively a Nash equilibrium corresponding to a solution of the considered pattern formation problem and providing at the same time the maximization of a global criterion that is related to the studied problem but not known to the agents.

The structure of the paper is the following. The next section states the pattern formation problem considered in the paper. Section 3 presents a game-theoretic multi-agent approach to the distributed solving of the pattern formation problem by self-organization. Section 4 presents results of the experimental study, and the last Section contains conclusions.

2 Pattern Formation Problem

We consider a 2D discrete space consisting of $m+2$ rows and $n+2$ columns, where rows i are enumerated as $0, 1, \ldots m, m+1$, and columns j as $0, 1, \ldots n, n+1$, with a total number of cells equal to $(m+2) \times (n+2)$ and $m, n \geq 1$. We split this space into two subareas: an active area consisting of $m \times n$ cells and a border area containing border cells corresponding to the rows 0 and $m+1$, and the

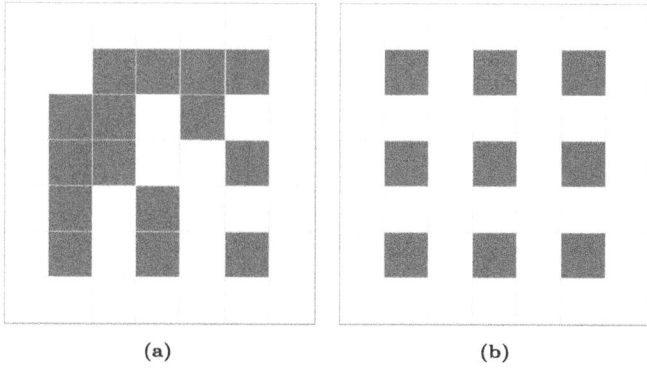

Fig. 1. Instance 5 × 5: (a) an inital configuration, (b) a requested final configuration.

columns 0 and $n + 1$. We further assume that cells of the active area can take the values $x_{i,j} \in \{0, 1\}$ while the border area cells will have values $x_{i,j} = 0$.

The purpose of the active area is to create, store and modify patterns. The border area has a supplementary character. Its main purpose is to provide the same homogenous conditions for all agents being a part of a game-theoretic framework which will be introduced, more exactly, for those agents from the active area who are adjacent to the border area. Functionalities of these both areas are provided by two classes of agents assigned to cells: border agents assigned to cells constituting the border area, and active agents assigned to cells representing the active area. We will see later that each active agent is composed of two subagents: an environmental agent controlling a local environment of an active agent, and an agent-player participating in games with agent-players from a neighborhood. An agent-player is a CA-based reinforcement learning machines recently proposed [9] which we shortly call *competing CA*.

Figure 1 shows 2 examples of a 2D discrete space for $m = n = 5$ with a total number of cells equal to 49. Some cells from the active area have values equal to 1 (in red). The remaining cells from both active and border areas have values equal to 0 (in white). Let us consider Fig. 1a which presents some initial randomly created pattern. Figure 1b presents a desired pattern which we wish to construct using a game-theoretic framework based on the SPD game and applying competing CA-based agents. We expect that from any initial pattern configuration the system will be able to construct a pattern similar or identical to the one presented in Fig. 1b.

The considered problem is a simplified version of the domino problem studied recently [8] which was solved by a construction of probabilistic CA rules and applying a methodology of matching pattern templates. In this paper we propose a novel methodology of solving the problem by the collective behavior of agents who participate in a SPD-like games and act in such a way to maximize their personal payoffs but at the same time when reaching a Nash equlibrium they maximize the global payoff criterion related to the considered problem which

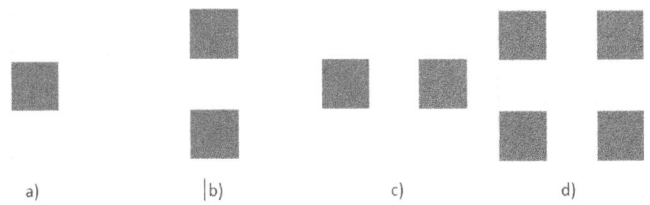

Fig. 2. Correct neighborhoods of: a) a cell in state 1, b) a cell in state 0 (type 1), c) correct state 0 (type 2), d) a cell in state 0 (type 3).

is not known for them. The basic question which must be solved under this approach is how to design the local goals for the agents to reflect by them the global optimization goal related to the problem.

Let us see at the problem from the point of view of a global optimization. Looking at Fig. 1b we can notice that the global optimization problem is very clear. We want to maximize the number of "1 s" with "a correct neighborhood". A correct neighborhood of a single "1" is a neighborhood consisting of 8 "0s", as it is shown in Fig. 2a. So, we can construct a function $f(X) = \sum_{i=1}^{m} \sum_{j=1}^{n} x_{ij}$, where $x_{ij} = 1$, if $s_{ij} = 1$ and all neighbors of A_{ij} are in state 0, otherwise $x_{ij} = 0$. The function counts a number of "correct 1s" in a considered pattern and should be maximized.

For the pattern from Fig. 1a the value $f(X) = 1$ while for Fig. 1b $f(X) = 9$. It is the maximal value of this global function and the corresponding pattern for the considered instance of the problem is optimal. How to construct an optimal pattern – a solution for a given instance? Because the problem is known to be NP-hard some metaheuristics can be applied. However, here we are not interested in global optimization algorithms. We want to deliver an optimal/suboptimal solutions in a fully distributed way, and below we present details of our approach.

3 Multi–agent CA System Generating Patterns

We interpret the 2D discrete space as a 2D CA consisting of $(m + 2) \times (n + 2)$ cells with the main cell states x_{ij} defining the pattern X. The border cell values are fixed to 0, and the cells of the active area of size $m \times n$ can take on the values $x_{ij} \in \{1, 0\}$. To each cell a CA-based agent is assigned, in particular to active cells agents with a CA rule that is space-dependent (non-uniform) and can also vary under time (second-order CA). The computation of the pattern $X(t+1)$ from the pattern $X(t)$, where t denotes the time-step or iteration ($iter$), is performed by the following steps:

- Test: The pattern $X(t)$ is checked by environmental agents for correct local neighborhoods. The result is an intermediate pattern $X'(t)$.
- Matrix of cooperators/defectors: Matrix of players actions C^*/D^* is computed on the base of $X'(t)$.

- Game: each agent-player uses his action C*/D* in games with neighbors and receives a cumulated payoff.
- Competition: A player compares its payoff with the payoffs of its neighbors. If a neighbor's payoff is higher, then its rule is adopted.
- Update: The next cell state is computed with the agent-player rule/strategy currently assigned to the cell. The result is an intermediate pattern $X'(t)$.
- Mutation: an agent-player rule/strategy and/or a cell's pattern state is a subject of mutation. The result is a pattern $X(t+1)$.

3.1 Agents and Their Actions

Agent-Players. An agent-player is one of two agents attached to each active cell of CA controlling a state of a corresponding cell. His main task is participation in a SPD-like game with agent-players from his neighborhood, participation in competition between his neighbors, and applying currently assigned to him rule (strategy) to make a discrete-time decision on its next state.

Applying the terminology used in the SPD game, a state 0 of a CA cell can also be called *Defect* (D) and a state 1 can be called *Cooperate* (C). We assume that the following set of rules (called also main rules) is available for the agents:

- *all C*: always cooperate (C), i.e. set the cell's own state to 1,
- *all D*: always defect (D), i.e. set the cell's state to 0,
- *k–D*: cooperate, i.e. set the state to 1 as long as not more than k neighbors defect (are in state 0), otherwise defect (set the state to 0),
- *k–C*: cooperate, i.e. set the state to 1 as long as not more than k neighbors cooperate, otherwise defect,
- *k–DC*: defect, i.e. set the state to 0 as long as not more than k neighbors defect, otherwise cooperate.

The presented rules are socially interpreted rules inspired by the SPD game and recently studied [9] from the point of view of their usefulness for the collective behavior of a *second order* CA (i.e. with non-uniform time-dependent rules) and also successively applied [10] to work out a self-optimizing algorithm oriented on solving a coverage problem in WSN.

Environmental agents. An environmental agent is responsible for controlling the local neighborhood of a given active cell. It is oriented on destroying either groups of 0s or groups of 1s around a cell applying the following rules:

- If the state $x_{i,j}$ of a cell (i,j) is equal to 0 and the states of all 8 neighbors are also equal to 0 then the state $x_{i,j}$ is changed into 1 with a predefined probability p *destroy 0s block*. This operation can be seen as an adjustment of the center cell to the correct value.
- if the state $x_{i,j}$ of a cell (i,j) is equal to 1 and at least one cell from the Moore neighborhood is equal to 1 then the state $x_{i,j}$ is changed into 0 with a predefined probability p *destroy 1s block*. This operation can be seen as noise-injection when the neighborhood is not a valid one.

Mutation. Some parameters of the system can be a subject of random modifications called mutation. Two parameters can be modified this way: mutation of a cell state happens with a predefined probability $p\ swap\ mut$ and mutation of a strategy with $p\ strat\ mut$.

3.2 Payoff Function of the SPD-Like Game

We assume that an agent-player A_{ij} corresponding to a cell (i,j) will participate in the SPD-like game with each of his 8 neighbors defined by a Moore neighborhood. Players from this neighborhood will be considered as opponents in the game. At a given discrete moment, each cell can be in one of two states: C or D. A standard approach in the SPD game assumes that a current state of a cell (i,j) is considered as an action of an agent-player A_{ij} in his games with his neighbors. However, in our version of the game which we use to describe the pattern formation problem we redefine an action of A_{ij} and we use for it the notation Cooperate* (C^*) or Defect* (D^*).

An action of a player A_{ij} against his opponent from his neighborhood is defined in the following way. When his cell is in the state 1 (or C) (see, Fig. 2a) and all cells from his neighborhood are in a state 0 then his action is C^*, i.e. he Cooperate*, because as we can see at Fig. 2a he has a correct neighborhood from the point of view of a global solution. Otherwise he Defect* and his action is D^*. When his cell is in a state 0 there are three possibilities (see, Fig. 2b,c,d) to consider his neighborhood as correct from the point of view of a global solution. These correct types of neighborhoods are labelled as a type 1, a type 2 or a type 3. If any of these three types of a neighborhood takes place then the action of the considered player is C^*, otherwise is D^*.

Active agents will play games with their 8 neighbors and among them can be also passive agents. So, the open remaining question is the issue of defining the actions C^*/D^* of passive agents. For this purpose we will make additional assumptions concerning passive agents. The first assumption is that passive agents formally participate in a game independent on games of active agents, and this is a game on the ring created by passive agents. A neighborhood of a given passive agent in this game is created by his two adjacent neighbors. As it was already stated each passive agent is equipped with *all D* rule what results in a state 0 of each passive cell, and this is a correct state of all these cells. Therefore, we can assume that an action of each passive agent is C^*. These actions will be the opponent actions met by active agents in their games with passive agents.

After redefining the values of the players actions C^*/D^* the payoff function of the game is presented in Table 1. An iterated game will consist of a user predefined number of rounds (iterations) not known for the players. In each round an active agent-player A_{ij} will participate in 8 single games with his neighbors and obtain in each game a payoff. His payoff in a single game with an opponent is defined by the values d, b, a and c from the payoff function. Let us assume that $d = 1$, $b = 1.2$, $a = 0.1$, $c = 0$ and assume a game of a player $A(i,j)$ with one of his opponent from a neighborhood. If a player $A(i,j)$ takes

Table 1. Payoff function assigning payoffs to player A_{ij} participating in a SPD-like game with a neighbor-opponent.

Player's A_{ij} action	Opponent's action	
	Cooperate* (C^*)	Defect* (D^*)
Cooperate* (C^*)	$d = 1$	$c = 0$
Defect* (D^*)	$b = 1.2$	$a = 0.1$

the action $s_{ij} = C^*$ and the opponent (i_k, j_k) from a neighborhood also takes the action $s_{i_k j_k} = C^*$, then the player receives a payoff $u_{ij}(s_{i,j}, s_{i_k j_k}) = 1$. If the player takes the action D^* and the opponent player still keeps the action C^*, the defecting player receives a payoff equal to $b = 1.2$. If the player takes the action C^* while the opponent takes the action D^*, the cooperating player receives the payoff equal to $c = 0$. When both players use the action D^*, then both receive a payoff equal to $a = 0.1$. A player $A(i,j)$ cumulates his payoffs obtained in games with the neighbors, so his cumulated payoff u_{ij}^{cumul} received in an iteration can be described as

$$u_{ij}^{cumul}() = \sum_{k=1}^{n_{ij}=8} u_{ij}(s_{ij}, s_{i_k j_k})/8. \quad (1)$$

We assume that players are rational and act in such a way as to maximize their cumulated payoffs. Game theory predicts that players' behavior in non-cooperating games is oriented towards achieving a Nash equilibrium (NE), where none of them can improve his own cumulated payoff by just changing his action. In our approach we will be interested in the collective behavior of players in the game measured by the average total payoff (ATP) $\bar{u}()$ of the whole set of players:

$$\bar{u}(s_{11}, s_{12}, ..., s_{mn}) = \frac{1}{mn} \sum_{j=1}^{m} \sum_{i=1}^{n} u_{ij}^{cumul}(). \quad (2)$$

The game can have many NE points with different values of ATP. We call NE with the highest ATP the *maximal price point* (MPP). We expect of such a behavior of players that, while they attempt to reach a NE, at the same time the ATP of the whole set of players is maximized, i.e. MPP is reached. Such a behavior depends on many factors, and one of them is the model of a player making decisions. In this paper, we examine the collective behavior of players modeled by competitive CA-based agents in an attempt to solve the pattern formation problem.

3.3 Competition of Agent-Players

The competition mechanism assumes that after completing games with neighbors in a given iteration each agent compares its cumulated payoff with the cumulated payoffs of its neighbors. If a more successful player exists in the neighborhood of the considered player, this player-winner with his rule replaces the rule of the considered player. In such a way, from the point of view of solving the considered problem, more prospective CA-rules achieve a higher frequency in the population of CA-based agents. The competition mechanism is a mechanism similar to evolutionary selection mechanisms applied in evolutionary algorithms. It converts a classical CA into a *second–order* CA, which can adapt in time.

4 Experimental Results

In this Section we report some experimental results with use of an instance of the problem called *Instance 31 × 31* with $m = n = 31$. Figure 3a presents this instance at $t = 0$ with initial states set with a probability equal to 0.5. Rules were assigned to the agent-players with probability 0.2. Figure 3b presents this initial assignment (rule *all C* in red color, *all D* in blue, *k–D* in green, *k–C* in cyan, *k–DC* in pink). Figure 3c shows initial actions C^* (in orange) or action D^* (in blue) used in the first game. The experiment has been conducted with the following values of parameters: *p destroy 0 s block*=0.3, *p destroy 1 s block*=0.1, *p strat mut*=0.0015 and *p swap mut*=0. It lasted 500 iterations (games). Figure 4 presents details of the experiment in a single run of the system.

The most important parameter showing the performance of our approach is the number of correct 1 s, as it was stated in Sect. 2. Figure 4a shows a plot (in red) of a fraction of correct 1 s *f C corr* for different iterations. For *iter* = 0 (corresponding to CA from Fig. 3a) *f C corr*=0.03, for *iter* = 50 (see, Fig. 3d) *f C corr*=0.23, and for *iter* = 500 (see, Fig. 3g) we have *f C corr*=0.98. It means that the system found a solution very close to the optimal one. In this figure we can also notice how changes the value of *av pay* (in blue) corresponding to *ATP* (see, Eq. 2) and reaching the value 0.97 at *iter* = 500.

Figure 4a shows also how important is the role of the environmental agents, controlling locally states of an environment by parameters *p destroy 0 s block* and *p destroy 1 s block*. Plot *f 0 s block* (in orange) shows how the fraction of destroyed 0 s blocks changes, and the plot *f 1 s block* (in green) shows how the fraction of destroyed 1 s blocks changes during the iterated game. We can see a strong activity of these two parameters during the first 100 iterations. The observed collective behavior is a result of interactions between both types of agents, and does not exist if only one type of agents is working.

Figures 3b,e,h show how the distribution of the rules assigned to the agents changes in the course of an iterated game. While at *iter* = 0 the rules are statistically equally distributed among agents, in the next iterations we can see significant changes in their redistribution. At iter=50 when we can see in Fig. 3d the emerging of the desired configuration of CA states, we can also see in the corresponding Fig. 3e that the two rules *k–D* and *k–DC* become dominating. At *iter* = 500 when a solution was reached (see, Fig. 3g), *k–DC* becomes the rule which is the winner of the rules competition (see, Fig. 3h and Fig. 4b). More exactly, it is the rule *6–DC* which is assigned to 99.9% of cells.

Figures 3c,f,i show how the agents' actions C^*/D^* evolve. One can see that at *iter* = 0 only 15 players offered an action C^* (in orange) for their opponents

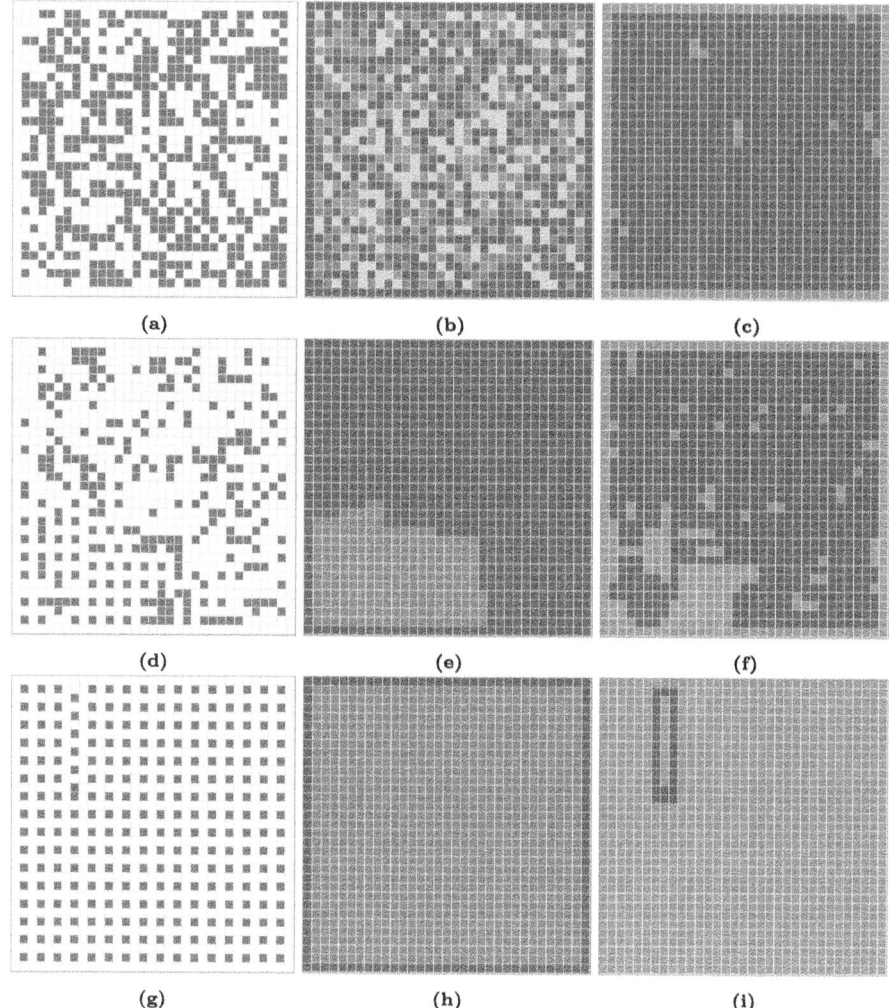

Fig. 3. Instance 31 × 31: single run. iter=0: (a) agents' states, (b) agents' rules, (c) agents' C^*/D^* actions; iter=50: (d) agents' states, (e) agents' rules, (f) agents' C^*/D^* actions; iter=500: (g) agents' states, (h) agents' rules, (i) agents' C^*/D^* actions.

while the remaining players offered an action D^* (in blue). We can see that the number of cooperating agents increases during the iterated game, and at $iter = 500$ near all players cooperate.

Figures 5a,b summarize the experimental results of this section obtained by averaging the results over 20 runs. Figure 5a presents the averaged results discussed for a single run and presented earlier in Fig. 4. One can see that below 500 iterations the number of correct 1 s (in red)—the main performance parameter, is equal in the average 0.86, with a low value of standard deviation (std) equal

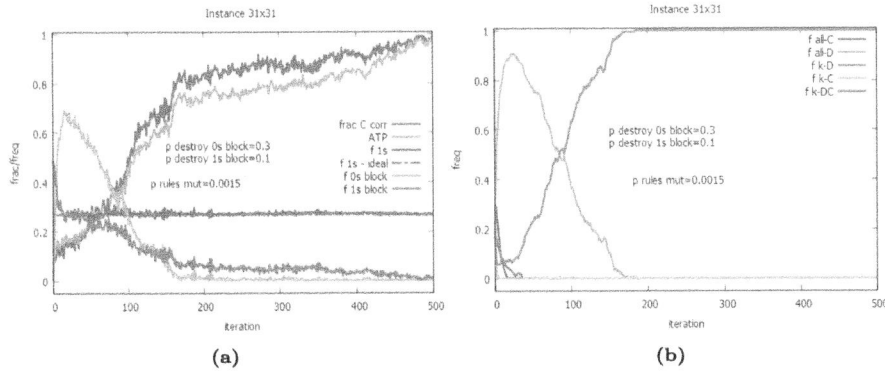

Fig. 4. Single run: (a) fraction of the correct number of 1 s (in red) and other experimental characteristics, (b) relative frequencies of rules. (Color figure online)

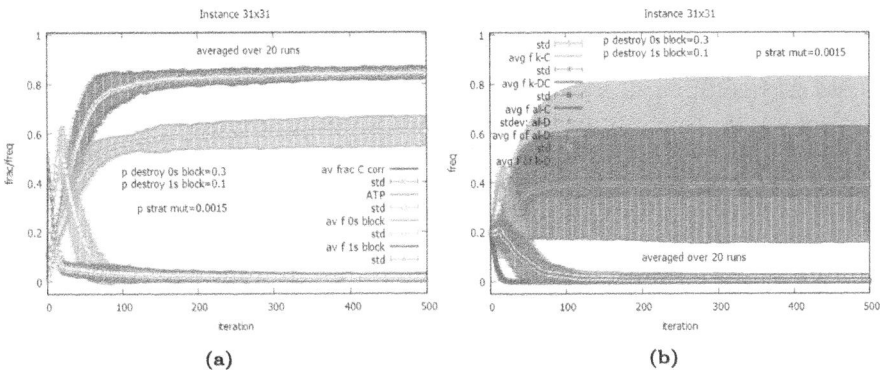

Fig. 5. Multiple run: averaged (a) fraction of the correct number of 1 s (in red) and other characteristics, (b) frequencies of rules. (Color figure online)

to around 0.02. This corresponds to the average value of ATP equal to around 0.65 with a slightly higher value of *std*. Figure 5b presents statistics concerning the evolution of the rules. What we can notice is that the two main rules providing collectively solutions have a relatively large *std*. This means that the system automatically selects at the end of each run either a single rule (k–C or k–DC) or pairs of two rules k–C and k–DC with different values of k, which are in some way related and provide a relatively good performance for solving the addressed pattern formation problem.

5 Conclusions

We have proposed a self-organizing system to solve the pattern formation problem by creating a desired 2D pattern. The approach is based on three components: the multi-agent interpretation of the problem, the application of a variant of the SPD game as a model of interaction between agent-players, and the using of evolutionary competing CA-based agents. We designed a payoff function reflecting the local goal of CA-based agents, and we show that the system of competing players is able to reach NE providing at the same time the maximization of a global criterion not known for the agents that is related to the considered pattern formation problem. Our preliminary experimental results show a good performance in solving the considered problem by the collective behavior of agents in an iterated game created for solving this problem. Our future work will be oriented toward a more detailed study of the proposed approach with the use of different instances of the problem, to consider issues of scalability and the influence of specific model parameters on the algorithm's performance and extending the model to solve more complex pattern formation problems.

References

1. Birgin, E.G., Lobato, R.D., Morabito, R.: An effective recursive partitioning approach for the packing of identical rectangles in a rectangle. J. Oper. Res. Soc. **61**, 303–320 (2010)
2. Chopard, B., Droz, M.: Cellular Automata Modeling of Physical Systems. Cambridge University Press, Cambridge (1998)
3. Deutsch, A., Dormann, S., Cellular automaton modeling of biological pattern formation. Birkäuser (2005)
4. Nagpal, R.: Programmable pattern-formation and scale-independence. In: Minai, A.A, Bar-Yam, Y. (eds.) Unifying Themes in Complex Sytems IV, pp. 275–282 (2008)
5. Hoffmann, R.: Cellular automata agents form path patterns effectively. Acta Phys. Pol. B Proc. Suppl. **9**(1), 63–75 (2016)
6. Hoffmann, R., Désérable, D.: Line patterns formed by cellular automata agents. In: El Yacoubi, S., Wąs, J., Bandini, S. (eds.) ACRI 2016. LNCS, vol. 9863, pp. 424–434. Springer, Cham (2016). https://doi.org/10.1007/978-3-319-44365-2_42
7. Hoffmann, R., Désérable, D.: Generating maximal domino patterns by cellular automata agents. In: Malyshkin, V. (ed.) PaCT 2017. LNCS, vol. 10421, pp. 18–31. Springer, Cham (2017). https://doi.org/10.1007/978-3-319-62932-2_2
8. Hoffmann, R., Désérable, D., Seredyński, F.: A cellular automata rule placing a maximal number of dominoes in the square and diamond. J. Supercomput. **77**, 9069–9087 (2021)
9. Seredyński, F., Kulpa, T., Hoffmann, R.: Evolutionary self-optimization of large CA-based multi-agent systems. J. Comput. Sci. **68**, 101994 (2023)
10. Seredyński, F., Kulpa, T., Hoffmann, R., Désérable, D.: Coverage and lifetime optimization by self-optimizing sensor networks. Sensors **23**(8), 3930 (2023)

Effects of a Vanishing Noise on Elementary Cellular Automata Phase-Space Structure

Franco Bagnoli[1,2,3]($^{\boxtimes}$)[iD], Michele Baia[1,2], and Tommaso Matteuzzi[1][iD]

[1] Department of Physics and Astronomy and CSDC, University of Florence,
via G. Sansone 1, 50019 Sesto Fiorentino, Italy
`{franco.bagnoli,michele.baia,tommaso.matteuzzi}@unifi.it`
[2] INFN, Florence sect., Florence, Italy
[3] UMR Espace-Dev, University of Perpignan,
via Domitia, Perpignan, France

Abstract. We investigate elementary cellular automata from the point of view of (discrete) dynamical systems. By studying small lattice sizes, we obtain the complete phase space of all minimal elementary cellular automata, and, starting from a maximal entropy distribution (all configurations equiprobable), we show how the dynamics affects this distribution. We then investigate how a vanishing noise alters this phase space, connecting attractors and modifying the asymptotic probability distribution. What is interesting is that this modification not always goes in the sense of decreasing the entropy.

Keywords: Elementary cellular automata · discrete dynamical systems · attractors · noise-induced transitions

1 Introduction

Cellular automata (CA) are interesting systems, see the series of proceedings of this conference to get a wide scenario of this subject [1–11].

Deterministic CA have been introduced as discrete dynamical systems by S. Wolfram [12] and then deeply studied by A. Wuensche [13–15]. One can easily see that many CA rules are equivalent upon reflection or zero-one exchange. Of the total of 256 elementary cellular automata (ECA), only 88 are independent (minimal).

Let us consider the space of all possible configurations on a finite lattice, with periodic boundary conditions (the phase-space). A given CA dynamics (a CA rule) connects configurations, originating trajectories. Since it is deterministic, we can have joining trajectories, but not separation. For finite lattices, trajectories, after a transient, always end into a cycle or a fixed point (which is a cycle with period equal to one), that constitutes the only attractors. Therefore, the phase-space of a given CA rule is partitioned into basins of attractors. This characterization is not new at all, it has been studied for instance in Ref. [13,14].

One of the investigations reported in this paper is that of evaluating the probability distribution of attractors starting from a maximum-entropy distribution (i.e., random initial conditions) for all the minimal ECA rules and lattice sizes up to 17 sites. This study is only preparatory for what follows, even if it is the first extensive study of all attractors of all minimal ECA of such size as far as we know.

CA rules can be classified according on the pattern they generate, which roughly corresponds to the kind and distribution of attractors and basins [12,16]; many other classification techniques have been defined [15,17].

Another way of looking at the dynamics properties of CA is that of measuring the stability of a trajectory with respect to some perturbation or noise, by observing the spreading of a damage [18] or defining indicators like Lyapunov exponents [19–21] or similar ones [22].

The idea of stability can be extended to stochastic CA, comparing two replicas of the same system evolving with the same realization of stochasticity (i.e., using the same random numbers) and examining how a difference between them evolves.

The fact that stochasticity can induce sudden transitions in the stability and predictability of the dynamics has also been investigated many times, see for instance Refs. [23–25], but this phenomena has been generally investigated for a finite noise level, mainly studying how this noise affects the sensitivity to a variation in the initial configuration.

The goal of this paper is to investigate how the phase space structure is affected by a vanishing noise, i.e., a noise that is only occasionally present. This topic is interesting since in nature even a deterministic system like cellular automata (which are robust to small noise amplitude due to their discrete dynamics) can occasionally be subject to a perturbation.

We shall show how a vanishing noise connects attractors, therefore determining a Markov process among them; the probability distribution of attractors is thus modified by this noise, and the corresponding entropy changes, in some cases even diminishing with the noise.

The outline of this paper is the following. In Sect. 2 we give some definition, then we present the phase space of ECA in Sect. 3 and the stability of their attractors in Sect. 4. In the following Sect. 5 we show how attractors are connected by a vanishing noise, and how the phase-space entropy if affected by noise and in Sect. 7 we furnish some algorithms to enumerate all attractors in small systems. Conclusions are drawn in the last section.

2 Definitions

Cellular automata are completely discrete dynamical or statistical systems, defined on a lattice of cells, which can take a finite number of states. The evolution of the state of the cells is given by a function of their neighborhood, i.e., of the state of the cells connected with an incoming link to the cell itself.

Table 1. Look-up table of some CA rules. Here s_-, s, s_+ stands for s_{i-1}, s_i, s_{i+1} and **R n** stands for rule n.

s_-,s,s_+	R 1	R 164	R 232	R 11	R 30	R 150	R 204	R 110
0,0,0	1	0	0	1	0	0	0	0
0,0,1	0	0	0	1	1	1	0	1
0,1,0	0	1	0	0	1	1	1	1
0,1,1	0	0	1	1	1	0	1	1
1,0,0	0	0	0	0	1	1	0	0
1,0,1	0	1	1	0	0	0	0	1
1,1,0	0	0	1	0	0	0	1	1
1,1,1	0	1	1	0	0	1	1	0

In order to be more specific, let us define a network though an adjacency matrix a, where $a_{ij} = 1$ if cell i is connected (takes information) from cell j and zero otherwise. In general, regular lattices with periodic boundary conditions are used, for which the adjacency matrix is shift-invariant (circulant). In particular, for ECA, a cell is connected to the cell itself and its two nearest neighbors, i.e., the matrix a is tri-diagonal.

Let us denote by $s_i(t)$ the state of cell i at time t. In order to be more concise, the index t will be considered implicit, and we shall write $s'_i \equiv s_i(t+1)$. Similarly, the state of cells in the neighbourhood osf cell i at time t will be denoted as $v_i \equiv v_i(t) = \{s_j : a_{ij} = 1\}$. ECA are Boolean CA, i.e., $s_i \in \{0,1\}$.

The state of the neighborhood of a cell i, $v_i = \{s_{i-1}, s_i, s_{i+1}\}$, can also be read as a number in base-two representation, i.e., $v_i = 4s_{i+1} + 2s_i + s_{i-1}; 0 \le v_i \le 7$.

The evolution of the state of a cell is given by a function of the neighborhood $f(v)$, which is applied in parallel to all cells, $s'_i = f(v_i)$.

Since the neighborhood v can take only a finite number of values, the function f can be seen as a look-up table. Therefore, there are $2^8 = 256$ possible Boolean functions of three inputs, and each function is specified by listing the 8 values corresponding to $v = 0, \ldots, 7$.

Reading again these lists of values as a number in base-2 representation, $(f(1,1,1), f(1,1,0), \ldots, f(0,0,0))$ we get the Wolfram notation for ECA, as illustrated for some rules in Table 1.

By exploiting left-right and 0–1 symmetries, one can reduce the number of independent rules to 88, called the "minimal" rules, listed in Table 2.

The time evolution of some rules is reported in Fig. 1. Rule 1, 164, 232, 204 are typical class-2 rules, quickly falling into a fixed point or a cycle of small period. This implies that their phase-space is partitioned into many attractors of short period. Also rule 11 is considered class-2 since it does not exhibit interesting patterns, although in this case the cycles are not so many and periods depend on lattice size. Rule 204 is the identity. Rule 232 is the majority rule.

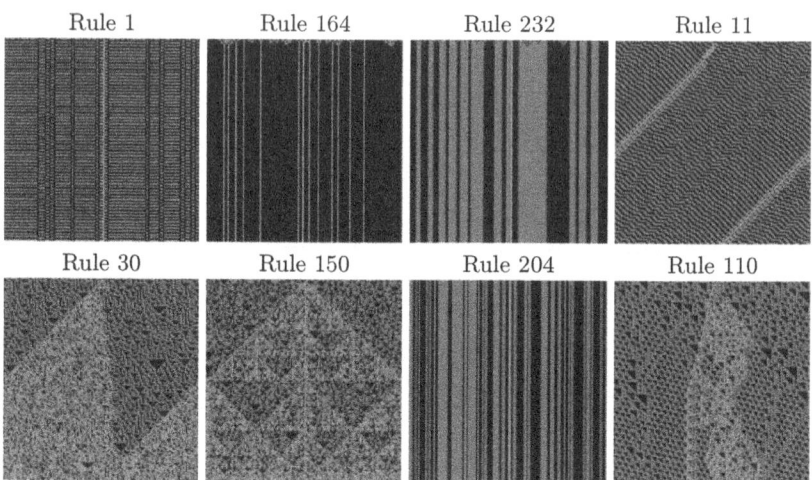

Fig. 1. Time evolution of some rules starting from a random initial configuration ($s_i(t)$) and the evolution of an initial single defect ($S_i(t)$). Time runs from top to down. Color code: black: $s_i(t) = S_i(t) = 0$; blue: $s_i(t) = 1$, $S_i(t) = 0$; blue: $s_i(t) = 1$, $S_i(t) = 0$; green: $s_i(t) = S_i(t) = 1$; red: $s_i(t) = 0$, $S_i(t) = 1$. (Color figure online)

Rules 22, 30 and 150 are typical class-3, i.e., chaotic, and indeed for these rules an initial damage tends to spread (sensitivity to initial configuration).

Rule 100 is class-4, capable of universal computing [26], exhibiting gliders and travelling structures.

3 Stability Properties of CA

It is possible to define the Boolean derivative of a Boolean function [27,28], which is similar to the usual definition: it takes value one if the change of a variable makes the function change, and zero otherwise.

Linear rule are those for which all derivatives are constant, *i.e.*, variables appear only alone, possibly with a constant. Let us provide some examples.

$$f_{150}(x,y,z) = x \oplus y \oplus z; \qquad f_{204}(x,y,z) = y$$

are linear function, while

$$f_0 = 0; \qquad f_{22}(x,y,z) = x \oplus y \oplus z \oplus xyz$$
$$f_{30}(x,y,z) = x \oplus y \oplus z \oplus yz; \qquad f_{35}(x,y,z) = 1 \oplus x \oplus y \oplus xy \oplus xz \oplus xyz$$
$$f_{76}(x,y,z) = x \oplus xyz; \qquad f_{110}(x,y,z) = x \oplus y \oplus xy \oplus xyz$$
$$f_{178}(x,y,z) = x \oplus z \oplus xy \oplus xzvyz; \qquad f_{232}(x,y,z) = xy \oplus yz \oplus xz$$

are not.

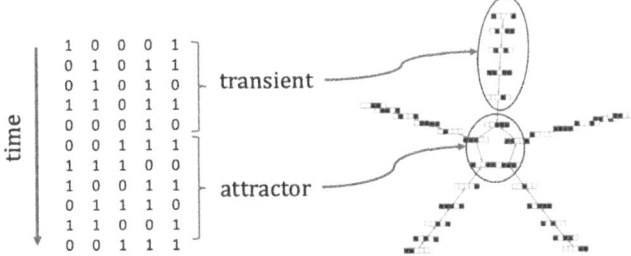

Fig. 2. Transients and attractors for a basin.

Exploiting the Boolean derivatives, it is also possible to define the Jacobian J for a given rule,

$$J_{ij} = \frac{\partial x'_i}{\partial s_j} = \frac{\partial f(v_i)}{\partial s_j}$$

which, for ECA, is a tri-diagonal circulant matrix (for periodic boundary conditions), that depends on the configuration $\{s_i(t)\}$, except for linear rules for which is constant.

By computing the maximum eigenvalue of the product of Jacobians over a trajectory it is possible to compute the maximum Lyapunov exponent of the trajectory [19], which corresponds to the spreading rate of a defect, in the limit of just one defect for each time step (i.e., taking replicas of the configuration at each time, and distributing the defects on each replica). In other words, this product gives all possible paths of a vanishing defects, i.e., supposing that only one defect can survive at each time step.

What is important for the present analysis is that even an unstable trajectory cannot be left, due to the Boolean nature of CA. For instance, the trajectory of rule 232 reported in Fig. 1 is unstable and shows a positive Lyapunov exponent, however for this rule any block of at least two cells with the same value gives origin to a permanent "stripe". By adding a vanishing noise (i.e., at most one cell state flip per time step), the boundaries of a stripe can move. Upon meeting, two boundaries fuse, making a strip disappear, while a new strip cannot appear (unless the nose is so high that two neighboring defects are created at the same time). Finally, all stripes coalesce (for periodic boundary conditions, if boundaries are fixed a frustration may be present). The stable configuration in the presence of a vanishing noise are those composed by all zeros and all ones, which have a negative (minus infinity) Lyapunov exponent.

These considerations illustrate the importance of examining the effects of a vanishing noise.

4 ECA Attractor Structure

We summarize here concepts introduced by A. Wuensche [13,14].

If we consider the whole lattice,

$$s(t) \equiv \{s_i(t), i = 1, \ldots, L\},$$

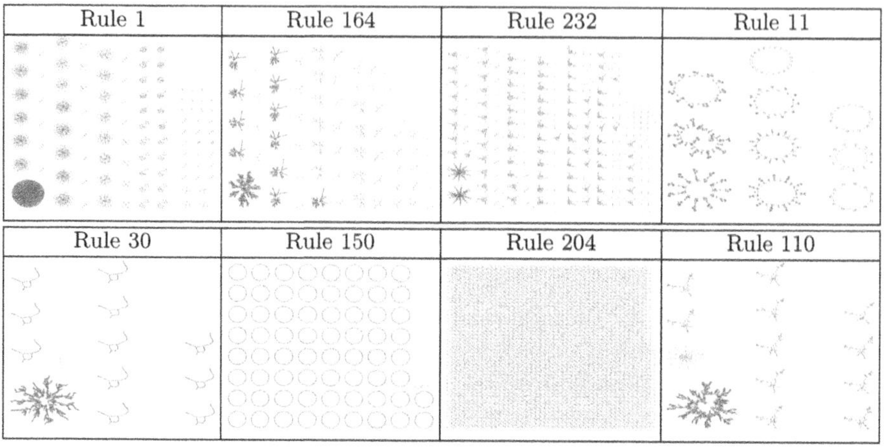

Rule 1	Rule 164	Rule 232	Rule 11
Rule 30	Rule 150	Rule 204	Rule 110

Fig. 3. Attractors of rules shown in Fig. 1. Depending on the rule the dynamics shows different behaviors; for example, fixed point attractors (e.g., bottom left of rule 1), cycles with/without a basin of attraction (e.g., rule 30 and 150).

with appropriate (for instance periodic) boundary conditions, we can define a global function $\boldsymbol{F}$ so that

$$\boldsymbol{s}(t+1) = \boldsymbol{F}(\boldsymbol{s}(t)).$$

and this defines a discrete dynamical system.

For a Boolean lattice of L sites, $\boldsymbol{s}$ can take 2^L possible values. Let us denote a trajectory γ the ordered set of configurations given by the dynamics,

$$\gamma \equiv \gamma(\boldsymbol{s}(0)) = \{\boldsymbol{s}(0), \boldsymbol{s}(1), \dots\}.$$

Since the evolution is deterministic and the number of states (on a finite lattice) is finite, a trajectory always ends in a cycle, possibly in a fixed point (which is cycle of length 1); these limit cycles are the only attractors for finite CA. A cycle α is a trajectory $\alpha = \{\boldsymbol{s}(1), \boldsymbol{s}(2), \dots, \boldsymbol{s}(k)\}$ such that $\boldsymbol{s}(1) = \boldsymbol{F}(\boldsymbol{s}(k))$. The length of the cycle is indicated by $c(\alpha)$.

A generic trajectory is represented by a (possible empty) transient τ followed by a cycle α: $\gamma(\boldsymbol{s}(0)) = \{\tau(\boldsymbol{s}(0)), \alpha, \alpha, \dots\}$ where τ is the minimal one, i.e., it has no overlap with α, see Fig. 2 for an illustration.

The basin $\beta(\alpha)$ of a cycle α is given by all configurations $\boldsymbol{s}$ such that $\gamma(\boldsymbol{s}) = \{\tau(\boldsymbol{s}), \alpha\}$. The size of the basin is indicated by $b(\alpha)$. We can also arbitrarily number the cycles with an index $k = 1, \dots, n$, so we can speak of length of cycle k, c_k, and size of its basin, b_k. Clearly, $\sum_k b_k = 2^L$.

Another approach to the same problem is the following. Let us define the connection matrix $M_{\boldsymbol{s}',\boldsymbol{s}} = M(\boldsymbol{s}'|\boldsymbol{s}) = 1$ if $\boldsymbol{s}' = \boldsymbol{F}(\boldsymbol{s})$ and zero otherwise (it is a $2^L \times 2^L$ matrix). M clearly depends on the boundary conditions, that we assume are periodic. It is a (degenerate) Markov matrix, with $\sum_{\boldsymbol{s}} M_{\boldsymbol{s}',\boldsymbol{s}} = 1$ (actually, only one entry per columns of M is one, all the rest is zero). If we start with a probability distribution of configurations $\boldsymbol{P}(\boldsymbol{s}, 0)$, we have

$$P(s', t+1) = \sum_S M(s'|s) P(s, t).$$

Starting with a delta, $P(s, 0) = [s = s(0)]$, we get a sequence of deltas corresponding to the trajectory starting from $s(0)$. After a transient time, the evolution enters an attractor cycle. The distributions corresponding to the states belonging to a cycle are eigenvectors of M.

We are interested in the maximally entropic starting state $P(s, 0) = 1/2^L$, for which the entropy is $S_{\text{MAX}} = \log_2(L)$. If we iterate the evolution for a long time in many cases we get a distribution $\overline{P}$ whose entries are zero for the configurations belonging to the transient part of cycle, and equal to $b_k/(2^L c_k)$ for configurations belonging to a cycle, if all of them are the entry points in the limit cycle of transient trajectories with the same number of configurations. Since in cycle k there are c_k configurations we have

$$S(\overline{P}) = \frac{-1}{S_{\text{MAX}}} \sum_s \overline{P}(s) \log_2(\overline{P}(s)) = \frac{-1}{2^L \log_2(L)} \sum_{k=1}^n b_i \log_2 \left(\frac{b_i}{c_i 2^L}\right).$$

It may happen that the asymptotic distribution is oscillating, for instance rule 1 maps all local configurations but $(0,0,0)$ to 0, and the local configuration $(0,0,0)$ to 1. So, starting from a homogeneous configuration, we have an alternation of deltas (corresponding to configurations $\mathbf{0} = (0, 0, \ldots, 0)$ and $\mathbf{1} = (1, 1, \ldots, 1)$).

The dynamics generally induces a production of information (reduction of entropy), which is related to the presence of transient states. One can have a visual representation of this contraction by looking at Fig. 3. One can see that in many cases there are transients, but this is not always true. For instance, rule 204 (the identity) only has fixed points, while rule 150 has many cycles and two fixed points ($\mathbf{0}$ and $\mathbf{1}$) with no transients.

The entropy S is reported in Table 2 for all minimal rules and $L = 17$. Let us illustrate some examples (see Fig. 1 and 3).

CA rule 0 (255) has a very simple structure: there is only one cycle, $\mathbf{0}$ ($\mathbf{1}$) whose basin includes all configurations ($K = 1$), which fall into this cycle after one step: $b_1 = 2^L$. The final entropy is zero.

The phase space of the identity CA rule 204 is composed by $n_{\text{MAX}} = 2^L$ fixed points (cycles of length $c_k = 1$), each of them coincides with its own basin ($b_k = 1$). The final relative entropy is one.

The phase space of the right shift CA 170 and of left-shift rule 240 is also composed by cycles which coincides with their basins, but they have different length: configuration $\mathbf{0}$ is a fixed point, configuration $(0, 1, 0, 1, 0, 1, \ldots)$ and $(1, 0, 1, 0, 1, 0, \ldots)$ form of a cycle of length 2 for even L, but they have period L if L is odd, as all other configurations if L is prime. In any case their asymptotic relative entropy is one.

The phase space of the majority CA 232 is composed by two fixed points $\mathbf{0}$ and $\mathbf{1}$ whose basin is composed by all configurations with isolated "other" values (i.e., of the type $(0, 0, \ldots, 1, 1 \ldots, 0, 0 \ldots)$, plus a number of other fixed points i.e., configuration with have at least two consecutive cells with the same value $(0, 0, \ldots, 1, 1, \ldots, 0, 0, \ldots)$. Rules 1 and 164 have similar phase spaces.

Finally, there are CA with very long cycles, which however depend on L. Since the evolution is parallel, any symmetry (reflection, inversion, translation) in the initial configuration cannot be forgotten, one cannot have a cycle which includes symmetric and non-symmetric configurations. Therefore, the largest cycles are for L prime, so that no symmetric configurations are possible.

However, there are always two configurations, **0** and **1** which are translationally invariant, and which therefore cannot be part of a larger cycle.

The phase-space of the chaotic rule 150 is composed only by cycles, and again in this case the final relative entropy is one. Not all chaotic CA have only cycles, for instance, for $L = 11$, CA rule 30 has 13 attractors: configuration **0** with a basin of 1, 11 cycles of length 17 and basin 45, and a long cycle of length 154 with a basin of 1551 configurations. Rule 110 has a similar phase space. A complete investigations of such attractors can be found in Ref. [13].

The computation of asymptotic probability distributions using the matrix M (iterating or finding eigenvectors) is however computationally quite expensive. We shall give in Sect. 7 a more effective procedure.

5 Stability of Attractors with Respect to a Vanishing Noise

Another important issue is the effect of a vanishing noise, i.e. the effects of an occasionally flip of the value of just one site, as illustrated before for the rule 232.

Since the noise is vanishing, one can assume that the automata already reached an attractor. The noise causes a jump from a configuration to the perturbed one, which can belong to the basin of the same attractor or to that of another attractor. Thus, the effect of the noise is that of connecting the attractors.

It is possible, exploiting the information obtained by Algorithm 1, to build a matrix $W_{ij} \equiv W(i|j)$ which counts, for all configurations belonging to cycle j, and for all possible one-site perturbations, how many of them will belong to the basin of attractor i. Clearly, $\sum_j W_{ij} = L$.

Normalizing it over columns, one gets a Markov matrix A_{ij} which gives the probability of going from attractor j to attractor i under the influence of a vanishing noise.

So, beyond the entropy reduction given by dynamics, we can have an entropy variation due to this noise. We shall denote the entropy on the probability distribution $\overline{Q}$, obtained by iterating $Q' = AQ$, by the symbol S^*.

What is interesting, is that this vanishing noise can act in different ways on CA rules. Results are reported on Table 2.

Table 2. Number of attractors n, fraction of them with respect to the number of possible configurations $f = n \cdot 2^{-L}$, entropy induced by dynamics S and entropy induced by a vanishing noise $S*$. Computations for all minimal rules **R**, lattice size $L = 17$, $n_{\text{MAX}} = 131072$.

R	n	S	S*	R	n	S	S*	R	n	S	S*	R	n	S	S*
0	1	0	0	26	148	0.792	0.809	56	218	0.613	0.605	132	3572	0.6	0.624
1	1786	0.486	0.634	27	422	0.801	0.807	57	15	0.343	0.299	134	49	0.551	0.578
2	40	0.485	0.538	28	630	0.501	0.299	58	20	0.407	0.24	136	2	0	0
3	422	0.736	0.766	29	15777	0.863	0.857	60	260	0.941	0.941	138	837	0.791	0.764
4	3571	0.508	0.612	30	7	0.799	0.799	62	1186	0.611	0.607	140	3572	0.678	0.687
5	7158	0.736	0.766	32	1	0	0	72	664	0.323	0	142	317	0.572	0.393
6	48	0.577	0.607	33	1786	0.638	0.664	73	1361	0.71	0.681	146	121	0.587	0.586
7	42	0.502	0.059	34	211	0.678	0.687	74	101	0.642	0.656	150	8740	1	1
8	1	0	0	35	429	0.689	0.683	76	31553	0.85	0.85	152	41	0.515	0.535
9	24	0.45	0.451	36	664	0.323	0.445	77	3571	0.6	0.299	154	1688	1	1
10	211	0.678	0.687	37	349	0.531	0.517	78	120	0.377	0.24	156	631	0.502	0.299
11	108	0.594	0.299	38	234	0.731	0.738	90	4404	0.941	0.941	160	2	0	0
12	3571	0.678	0.687	40	8	0.001	0	94	1191	0.564	0.526	162	212	0.667	0.694
13	120	0.377	0.24	41	16	0.458	0.467	104	239	0.208	0	164	665	0.391	0.487
14	120	0.507	0.349	42	1857	0.839	0.862	105	4370	1	1	168	212	0.03	0
15	3856	1	1	43	316	0.572	0.393	106	214	0.699	0.701	170	7712	1	1
18	120	0.578	0.586	44	664	0.528	0.538	108	10966	0.776	0.764	172	704	0.5	0.538
19	1786	0.63	0.059	45	22	1	1	110	20	0.495	0.493	178	1787	0.6	0.299
22	52	0.44	0.46	46	40	0.544	0.533	122	120	0.588	0.588	184	422	0.572	0.567
23	1786	0.6	0.059	50	1786	0.6	0.299	126	120	0.578	0.588	200	14197	0.699	0
24	40	0.544	0.535	51	65536	1	1	128	2	0	0	204	131072	1	1
25	27	0.453	0.441	54	124	0.501	0.52	130	41	0.525	0.538	232	3572	0.6	0

6 Noise Effect

The effect of noise depends on the stability of attractor trajectories. This effect is clearly zero for "contracting" rules like rule 0, 8, etc. For these rules there is only one attractor, a stable fixed point with Jacobian equal to zero, so nothing happens and $S* = S = 0$.

The effect of noise is also zero for "flat" rules like the identity 204 or the shift 170. In these cases, the Jacobian is diagonal (eventually circularly shifted by one) and therefore the perturbation is simply maintained, so $S* = S = S_{\text{MAX}}$, irrespective of noise.

A similar result occurs for the linear rule 150. In this case the effect of noise propagates on all the lattice, but, since there is no preferred attractor, it has no influence on the probability distribution.

In general, the statistical properties of "chaotic" rules like 30, 110 are not affected by noise, except for rule 22 which shows a slight increase of entropy.

In many case there is an entropy reduction by noise. This is particularly evident for rule 232. In this case, most of attractors are composed by configuration made by clusters of zeros and ones of at least width 2. These configurations are

```
global A;
A(0 : 2^L − 1) ← 0;
k ← 1;                                    /* attractor index */
s_0 ← 0;                                  /* first configuration */
while s_0 < 2^L do
    s ← s_0;                              /* working configuration */
    while A(s) = 0 do
        A(s) ← −k                         /* negative indices denote transient conf. */
        s ← F(s);                         /* iteration */
    end
    if A(s) = −k then                     /* entered the cycle */
        while A(s) = −k do
            A(s) ← k;                     /* positive indices denote cycle */
            s ← F(s);
        end
        k ← k + 1;                        /* increase the attractor number */
    else                                  /* encountered another cycle */
        m ← |A(s)|;                       /* index of the other cycle */
        s ← s_0;                          /* restarting */
        relabel(s, m);
    end
    s_0 ← s_0 + 1;                        /* next configuration */
end
```

Algorithm 1: Finding all attractors and basins

stable with respect to single perturbations inside a cluster but are "connected" by perturbations at the boundary of a cluster.

For vanishing noise, these boundaries perform a self-annihilated random walk, so that the effect of the noise is to drive the system to the stable configurations 0 and 1.

This is also the case of rule 13 and 77, for which the asymptotic configuration is an alternation of zeros and ones, with occasionally clusters of double zeros which however can move and self-annihilate in the presence of noise. The same happens for instance to rule 58 for which a stable (translating) configuration is a repetition of $\{1, 1, 0, \}$, and the noise serves to remove the $\{1, 0, 1, 0\}$ defects.

For a few rules, entropy increases by noise, but not for the trivial effect of introducing disturbances, since the measure is always performed on the distribution of attractors. Let us take as an example rule 1.

In this rule, the local pattern $\{0, 0, 0\}$ gives 1, all other patterns give zero. All isolated zeros in the initial configuration are removed, while isolated 1 maintains (every other step). So, the effect of noise is that of creating cluster of isolated ones in a greater number with respect to the random initialization process, and entropy increases.

A similar increase in entropy is observed for rule 164. In this case, too, the effect of noise is that of connecting basins with few states (with low statistical weight starting from a random configuration) to configurations belonging to the largest basin.

```
global A;
relabel (s, m)
    while |A(s)| ≠ m do
        A(s) ← −m;
        s ← F(s);
    end
end
```
Algorithm 2: Relabeling transients

7 Numerical Determination of Attractors for Small Configurations

The problem of finding the attractor and its basin starting from a given configuration was investigated using a preimage algorithm in Ref. [13], which is much faster than investigating all the 2^L configurations for a lattice size L. However, since we are interested in enumerating all the attractors, we cannot avoid it.

We can find all attractors and their characteristics for a small-size cellular automata, by enumerating all of them, following their evolution until entering a limit cycle, and numbering these cycles (see Algorithms 1 and 2).

Let L be the size of CA. There are 2^L possible configurations s, which can be read as a number s in base-two representation from 0 to $2^L - 1$. Let A be an array of size 2^L set to zero. The idea is to label each configuration in a trajectory $\{s(0), s(1), \ldots\}$ with the same index $-k$, until we find a configuration S which is already marked (i.e., $A(S) \neq 0$. If $A(S) = -k$ then we proceed by inverting the sign of all configurations in the trajectory until we find a positive entry, i.e., we have explored all the cycle. We then increment k.

If however $|A(S)| \neq k$, it means that the trajectory belongs to the basin of an already encountered attractor, so we restart from $s(0)$ and we change all k with $|A(S)|$. We do not increase k after this phase.

When there are no more configurations we have found all attractors and basins and we set their number to $n = k - 1$. Then we can compute, for each attractor $k \in \{1, \ldots, n\}$, the length of its cycle (the number of entries $A(s) = k$) and its size (the number of entries $|A(s)| = k$).

8 Conclusions

We investigated elementary cellular automata as discrete dynamical systems. We obtained the complete phase-space (i.e. accessible states, attractors, basins of attraction) of all minimal ECA for small lattice sizes (up to $L = 17$), and, starting from a maximal entropy distribution (i.e., all configurations equiprobable), we have shown how the dynamics affects this distribution, implying in general a reduction of entropy.

We then investigated how a vanishing noise alters the phase-space landscape, connecting attractors and modifying the asymptotic probability distribution over configurations. For chaotic rules and in general for those rules for which the dynamics does not reduce the entropy, noise has no effect on it.

In many of the other cases, the noise decreases the entropy, since it connects unstable limit cycles or fixed points to the basin of stable one.

In a few case, the opposite happens. This, is probably due to the instability of the limit cycle of the largest attractor. The connections between stability (Lyapunov exponents) and attractors, and the application of statistical techniques for dealing with larger lattices will be the subject of a future work.

Acknowledgments. This publication was produced with the co-funding of European Union - Next Generation EU, in the context of The National Recovery and Resilience Plan, Investment 1.5 Ecosystems of Innovation, Project Tuscany Health Ecosystem (THE), CUP: B83C22003920001.

Disclosure of Interests. The authors declare no conflict of interest.

References

1. Bandini, S., Chopard, B., Tomassini, M. (eds.): ACRI 2002. LNCS, vol. 2493. Springer, Heidelberg (2002). https://doi.org/10.1007/3-540-45830-1. isbn: 9783540458302
2. Sloot, P.M.A., Chopard, B., Hoekstra, A.G. (eds.): ACRI 2004. LNCS, vol. 3305. Springer, Heidelberg (2004). https://doi.org/10.1007/b102055. isbn: 9783540304791
3. El Yacoubi, S., Chopard, B., Bandini, S. (eds.): ACRI 2006. LNCS, vol. 4173. Springer, Heidelberg (2006). https://doi.org/10.1007/11861201. isbn: 9783540409328
4. Umeo, H., Morishita, S., Nishinari, K., Komatsuzaki, T., Bandini, S. (eds.): ACRI 2008. LNCS, vol. 5191. Springer, Heidelberg (2008). https://doi.org/10.1007/978-3-540-79992-4. isbn: 9783540799924
5. Bandini, S., Manzoni, S., Umeo, H., Vizzari, G. (eds.): ACRI 2010. LNCS, vol. 6350. Springer, Heidelberg (2010). https://doi.org/10.1007/978-3-642-15979-4. isbn: 9783642159794
6. Sirakoulis, G.C., Bandini, S. (eds.): ACRI 2012. LNCS, vol. 7495. Springer, Heidelberg (2012). https://doi.org/10.1007/978-3-642-33350-7. isbn: 9783642333507
7. Was, J., Sirakoulis, G.C., Bandini, S. (eds.): ACRI 2014. LNCS, vol. 8751. Springer, Cham (2014). https://doi.org/10.1007/978-3-319-11520-7. isbn: 9783319115207
8. El Yacoubi, S., Was, J., Bandini, S. (eds.): ACRI 2016. LNCS, vol. 9863. Springer, Cham (2016). https://doi.org/10.1007/978-3-319-44365-2. isbn: 9783319443652
9. Mauri, G., El Yacoubi, S., Dennunzio, A., Nishinari, K., Manzoni, L. (eds.): ACRI 2018. LNCS, vol. 11115. Springer, Cham (2018). https://doi.org/10.1007/978-3-319-99813-8. isbn: 9783319998138
10. Gwizdaa, T.M., et al. (eds.): ACRI 2020. LNCS, Springer, Heidelberg (2021). https://doi.org/10.1007/978-3-030-69480-7. isbn: 9783030694807
11. Chopard, B., et al. (eds.): ACRI 2022. LNCS, Springer, Heidelberg (2022). https://doi.org/10.1007/978-3-031-14926-9. isbn: 9783031149269
12. Wolfram, S.: Statistical mechanics of cellular automata. Rev. Mod. Phys. **55**(3), 601 (1983). https://doi.org/10.1103/revmodphys.55.601

13. Wuensche, A., Lesser, M.: Global Dynamics of Cellular Automata: An Atlas of Basin of Attraction Fields of One-Dimensional Cellular Automata. CRC Press, Boca Raton (1992)
14. Wuensche, A.: Complexity in one-D cellular automata: Gliders, basins of attraction and the Z parameter. University of Sussex, School of Cognitive and Computing Sciences (1994)
15. Wuensche, A.: Classifying cellular automata automatically: Finding gliders, filtering, and relating space-time patterns, attractor basins, and the Z parameter. Complexity **4**(3), 47–66 (1999). https://doi.org/10.1002/(SICI)1099-0526(199901/02)4:3⟨47::AID-CPLX9⟩3.0.CO;2-V. issn: 1099-0526
16. Li, W., Packard, N.: The structure of the elementary cellular automata rule space. Complex Syst. **4**(3), 281–297 (1990)
17. Martinez, G.: A note on elementary cellular automata classification. J. Cellular Automata **8**(3–4), 233–259 (2013)
18. Stauffer, D.: Dynamics and damage spreading in cooperative systems: a numerical search for universality. In: Universalities in Condensed Matter, pp. 246–249. Springer, Heidelberg (1988). https://doi.org/10.1007/978-3-642-51005-2_50. isbn: 9783642510052
19. Bagnoli, F., Rechtman, R., Ruffo, S.: Damage spreading and Lyapunov exponents in cellular automata. Phys. Lett. A **172**(1–2), 34–38 (1992). https://doi.org/10.1016/0375-9601(92)90185-o. issn: 0375-9601
20. Shereshevsky, M.A.: Lyapunov exponents for one-dimensional cellular automata. J. Nonlinear Sci. **2**, 1–8 (1992). https://doi.org/10.1007/bf02429850. issn: 1432-1467
21. Vispoel, M., Daly, A.J., Baetens, J.M.: Lyapunov exponents of multi-state cellular automata. Chaos Interdisc. J. Nonlinear Sci. **33**(4), 043108 (2023). https://doi.org/10.1063/5.0139849. issn: 1089-7682
22. Baetens, J., Gravner, J.: Introducing Lyapunov profiles of cellular automata. J. Cellular Automata **13**(3), 267–286 (2018). issn: 1557-5969
23. Martins, M.L., de Resende, H.V., Tsallis, C., de Magalhes, A.C.N.: Evidence for a new phase in the Domany-Kinzel cellular automaton. Phys. Rev. Lett. **66**(15), 2045 (1991). https://doi.org/10.1103/physrevlett.66.2045. issn: 0031-9007
24. Bagnoli, F.: On damage-spreading transitions. J. Stat. Phys. **85**, 151–164 (1996). https://doi.org/10.1007/bf02175559
25. Baetens, J., Van der Meeren, W., De Baets, B.: On the dynamics of stochastic elementary cellular automata. J. Cellular Automata **12**(1–2), 63–80 (2017)
26. Cook, M.: Universality in elementary cellular automata. Complex Syst. **15**(1), 1–40 (2004)
27. Vichniac, G.Y.: Boolean derivatives on cellular automata. Physica D **45**(1–3), 63–74 (1990). https://doi.org/10.1016/0167-2789(90)90174-n. issn: 0167-2789
28. Bagnoli, F.: Boolean derivatives and computation of cellular automata. Int. J. Mod. Phys. C **3**(02), 307–320 (1992). https://doi.org/10.1142/s0129183192000257

A New Class of the Smallest 4-State Semi-symmetric FSSP Partial Solutions for 1D Arrays

Hiroshi Umeo(✉), Naoki Kamikawa, and Gen Fujita

University of Osaka Electro-Communication, Neyagawa-shi, Hastu-cho, 18-8, Osaka 572-8530, Japan
umeo@cyt.osakac.ac.jp

Abstract. A synchronization problem in cellular automata has been known as the Firing Squad Synchronization Problem (FSSP), where the FSSP gives a finite-state protocol for synchronizing a large scale of cellular automata. A quest for smaller state FSSP solutions has been an interesting problem for a long time. It has been shown by Balzer [1967], Sanders [1994], Berthiaume et al. [2004], and Ng [2011] that there exists no 4-state FSSP solution to one-dimensional (1D) arrays and rings. The number four is the state lower bound in the class of FSSP protocols. Umeo, Kamikawa and Yunès [2009], by introducing a notion of *full* versus *partial* FSSP solutions, provided a list of the smallest 4-state *symmetric* powers-of-2 FSSP solutions that can synchronize any 1D ring cellular automata of length $n = 2^k$ for any positive integer $k \geq 1$. Afterwards, Ng [2011] also added a list of *asymmetric* FSSP partial solutions, thus completing the 4-state powers-of-2 FSSP partial solutions. On the other hand, nothing has been explored for the smaller-state 1D array synchronizers. A question whether how many 4-state partial solutions there are for 1D arrays has been remained open. In this paper, we answer the question by providing a new class of the smallest 4-state FSSP partial solutions that can synchronize any 1D arrays of length $n = 2^k - 1$, 2^k, and $2^k + 1$ for any positive integer $k \geq 2$. We present a class of the smallest 4-state semi-symmetric array synchronizers: **4** solutions for 1D arrays of length $n = 2^k - 1$, **415** solutions for length $n = 2^k$, and **41** solutions for length $n = 2^k + 1$.

1 Introduction

We study a synchronization problem that gives a finite-state protocol for synchronizing a large scale of cellular automata. A synchronization problem in cellular automata has been known as the Firing Squad Synchronization Problem (FSSP) since its development, in which it was originally proposed by J. Myhill in Moore [8] to synchronize some/all parts of self-reproducing cellular automata. The FSSP has been studied extensively for more than sixty years in [1–20].

The minimum-time (i.e., $(2n-2)$-step) FSSP algorithm was developed first by Goto [5] for synchronizing any 1D array of length $n \geq 2$. The algorithm required many thousands of internal states for its finite-state realization. Afterwards, Waksman [17], Balzer [1], Gerken [4], Mazoyer [7], and Clergue, Verel, and Formenti [3] also developed a minimum-time FSSP algorithm and reduced the number of states required, each with 16, 8, 7 and 6 states, respectively. On the other hand, Balzer [1], Sanders [10], Berthiaume et al. [2], and Ng [9] have shown that there exists no 4-state synchronization algorithm. Thus, an existence or non-existence of 5-state FSSP protocol has been an open problem for a long time.

Umeo, Kamikawa and Yunès [12] answered partially by introducing a notion of *partial* versus *full* FSSP solutions and proposed a full list of the smallest 4-state symmetric powers-of-2 FSSP partial protocols that can synchronize any 1D ring cellular automata of length $n = 2^k$ for any positive integer $k \geq 1$. Afterwards, Ng [9] also added a list of asymmetric FSSP partial solutions for rings. Thus, the number four is the state lower bound in the class of FSSP protocols for 1D rings. A question: what is the class of the smallest FSSP protocols for 1D arrays? has been remained open.

In this paper, we answer the question by proposing a new class of the smallest 4-state FSSP protocols that can synchronize any 1D arrays of length $n = 2^k - 1, 2^k$, and $2^k + 1$ for any positive integer $k \geq 2$. We present a class of the smallest 4-state semi-symmetric 1D array synchronizers: that is, 4 solutions for arrays of length $n = 2^k - 1$, 415 solutions for length $n = 2^k$, and 41 solutions for length $n = 2^k + 1$. Throughout this paper, we only consider 1D arrays and rings. In Sect. 2, we give a description of the FSSP and review some basic results on ring and array FSSP algorithms. Section 3 presents a new class the semi-symmetric partial solutions for arrays. Due to the space of available, we only give an overview of those solutions. Details can be found in Kamikawa, Umeo, and Fujita [6]. Section 4 gives a summary and discussions of the paper.

2 Firing Squad Synchronization Problem on Arrays

2.1 Definition of the FSSP on Arrays

The FSSP on 1D arrays is formalized in terms of the model of cellular automata. Figure 1 shows a 1D array cellular automaton consisting of n cells, denoted by C_i, where $1 \leq i \leq n$. All cells are identical finite state automata. The array cellular automaton operates in lock-step mode such that the next state of each cell is determined by both its own present state and the present states of its right and left neighbors. All cells (*soldiers*), except one cell, are initially in the *quiescent* state at time $t = 0$ and have the property whereby the next state of a quiescent cell having quiescent neighbors is the quiescent state. At time $t = 0$ the cell C_1 (*general*) is in the *fire-when-ready* state, which is an initiation signal to the array.

The FSSP is stated as follows: given an array of n identical cellular automata, including a *general* cell which is activated at time $t = 0$, we want to give a description (state set and next-state transition function) of the automata so that, *at some future time*, all of the cells will *simultaneously* and, *for the first*

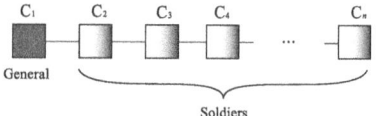

Fig. 1. One-dimensional array cellular automaton

time, enter a special *firing* state. The set of states and the next-state transition function must be independent of n. Without loss of generality, we assume $n \geq 2$. The tricky part of the problem is that the same kind of soldier having a fixed number of states must be synchronized, regardless of the length n of the array.

A formal definition of the FSSP on array is as follows: a cellular automaton $\mathcal{M}$ is a pair $\mathcal{M} = (\mathcal{Q}, \delta)$, where

1. $\mathcal{Q}$ is a finite set of states with three distinguished states G, Q, and F. G is an initial general state, Q is a quiescent state, and F is a firing state, respectively.
2. δ is a next-state function such that $\delta : \mathcal{Q} \cup \{*\} \times \mathcal{Q} \times \mathcal{Q} \cup \{*\} \to \mathcal{Q}$. The state $* \notin \mathcal{Q}$ is a pseudo state of the border of the array.
3. The quiescent state Q must satisfy the following conditions: $\delta(\text{Q},\text{Q},\text{Q}) = \delta(*,\text{Q},\text{Q}) = \delta(\text{Q},\text{Q},*) = \text{Q}$.

A cellular automaton of length n, $\mathcal{M}_n$ consisting of n copies of $\mathcal{M}$, is a 1D array of $\mathcal{M}$, numbered from 1 to n. Each $\mathcal{M}$ is referred to as a cell and denoted by C_i, where $1 \leq i \leq n$. We denote a state of C_i at time (step) t by S_i^t, where $t \geq 0$ and $1 \leq i \leq n$. A *configuration* of $\mathcal{M}_n$ at time t is a function $\mathcal{C}^t : [1, n] \to \mathcal{Q}$ and denoted as $\text{S}_1^t \text{S}_2^t .... \text{S}_n^t$. A *computation* of $\mathcal{M}_n$ is a sequence of configurations of $\mathcal{M}_n$, $\mathcal{C}^0, \mathcal{C}^1, \mathcal{C}^2, ..., \mathcal{C}^t, ...$, where $\mathcal{C}^0$ is a given initial configuration. The configuration at time $t + 1$, $\mathcal{C}^{t+1}$ is computed by synchronous applications of the next-state function δ to each cell of $\mathcal{M}_n$ in $\mathcal{C}^t$ such that:

$$\text{S}_1^{t+1} = \delta(*, \text{S}_1^t, \text{S}_2^t), \text{S}_i^{t+1} = \delta(\text{S}_{i-1}^t, \text{S}_i^t, \text{S}_{i+1}^t), 2 \leq i \leq n-1, \text{ and}$$
$$\text{S}_n^{t+1} = \delta(\text{S}_{n-1}^t, \text{S}_n^t, *).$$

A *synchronized configuration* of $\mathcal{M}_n$ at time t is a configuration $\mathcal{C}^t$, $\text{S}_i^t = \text{F}$, for all $1 \leq i \leq n$.

The FSSP is to obtain an $\mathcal{M}$ such that, for all $n \geq 2$,

1. A synchronized configuration at time $t = T(n)$, $\mathcal{C}^{T(n)} = \overbrace{\text{F} \cdots \text{F}}^{n}$ can be computed from an initial configuration $\mathcal{C}^0 = \text{G} \overbrace{\text{Q} \cdots \text{Q}}^{n-1}$.
2. For every t, i such that $1 \leq t \leq T(n) - 1$, $1 \leq i \leq n$, $\text{S}_i^t \neq \text{F}$.

No cells fire before time $t = T(n)$. We say that $\mathcal{M}_n$ is synchronized at time $t = T(n)$ and the function $T(n)$ is a time complexity for the synchronization. We denote the function as $T_G(n)$, if the general state G must be specified. According to conventions in FSSP studies, the right and left border state * is not counted as an internal state.

2.2 Full vs. Partial Solutions

One has to note that any solution in the original FSSP problem is to synchronize any array of length $n \geq 2$. We call it **full** solution. A solution that can synchronize some infinite set of arrays, but not all, is called **partial** solution. The notion of *partial* versus *full* FSSP solution was first introduced in Umeo, Kamikawa, and Yunès [12].

2.3 Recent Developments on Array and Ring FSSP Solutions

Here, we summarize recent developments in the array and ring FSSP solutions in Table 1. The table also shows the standpoint of the solutions given in Sect. 3.

Table 1. Comparison between array and ring FSSP solutions.

	Arrays	Rings
Lower Bounds in Synchronization Steps	Balzer [1], Goto [5]: $T(n) = 2n - 2$	Berthiaume et al. [2]: $T(n) = n$
State Lower Bounds in Full Solutions	Balzer [1], Sanders [10]: **No 4-state Full Solution**	Ng [9], Berthiaume et al. [2]: **No 4-state Full Solution**
State Lower Bounds in Partial Solutions	Yunès [20]: **No 3-state Partial Solution**	Umeo, Kamikawa, and Yunès [12]: **No 3-state Partial Solution**
Minimum- and Non-mimimum-time Full Solutions	Waksman [17]: 16 states Balzer [1] : 8 states Gerken [4]: 7 states Mazoyer [7]: 6 states Umeo et al. [14]: 6 states Clergue, Verel, and Formenti [3]: 6 states	Settle and Simon [11]: 8 states Umeo et al. [15]: 6 states
Partial Solutions for $n = 2^k - 1$	**Semi-symmetric 4 Partial Solutions (This paper)**	Umeo, Kamikawa, and Fujita [13]: **Full-symmetric 39 Solutions Asymmetric 132 Solutions**
Partial Solutions for $n = 2^k$	**Semi-symmetric 415 Partial Solutions (This paper)**	Umeo, Kamikawa, and Yunès [12]: **Full-symmetric 17 Partial Solutions** Ng [9]: **Asymmetric 80 Partial Solutions**
Partial Solutions for $n = 2^k - 1$	**Semi-symmetric 41 Partial Solutions (This paper)**	Umeo, Kamikawa, and Fujita [13]: **Shown to be absent**

62   H. Umeo et al.

	Right State			
Q	Q	G	A	*
Q	Q	•	•	Q
G		•	•	•
A			•	•
*				–

(Left State)

	Right State			
G	Q	G	A	*
Q	•	•	•	•
G		•	•	•
A			•	•
*				–

(Left State)

	Right State			
A	Q	G	A	*
Q	•	•	•	•
G		•	•	•
A			•	•
*				–

(Left State)

Fig. 2. Structure of 4-state symmetric transition tables for arrays. The symbol • indicates either any state in $\mathcal{Q}$ or - ($\notin \mathcal{Q}$) showing that there exists no state-transition at this point. The state F is only needed inside the 4 × 4 squares, but does not appear as a neighboring state. The state $* \notin \mathcal{Q}$ is a pseudo state of the right and left borders of the array. Half of the entry points in each square, including its diagonal points, must be specified to meet the symmetric property.

2.4 A Quest for 4-State Partial Solutions for Arrays

– **4-state 1D array cellular automata**
  Let $\mathcal{M}$ be a 4-state 1D array cellular automaton $\mathcal{M} = \{\mathcal{Q}, \delta\}$, where $\mathcal{Q}$ is an internal state set $\mathcal{Q} = \{\text{A}, \text{F}, \text{G}, \text{Q}\}$ and δ is a transition function such that $\delta : \mathcal{Q}^3 \cup \{*\} \to \mathcal{Q} \cup \{-\}$, where the symbol - means that there exists no state transition at this point. When we describe a solution, the symbol - is usually expressed as a blank square in the solution. Without loss of generality, we assume that Q is a quiescent state with a property $\delta(\text{Q}, \text{Q}, \text{Q}) = \delta(\text{Q}, \text{Q}, *) = \text{Q}$, G is a general state, A is an auxiliary state and F is the *firing* state, respectively. The initial configuration is $\text{G}\overbrace{\text{QQ}, ..., \text{Q}}^{n-1}$ for $n \geq 2$.
  Note that any FSSP solution can synchronize arrays even if it includes some redundant state-transitions, where the redundant means that the state-transition is not used in the synchronization process. In this paper we focus our attention to partial solutions with minimum number of state-transitions. The partial solutions given in Sect. 3 include no redundant state-transitions, thus, they consist of minimum number of state-transitions. Yunès [19] introduced a notion of Kolmogorov complexity of FSSP solution as the minimum-number of state-transitions used by the solution and Clergue, Verel, and Formenti [3] referenced the complexity in their 6-state solutions.

– **A computer investigation into 4-state FSSP solutions for arrays**
  Figure 2 is a symmetric 4-state transition table, where a symbol • in the table shows a possible state in $\mathcal{Q} = \{\text{A}, \text{F}, \text{G}, \text{Q}, -\}$. A blank square is uniquely determined by the symmetrical property such that $\delta(a, b, c) = \delta(c, b, a) = x$, where $a, c \in \mathcal{Q} \cup \{*\}$, $b \in \mathcal{Q} - \{F\}$ and $x \in \mathcal{Q} \cup \{-\}$. Note that we have totally 5^{25} possible state-transition tables. It is noted that the searching space for arrays is larger than rings.
  We make a computer investigation into the state-transition tables that might yield possible FSSP solutions. The outline of the searching procedures is as follows: We predicted that the partial solutions would exist for arrays of length around $n = 2^k, k = 1, 2, \ldots$, based on our experiments obtained in the earlier

studies of ring partial solutions in Umeo, Kamikawa and Yunès [12]. Most of the ring solutions obtained in Umeo, Kamikawa and Yunès [12] and Umeo, Kamikawa and Fujita [13] were more or less based on some kind of dichotomy and they were found for rings of length $n = 2^k$ and $n = 2^k - 1$.

Here the computer investigation consists of simulation and backtrack searching for state-transition tables. We compute the configuration together with the backtrack searching for small-size cellular space, say $n = 2, 3, 4, 5, 7, 8,$ and 9, depending on the search spaces of $n = 2^k - 1, 2^k$ and $2^k + 1$, respectively. The simulation was done starting from time $t = 0$ and then stopped at time $4n$, for an array of length n. The cut off time of the simulation was set to twice the upperbound of synchronization steps for ring solutions plus n steps. Our strategy of the searching for the state-transition tables is based on a backtracking. After getting a huge number of candidate solutions, we checked the validity of those solutions through computer for arrays of length $n = 2^k - 1, 2^k$ and $2^k + 1, 4 \leq k \leq 10$, respectively. Note that the solutions obtained are candidate solutions, since no formal proofs are given. Due to the limited space available, we omit the details of the simulation and backtracking searching strategy. The final solutions obtained and detailed computer procedures for the simulation and searching can be found in Kamikawa, Umeo and Fujita [6]. The backtracking searching technique was employed first successfully in Ng [9].

The computer we used for the quest is MacBook pro (16-inch, 2019) with 2.6 GHz, 6-core Intel Core i7 CPU, 16GB 2667 MHz DDR4 memory, and Mac Catalina 10.15.5 operating system. We implemented the backtrack search in C language using Apple clang version 11.0.3 C++ compiler with an optimized option -O2.

- **Full-symmetry, semi-symmetry and asymmetry in 4-state transition table**

  Here we introduce a property on symmetry and asymmetry in transition tables. A notion of semi-symmetry is a new one.
  - **Full-symmetry**: Consider a pair of two state-transitions such that $\delta(a, b, c) = \delta(c, b, a) = x$, where $a, c \in \mathcal{Q} \cup \{*\}$, $b \in \mathcal{Q} - \{F\}$ and $x \in \mathcal{Q} \cup \{-\}$. The pair of the state-transitions is said to be *full-symmetric*. A state-transition $\delta(a, b, a) = x$, where $a \in \mathcal{Q}$, $b \in \mathcal{Q} - \{F\}$ and $x \in \mathcal{Q} \cup \{-\}$, is said to be *self full-symmetric*. Any state-transition table consisting of full-symmetric and self full-symmetric state-transitions is said to be *full-symmetric*.
  - **Semi-symmetry**: The following two pairs of state-transitions such that $\delta(a, b, c) = d, \delta(c, b, a) = -$ and $\delta(a, b, c) = -, \delta(c, b, a) = d$, where $a, c \in \mathcal{Q} \cup \{*\}$, $b \in \mathcal{Q} - \{F\}$ and $d \in \mathcal{Q}$ are said to be *semi-symmetric*. A state-transition table consisting of full- and semi-symmetric state-transitions is said to be *semi-symmetric*.
  - **Asymmetry**: Consider a pair of two state-transitions such that $\delta(a, b, c) = x, \delta(c, b, a) = y, x \neq y$, where $a, c \in \mathcal{Q} \cup \{*\}$, $x, y \in \mathcal{Q}$ and $b \in \mathcal{Q} - \{F\}$. These two state-transitions are said to be *asymmetric state-*

Full-symmetric table

Semi-symmetric table

Asymmetric table

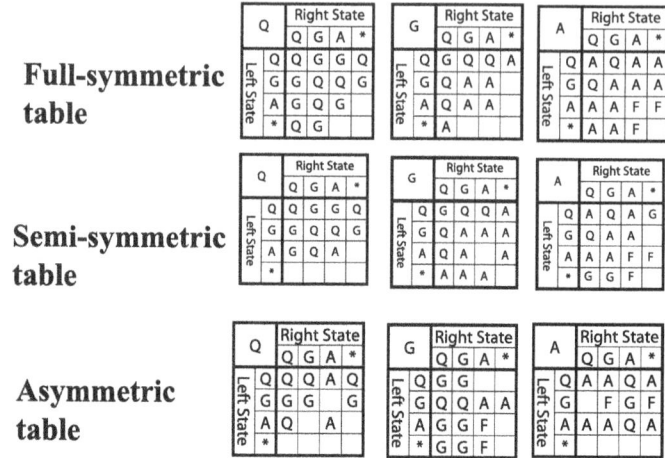

Fig. 3. An example of full-symmetric (top row), semi-symmetric (middle row) and asymmetric (bottom row) state-transition table, respectively. The blank square means there exist no state-transition at that point.

transition. A state-transition table including asymmetric state-transitions is called *asymmetric*.

Figure 3 shows an example of full-symmetric, semi-symmetric and asymmetric state transition tables.

3 Semi-symmetric Partial Solutions for 1D Arrays

In order to denote the ith partial solution for an infinite arrays of length n, we take the following notation: S_{n_i}.

3.1 Semi-Symmetric Partial Solutions for Arrays of Length $n = 2^k - 1$

With the help of a computer program we found **4** semi-symmetric solutions for arrays of length $n = 2^k - 1, k \geq 2$.

Theorem 1. *There exist* **4** *semi − symmetric 4-state* **partial** *solutions to the array FSSP for arrays of length* $n = 2^k - 1$ *for any positive integer* $k \geq 2$.

Figure 4 presents a state-transition table (leftmost) and snapshots (middle and rightmost) on an array of length $n = 15$ for the solution $S_{2^k-1\_001}$. Table 2 in Sect. 3.3 presents the number of state-transitions and synchronization steps of the solution $S_{2^k-1\_001}$. Details of the solutions $S_{2^k-1\_i}, 001 \leq i \leq 004$, can be found in Kamikawa, Umeo, and Fujita [6].

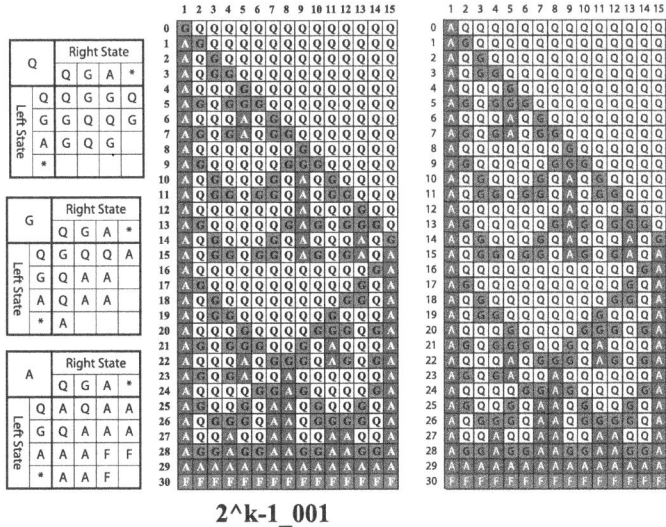

Fig. 4. State-transition table (leftmost) and snapshots (middle and rightmost) for semi-symmetric solution $S_{2^k-1\_001}$ on an array of length $n = 15$, each configuration starting from the general state G and the auxiliary state A.

3.2 Semi-symmetric Partial Solutions for Arrays of Length $n = 2^k$

With the help of a computer program we found **415** semi-symmetric solutions.

Theorem 2. *There exist **415 semi-symmetric** 4-state **partial** solutions to the **array** FSSP for arrays of length $n = 2^k$ for any positive integer $k \geq 1$.*

We classify those 415 solutions into 4 categories based on their synchronization steps as a multiplicative factor of n, for sufficiently large n. The breakdown of these solutions is as follows:

- $T_G(n) = \begin{cases} O(1), & k = 1 \\ 2n - 1, & k \geq 2 \end{cases}$ **Solutions:**
  We have got **343** $(2n-1)$-step semi-symmetric partial solutions $S_{2^k\_i}, 001 \leq i \leq 343$.
- $T_G(n) = \begin{cases} O(1), & k = 1 \\ 2n, & k \geq 2 \end{cases}$ **Solutions:**
  We have got **68** $2n$-step semi-symmetric partial solutions $S_{2^k\_i}, 344 \leq i \leq 411$.
- $T_G(n) = \begin{cases} O(1), & k = 1 \\ 2n + 1, & k \geq 2 \end{cases}$ **Solutions:**
  We have got **3** $(2n+1)$-step semi-symmetric partial solutions $S_{2^k\_i}, 412 \leq i \leq 414$.
- $T_G(n) = \begin{cases} 3n - 1, & k : \text{odd} \\ 4n - 1, & k : \text{even} \end{cases}$ **Solutions:**

One solution $S_{2^k\_415}$ obtained synchronizes any array of length $n = 2^k$ in
$$T_G(n) = \begin{cases} 3n-1, & k : \text{odd} \\ 4n-1, & k : \text{even} \end{cases}.$$

Figure 5 shows the state-transition table and snapshots on an array of length $n = 8$ and 16 for the solution $S_{2^k\_001}$ and $S_{2^k\_415}$. Table 2 in Sect. 3.3 presents the number of state-transitions and synchronization steps of these solutions. Details of the solutions $S_{2^k\_i}, 001 \leq i \leq 415$, can be found in Kamikawa, Umeo, and Fujita [6].

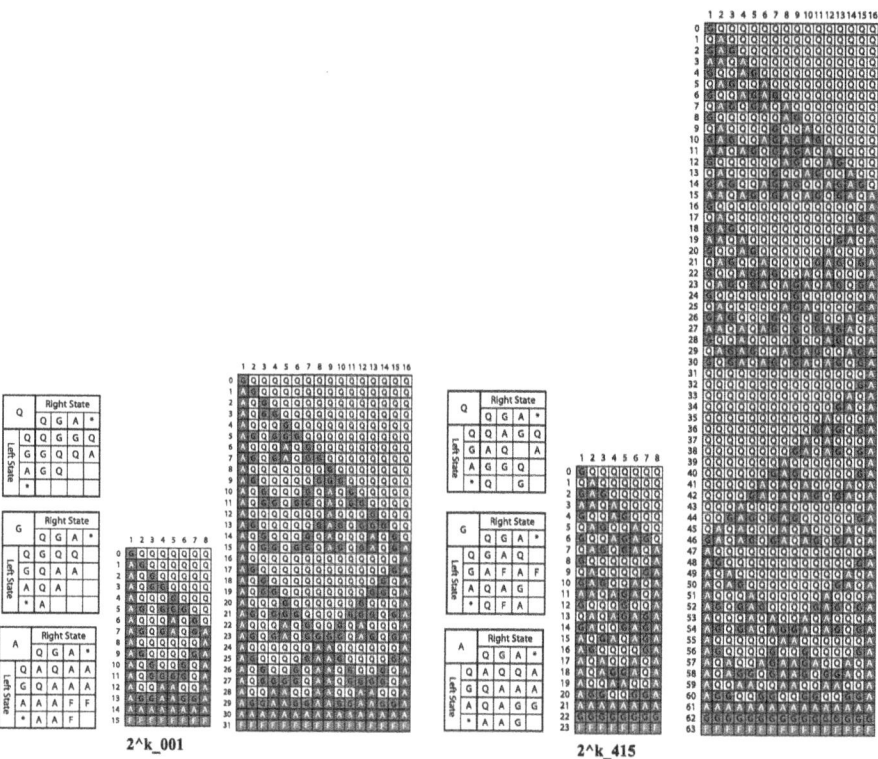

Fig. 5. State-transition tables and snapshots for semi-symmetric solutions $S_{2^k\_001}$ and $S_{2^k\_415}$ on an array of length $n = 8$ and 16.

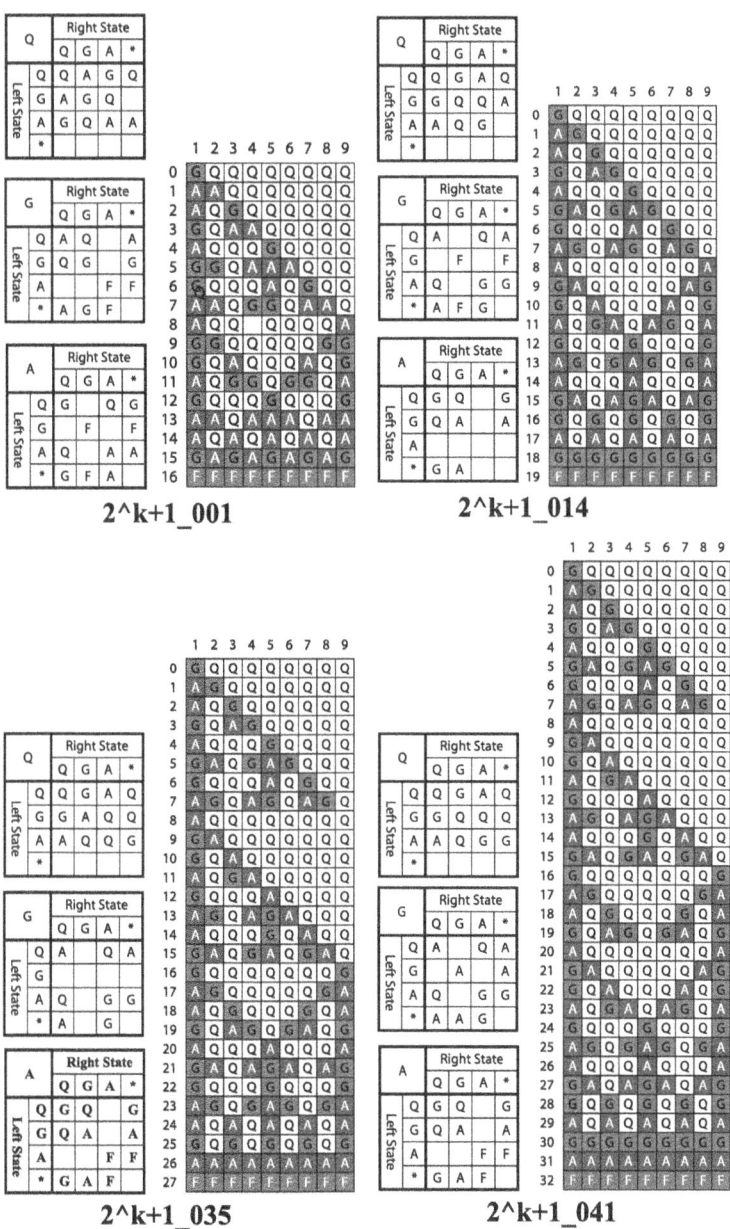

Fig. 6. State-transition tables and snapshots of 4 semi-symmetric solutions $S_{2^k+1\_001}$, $S_{2^k+1\_014}$, $S_{2^k+1\_035}$, and $S_{2^k+1\_041}$ on an array of length $n = 9$.

3.3 Semi-symmetric Partial Solutions for Arrays of Length $n = 2^k + 1$

With the help of a computer program we found **41** semi-symmetric solutions.

Theorem 3. *There exist **41 semi-symmetric** 4-state **partial** solutions to the **array** FSSP for arrays of length $n = 2^k + 1$ for any positive integer $k \geq 1$.*

We classify the 41 solutions into 7 categories based on their synchronization steps as a multiplicative factor of n, for sufficiently large n. The breakdown of these solutions is as follows:

- **Minimum-time Solutions:**
  We have got **4** minimum-time semi-symmetric partial solutions $S_{2^k+1\_i}$, $001 \leq i \leq 004$ operating in exactly $T_G(n) = 2n - 2, k \geq 1$ steps.
- **Nearly Minimum-time Solutions:** We have got **8** nearly minimum-time semi-symmetric partial solutions $S_{2^k+1\_i}$, $005 \leq i \leq 012$.
- $T_G(n) = 5n/2 - O(1)$ **Solutions:**
  We have got **3** $(5n/2 - O(1))$-step semi-symmetric partial solutions $S_{2^k+1\_i}$, $013 \leq i \leq 015$.
- $T_G(n) = 3n - O(1)$ **Solutions:**
  We have got **14** $3n - O(1)$-step semi-symmetric partial solutions $S_{2^k+1\_i}$, $016 \leq i \leq 029$.
- $T_G(n) = (3n - 3), (4n - 4)$ **Solutions:**
  We have got **4** $(3n - 4)$-, $(3n - 3)$-step semi-symmetric partial solutions $S_{2^k+1\_i}$, $031 \leq i \leq 033$.
- $T_G(n) = 7n/2 - O(1)$ **Solutions:**
  We have got **3** $(7n/2 - O(1))$-step semi-symmetric partial solutions $S_{2^k+1\_i}$, $031 \leq i \leq 033$.
- $T_G(n) = 4n - O(1)$ **Solutions:**
  We have got **5** $(4n - O(1))$-step semi-symmetric partial solutions $S_{2^k+1\_i}$, $037 \leq i \leq 041$.

Figure 6 shows the state-transition table and snapshots on an array of length $n = 9$ for the solutions $S_{2^k+1\_001}$, $S_{2^k+1\_014}$, $S_{2^k+1\_035}$, and $S_{2^k+1\_041}$. Table 2 presents the number of state-transitions and synchronization steps of these solutions. Details of the solutions $S_{2^k+1\_i}$, $001 \leq i \leq 041$, can be found in Kamikawa, Umeo, and Fujita [6].

Observation 3.1 (Swapping General States)

It is noted that the solution $S_{2^k-1\_001}$ has a property that both of the states G and A can be an initial general state and the solution presents successful synchronizations from each general state. See Fig. 4 (middle and rightmost). It can synchronize any array of length $n = 2^k - 1, k \geq 2$ in $T(n) = 2n$ steps from an initial configuration G $\overbrace{\text{Q} \cdots \text{Q}}^{n-1}$ and A $\overbrace{\text{Q} \cdots \text{Q}}^{n-1}$, respectively.

Table 2. Synchronization steps and number of state-transitions for semi-symmetric partial solutions for arrays of length $n = 2^k - 1, 2^k$, and $2^k + 1$. The symbol * denotes that the solutions never yield synchronized configurations from the state A. The symbol † means that the synchronization from the state A is successful for arrays of length $n = 2^k, k \geq 2$.

Semi-symmetric Solutions	Synchronization Steps	# of State-Transitions	Remarks
$S_{2^k-1\_001}$	$T_G(n) = 2n, k \geq 2$	37	$T_A(n) = 2n, k \geq 2$
$S_{2^k\_001}$	$T_G(n) = \begin{cases} 2, & k = 1 \\ 2n - 1, & k \geq 2 \end{cases}$	34	$T_A(n) = 2n - 1^\dagger, k \geq 2$
$S_{2^k\_415}$	$T_G(n) = \begin{cases} 3n - 1, & k = 1 \\ 4n - 1, & k \geq 2 \end{cases}$	40	$T_A(n) = \begin{cases} 3n - 1, & k = 1 \\ 4n - 1, & k \geq 2 \end{cases}$
$S_{2^k+1\_001}$	$T_G(n) = 2n - 2, k \geq 1$	33	*
$S_{2^k+1\_014}$	$T_G(n) = (5n - 7)/2, k \geq 1$	30	*
$S_{2^k+1\_035}$	$T_G(n) = (7n - 9)/2, k \geq 1$	31	$T_A(n) = (5n - 7)/2, k \geq 1$
$S_{2^k+1\_041}$	$T_G(n) = 4n - 4, k \geq 1$	34	$T_A(n) = 3n - 3, k \geq 1$

A similar observation can be made for some solutions $S_{2^k\_001}$, $S_{2^k\_415}$, $S_{2^k+1\_035}$, and $S_{2^k+1\_041}$. Table 2 shows the synchronization steps and number of state-transitions for the solutions $S_{2^k-1\_001}$, $S_{2^k\_001}$, $S_{2^k\_415}$, $S_{2^k+1\_001}$, $S_{2^k+1\_014}$, $S_{2^k+1\_035}$, and $S_{2^k+1\_041}$.

4 Summary and Discussions

A quest for the smaller-state FSSP solutions has been an interesting problem for a long time. At the first stage of the quest we have provided a new class of the smallest 4-state semi-symmetric FSSP protocols that can synchronize any 1D array of length $n = 2^k - 1$, 2^k, and $2^k + 1$, respectively, for any positive integer $k \geq 2$. That is, 4 solutions for arrays of length $n = 2^k - 1$, 415 solutions for length $n = 2^k$, and 41 solutions for length $n = 2^k + 1$. We conjecture that there exist no full-symmetric 4-state partial solutions for arrays. We haven't given any formal proof for the correctness of the solutions, thus the solutions presented are candidate solutions. Ng [9] gave a unified technique for showing the correctness of ring partial solutions. At the next stage a similar technique would be expected to employ for showing the correctness of array solutions.

The candidate solutions in this paper are found in arrays around length $n = 2^k$. All solutions found seem to be based on a kind of bisection divide-and-conquer algorithmic technique, yielding that no solution were obtained for arrays of length such as $n = 3^k, 5^k, 6^k, 7^k, k = 1, 2, 3, \ldots,$.

In the design of conventional FSSP algorithms, such as Goto [5], Balzer [1], Waksman [17], Umeo et al. [15], and Gerken [4], geometric design techniques based on signal propagations and interactions in space-time diagram are well-known ones. The solutions presented in this paper, however, seem to be based on some algebraic ones and they are related with Wolfram rule 60, 150 etc. How can we characterize those algebraic solutions in terms of Elementary cellular automata?

A question: how many 4-state asymmetric partial solutions exist for arrays? remains open. We think that there would be a large number of 4-state asymmetric partial solutions for arrays.

Acknowledgment. The authors would like to thank anonymous reviewers who helped in improving the final version.

References

1. Balzer, R.: An 8-state minimal time solution to the firing squad synchronization problem. Inf. Control **10**, 22–42 (1967)
2. Berthiaume, A., Bittner, T., Perković, L., Settle, A., Simon, J.: Bounding the firing synchronization problem on a ring. Theor. Comput. Sci. **320**, 213–228 (2004)
3. Clergue, M., Verel, S., Formenti, E.: An iterated local search to find many solutions of the 6-states firing squad synchronization problem. Appl. Soft Comput. **66**, 449–461 (2018)
4. Gerken, H.D.: Über Synchronisationsprobleme bei Zellularautomaten. Diplomarbeit, Institut für Theoretische Informatik, Technische Universität Braunschweig, pp. 1–50 (1987)
5. Goto, E.: A minimal time solution of the firing squad problem. In: Dittoed Course Notes for Applied Mathematics, vol. 298 (with an illustration in color), Harvard University, pp. 52–59 (1962)
6. Kamikawa, N., Umeo, H., Fujita, G.: The smallest 4-state partial synchronizers for 1D arrays. In: Annual Report for Institute of Informatics, University of Osaka Electro-Communication, Draft Version (2024)
7. Mazoyer, J.: A six-state minimal time solution to the firing squad synchronization problem. Theor. Comput. Sci. **50**, 183–238 (1987)
8. Moore, E.F.: The firing squad synchronization problem. In: Moore, E.F. (ED.) Sequential Machines, Selected Papers, pp. 213–214. Addison-Wesley, Reading (1964)
9. Ng, W.L.: Partial Solutions for the Firing Squad Synchronization Problem on Rings, pp. 1–363. ProQuest publications, Ann Arbor (2011)
10. Sanders, P.: Massively parallel search for transition-tables of polyautomata. In: Jesshope, C., Jossifov, V., Wilhelmi, W. (eds.) Proceedings of the VI International Workshop on Parallel Processing by Cellular Automata and Arrays, Akademie, pp. 99–108 (1994)
11. Settle, A., Simon, J.: Smaller solutions for the firing squad. Theor. Comput. Sci. **276**, 83–109 (2002)
12. Umeo, H., Kamikawa, N., Yunès, J.-B.: A family of smallest symmetrical four-state firing squad synchronization protocols for ring arrays. Parallel Process. Lett. **19**(2), 299–313 (2009)
13. Umeo, H., Kamikawa, N., Fujita, G.: A new class of the smallest FSSP partial solutions – symmetric and asymmetric synchronizers for 1D rings of length $n = 2^k - 1$. Acta Informaticax **58**, 427–450 (2021). https://doi.org/10.1007/s00236-020-00391-6
14. Umeo, H., Maeda, M., Hongyo, K.: A design of symmetrical six-state $3n$-step firing squad synchronization algorithms and their implementations. In: El Yacoubi, S., Chopard, B., Bandini, S. (eds.) ACRI 2006. LNCS, vol. 4173, pp. 157–168. Springer, Heidelberg (2006). https://doi.org/10.1007/11861201_21

15. Umeo, H., Yunès, J.-B., Kamikawa, N., Kurashiki, J.: Small non-optimum-time firing squad synchronization protocols for one-dimensional rings. In: Proceedings of the 2009 International Symposium on Nonlinear Theory and its Applications, NOLTA 2009, pp. 479–482 (2009)
16. Umeo, H., Yunès, J.-B., Kamikawa, N.: About 4-states solutions to the firing squad synchronization problem. In: Umeo, H., Morishita, S., Nishinari, K., Komatsuzaki, T., Bandini, S. (eds.) ACRI 2008. LNCS, vol. 5191, pp. 108–113. Springer, Heidelberg (2008). https://doi.org/10.1007/978-3-540-79992-4_14
17. Waksman, A.: An optimum solution to the firing squad synchronization problem. Inf. Control **9**, 66–78 (1966)
18. Wolfram, S.: A New Kind of Science, pp. 1280. Wolfram Media Inc. (2002)
19. Yunès, J. B.: A propos d'automates cellulaires, suivi par des fovtions booléennes. HDR (2007)
20. Yunès, J.B.: A 4-states algebraic solution to linear cellular automata synchronization. Inf. Process. Lett. **19**(2), 71–75 (2008)

Synchronization of Chains of Logistic Maps

Franco Bagnoli[1,2](✉), Michele Baia[1,2], Tommaso Matteuzzi[1], and Arkady Pikovsky[3]

[1] Department of Physics and Astronomy and CSDC, University of Florence, via G. Sansone 1, 50019 Sesto Fiorentino, Italy
{franco.bagnoli,michele.baia,tommaso.matteuzzi}@unifi.it
[2] INFN, Florence sect., Florence, Italy
[3] Institute of Physics and Astronomy, University of Potsdam, Karl-Liebknecht-Str. 24/25, 14476 Potsdam-Golm, Germany
pikovsky@uni-potsdam.de

Abstract. We study the synchronisation dynamics of a chain of coupled chaotic maps arranged in a parent-children configuration, with each parent node connected to two children nodes, one of which is also the parent of the next node. We analyse two distinct phenomena: parent-child synchronisation, characterised by the vanishing distance between consecutive nodes, and siblings synchronisation, for which the states of two children coalesce. Our investigation reveals strong differences in the synchronisation mechanisms between these two phenomena, which can be directly linked to the probability distribution of the parent. Theoretical analysis and simulations using the logistic map support our findings. Furthermore, we explore the propagation of perturbations in a synchronised chain by studying the rate of reabsorption of the perturbation.

Keywords: Master-slave (parent-child) synchronization · child-child synchronization · logistic map · machine-precision synchronization · chaotic critical slowing down

1 Introduction

Synchronization of chaotic systems can be interpreted in two ways. If the system is spatially extended, like, for instance, coupled map lattices [1], synchronization means spatially homogeneous systems. In this kind of systems, spatial coupling is generally symmetric, although also asymmetric coupling has been studied [2].

An extreme case of the asymmetric coupling is unidirectional coupling, which we will call parent-child (also called master-slave) coupling. It consists in feeding a part of the parent signal to the child system [3,4]. In this kind of coupling, the synchronization threshold is related to the chaotic properties of the parent system. When the coupling "strength" is large enough, the two systems synchronize (i.e., their states become identical), provided they have the same

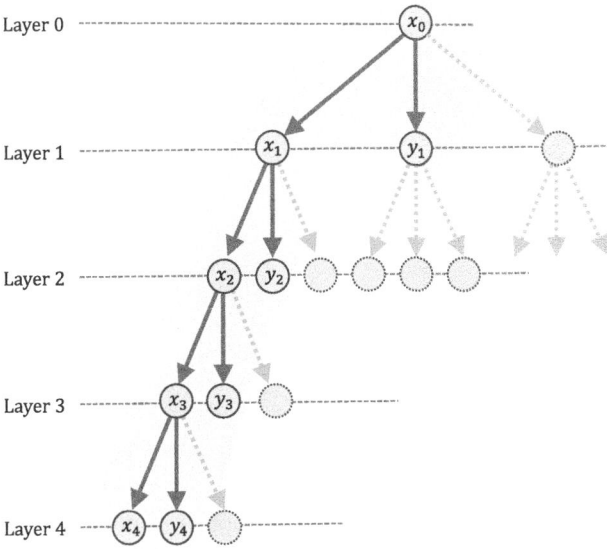

Fig. 1. The selection of chain. We start with a full Cayley tree (with levels denoted as "layers") with unidirectional coupling. For symmetry reasons, the behavior of all nodes in the same layer is the same, so we consider only two nodes ("siblings") for each layer except for the top one (layer 0).

parameters. According to the type of coupling, there can be a relation between the maximum Lyapunov exponent of the unperturbed system (the parent) and the critical strength for onset of synchronization.

It is also possible to feed the same parent signal to more than one child system, and study their mutual synchronization. In this case, the parent signal may act as a sort of common noise (although correlated with the children's state), and indeed synchronization by noise is a topic which has been studied [5], with even counter-intuitive results [6] showing that randomness may facilitate synchronization in the presence of chaos.

Beyond these foundational contributions, recent studies have pursued the generalization of synchronization principles. Investigations into phase synchronization [7] and generalized synchronization [8] with parameter mismatching have expanded the theoretical repertoire, while explorations into spatiotemporal chaos have uncovered synchronization phenomena in a variety of complex systems, including coupled maps lattices [9], coupled systems exhibiting spatiotemporal chaos [10], and Cellular Automata [11]. Notably, these attempts seek to establish connections between synchronization behaviors and universal classes such as directed percolation and the KPZ class [12–14]. Parent-child synchronization can be quite useful for data assimilation, i.e., extracting the parameters of a model from a time series [15].

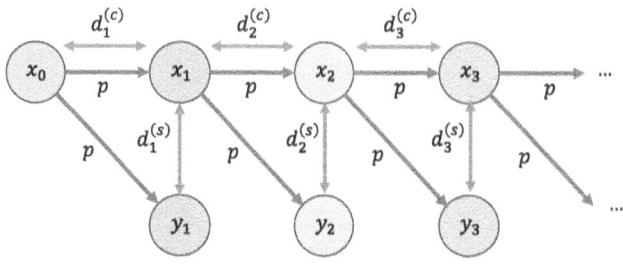

Fig. 2. Coupling and distances scheme. The sub-index i denotes layers. The arrow marked p denotes the one-directional coupling (see Eq. (1)). The two-directional arrows denotes distances. Node labeled x_0 is the first parent. It has two "children", x_1 and y_1, and we measure the distance $d_1^{(c)} = |x_0 - x_1|$ between the parent and one of its children, and $d_1^{(s)} = |x_1 - y_1|$ between the two "siblings". This scheme repeats itself at each layer i.

In the present work, we investigate the problem of parent-children synchronization in a unidirectional network represented by a Cayley tree (see Fig. 1). Since for symmetry reasons we can assume that the behavior of all nodes at the same level (denoted "layer" in the following) is the same, we take into consideration only two nodes (denoted "siblings") for each layer (except for the zero one).

In this way, we limit to the study of a chain of chaotic maps, see Fig. 2. We not only concentrate on the system formed by a parent and a child, but we also consider what happens to siblings that receive the signal from the same parent and how siblings synchronize among themselves and with the parent. As a particular example we examine the case of a chain of logistic maps. This chain exhibits quite interesting synchronization patterns.

After describing the general setup in Sect. 2, we study the case of logistic maps in Sect. 3. We shall show that quite unexpected phenomena appears, in particular that siblings synchronize among them for values of the coupling parameter much smaller than that required for the parent-child synchronization, and that this is related to the phenomenon of synchronization by common noise [5,6,12].

2 The Model

We consider one-dimensional maps, where the variable x evolves in discrete time steps, indexed by the time index t. In the following, we shall indicate by $x \equiv x(t)$ and $x' \equiv x(t+1)$.

Let us consider a chain scheme, in which the first map is an autonomous "pacemaker"

$$x_0' = f(x_0),$$

and all other maps are driven by their "parent map"

$$x_i' = (1-p)f(x_i) + pf(x_{i-1});$$
$$y_i' = (1-p)f(y_i) + pf(x_{i-1}). \tag{1}$$

as shown in Fig. 2. One can view this setup as the minimum sampling of a tree structure, see Fig. 1, where site 0 is the root, and on each layer $i > 0$ we consider just two elements x_i and y_i. This scheme repeats itself at the next layer. In essence, it is like considering a particular path in a full Cayley tree.

The elements x_i are coupled as in the Pecora-Carrol scheme [3,4] (when applied to maps) with parent x_{i-1}. At each layer two siblings x_i and y_i are driven by the same parent x_{i-1}; such a setup is usual in the theory of generalized synchronization, where a "clone" of a driven element is introduced [8].

Correspondingly, we will distinguish two types of synchronization in the chain. The *complete synchronization within a layer* occurs when $y_i(t) = x_i(t)$. We shall characterize it by computing the average difference

$$s_i = \langle d_i^{(s)} \rangle = \langle |x_i - y_i| \rangle = \frac{1}{T} \sum_{t=\tau}^{\tau+T} |x_i(t) - y_i(t)|,$$

where τ is a transient interval.

When s_i vanishes it means that both the "siblings" x_i and y_i follow the common drive x_{i-1} in the same way.

Another type of synchrony is the *complete synchronization across the layers* which means that $x_i(t) = x_{i-1}(t) = \ldots = x_0(t)$ for all t. In this case, all the elements exactly follow the driving one $x_0(t)$. One can characterize this regime by calculating the average in time of the difference across the layers

$$c_i = \langle d_i^{(c)} \rangle = \langle |x_i - x_{i-1}| \rangle = \frac{1}{T} \sum_{t=\tau}^{\tau+T} |x_i(t) - x_{i-1}(t)|,$$

checking if this quantity vanishes. One expects to observe a transition to the complete synchronization across the layers at some critical value of the coupling parameter p, since it is evident that $c_i > 0$ for $p = 0$ and $c_i = 0$ for $p = 1$. We denote this critical threshold (which does not depend on the index i) as π^c.

It is clear that from the complete synchronization across the layers it follows that the elements in one layer are also completely synchronized. However, as we shall see below, the opposite is not true and one can observe synchrony in a layer for a coupling p less than the threshold π^c. Moreover, synchronization in a layer can be observed for some layers and not observed for layers with higher index i.

As usual in the theory of complete synchronization, a criterion can be derived based on the analysis of the evolution of linear perturbations close to the synchronous state. Let us consider the synchronization of the first children x_1 with the parent x_0. Near the inter-layer synchronized state, $c_1 \simeq 0$, one can write $x_1(t) = x_0(t) + \delta x_1(t)$. Substitution of this in Eq. (1) yields

$$\delta x_1' = (1-p) \frac{\mathrm{d}f}{\mathrm{d}x}(x_0) \delta x_1.$$

Thus, the exponential growth rate of the perturbations in the correspondence of the complete synchrony is

$$\lambda_1^{(c)} = \ln(1-p) + \left\langle \ln \left| \frac{\mathrm{d}f}{\mathrm{d}x}(x_0) \right| \right\rangle = \ln(1-p) + \lambda_0, \qquad (2)$$

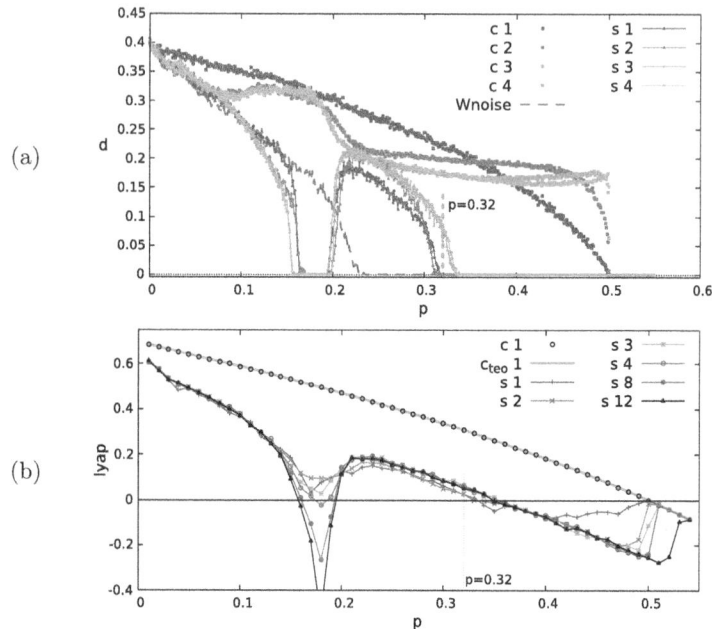

Fig. 3. (a) The distances between parent and child and between the two siblings for the first 5 layers of the logistic chain averaged over $R = 100$ repetitions. (b) The computed Lyapunov exponent $\lambda_1^{(c)}$ (c 1 in legend) and the Lyapunov exponent $\lambda_i^{(s)}$ of the chain for a single repetition. We also show the theoretical value of $\lambda_1^{(c)}$ (c_{teo} 1) estimated using Eq. 2. Other parameters $T = 10^4$, $\tau = 10^5$, $a = 4$.

where $\lambda_0 = \left\langle \ln \left| \frac{df}{dx}(x_0) \right| \right\rangle$ is the standard Lyapunov exponent of the driving map at level 0. Thus, the critical value of the coupling can be expressed as

$$\pi_1^{(c)} = 1 - \exp(-\lambda_0). \tag{3}$$

Assuming that the parent-children synchronization starts from the top layers (first x_1 with x_0, then x_2 with x_1, etc.), this relation is valid for the first non-synchronized layer, and thus in principle all layers should synchronize for the same value of $\pi^{(c)}$.

Consider now the synchronization within siblings in a layer. Assuming that the difference $\delta y_i = y_i(t) - x_i(t)$ is small, we get

$$\delta y_i' = y_i' - x_i' = (1-p)\left(\frac{df}{dx}(x_i)\right)(y_i - x_i) = (1-p)\left(\frac{df}{dx}(x_i)\right)\delta y_i.$$

With the same arguments as above, we obtain the exponential growth rate of the perturbation as

$$\lambda_i^{(s)} = \ln(1-p) + \left\langle \ln \left| \frac{df}{dx}(x_i) \right| \right\rangle = \ln(1-p) + \lambda_i, \tag{4}$$

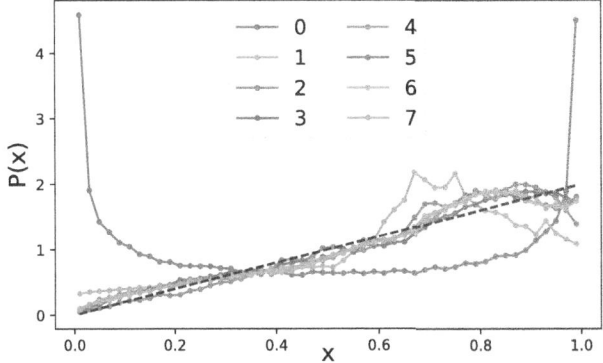

Fig. 4. Probability distribution $P(x_i)$ of maps along the chain for $p = 0.32$, and the linear approximation (straight black dashed line).

and therefore

$$\pi_i^{(s)} = 1 - \exp(-\lambda_i). \tag{5}$$

Notice that it is possible to obtain the value of the Lyapunov exponents of a map staring from the probability distribution $P(x)$ of the map itself,

$$\lambda = \int_0^1 \ln\left(\left|\frac{df}{dx}(x)\right|\right) P(x) dx. \tag{6}$$

We shall make use of this relation in the following.

3 Chain of Logistic Maps

The logistic map is defined by the recurrence equation

$$x' = ax(1-x),$$

and in the following we shall consider the case $a = 4$, for which the stationary probability distribution of the state variable x is $P_0(x) = \left(\pi\sqrt{x(1-x)}\right)^{-1}$. The corresponding Lyapunov exponent is

$$\lambda_0 = \int_0^1 \frac{\ln(4|1-2x|)}{\pi\sqrt{x(1-x)}} dx = \ln(2). \tag{7}$$

In Fig. 3 we show the main results of our simulations: in Fig. 3(a) we plot the average distances c_i and s_i. In Fig. 3(b) we show the computed sibling Lyapunov exponents, $\lambda_i^{(s)}$, Eq. (4), of the maps and the one estimated using Eq. (2), $\lambda^{(c)}$.

We can notice that the parent-children synchronization occurs for all layers around $\pi^{(c)} \simeq 0.5$, although with quite a different behavior near the transition

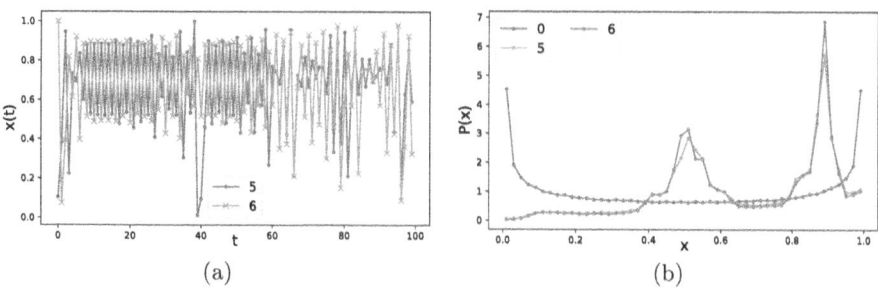

Fig. 5. (a) Time evolution of maps $x_5(t)$ and $x_6(t)$ for $p = 0.18$. Notice the period of alternating behavior, $\tau = 10^4$. (b) Probability distribution of the same maps and that of map 0, computed for $T = 10^5$, $\tau = 10^4$.

threshold, and that siblings synchronize among them for a value of $\pi_i^{(s)} \simeq 0.32 \ll \pi^{(c)}$. There is also a siblings synchronization window near $p = 0.18$.

We can get some theoretical estimation of the synchronized thresholds $\pi_i^{(c)}$ and $\pi^{(s)}$, Eqs. (3) and (5), exploiting Eq. (6).

For $\pi_i^{(c)}$ we can use the probability distribution of the unperturbed logistic map, Eq. (7), obtaining

$$\pi^{(c)} = \frac{1}{2},$$

which is in excellent agreement with simulations.

The siblings synchronization occurs quite far from the previous one, so we cannot assume that the probability distribution of the sibling x_i is the same of the unperturbed map, not even for $i = 1$. If we assume that the invariant probability distribution is flat, $P(x) = 1$, we get

$$\lambda_i(s) \simeq \int_0^1 \ln(4|1 - 2x|)\, dx = \ln(4) - 1. \tag{8}$$

so that the predicted threshold $\pi_i^{(s)}$ for which $\lambda_1^{(s)} = 0$ is

$$\pi_1^{(s)} = \frac{4 - e}{4} \simeq 0.32.$$

From Fig. 3(a) we see that these value is indeed close to the numerical results.

This result holds for all linear distributions $P(x) = ax + b$ due to the symmetry of the function $\ln(4|1 - 2x|)$ with respect to $x = 1/2$. Indeed, the probability distribution of maps near $p = 0.32$, Fig. 4, can be roughly approximated by $P(x) = 2x$ except for the unperturbed one, $i = 0$.

Going more in details, the actual siblings synchronization threshold $\pi_i^{(s)}$ shows a dependence on the layer index i. The probability distribution of the first layers $i = 1, 2, \ldots$ is below the linear approximation where the derivative $\ln(4|1 - 2x|)$ is larger, i.e., near $x = 1$, and this is consistent with the experimental observation that the corresponding synchronization threshold is smaller than that of maps at a deeper layer, Fig. 3(a).

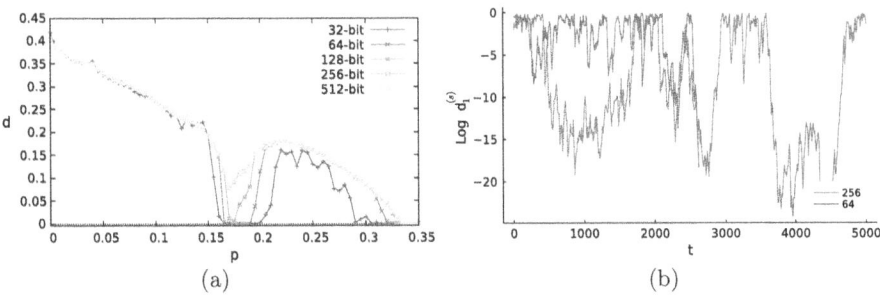

Fig. 6. (a) Siblings distances of the first node (s_1) computed with different machine precision ε (labeled as "i-bit", for $i = 32, 64, 128, 256$). Only for precision $\varepsilon \geq 256$ bit the synchronization window disappear. Here $T = 10^4$, $\tau = 10^5$, $R = 100$. (b) Plot of the time evolution of the siblings distance of the first nodes ($\log_{10} d_1^{(s)}$) computed with $\varepsilon = 64$ and $\varepsilon = 256$ machine precision at coupling parameter $p = 0.180$. Here $T = 5000$.

Notice that the corresponding Lyapunov exponent is still positive at the transition, Fig. 3(b). We shall discuss this aspect in the following subsection.

As we discussed above, we expect that the statistical properties along the layers for $p < \pi^c$ do not depend on i for large enough i, and correspondingly for these layers the Lyapunov exponent, Eq. (4), should also not depend on i. Comparing different curves in Fig. 3(b) one can conclude, that this is true for large domains of the parameter p for $i \geq 3$.

However, this is not true in an interval around $p = 0.18$, where we observe siblings synchronization, and close to the threshold $p = \pi^c = 0.5$, where also the value of the Lyapunov exponent is different for the various layers.

Let us first examine the synchronized region around $p \approx 0.18$, see Fig. 3.

Siblings Synchronization and Machine Precision Effects. Numerically, we observe that for p near 0.18, there are almost stable alternating patterns along the chain such that

$$x_{2m+1}(2k+1) = x_{2m}(2k) \simeq 0.5,$$
$$x_{2m}(2k+1) = x_{2m+1}(2k) \simeq 0.89,$$

see Fig. 5. This pattern dominates the dynamics at large i and makes the Lyapunov exponent negative. We shall report our investigations on this dynamics elsewhere.

While for large i the synchronized stability of this region can be explained by negative values of the transversal Lyapunov exponent $\lambda_i^{(s)}$, such an explanation is not valid for small i, where $\lambda_i^{(s)}$ is positive although close to zero. This "anomalous" synchronization can be explained by finite-precision effects (cf. Ref. [16]). Even when the transversal Lyapunov exponent is positive, the fluctuations of the distance between the systems x_i and y_i can make it so small, that the computer

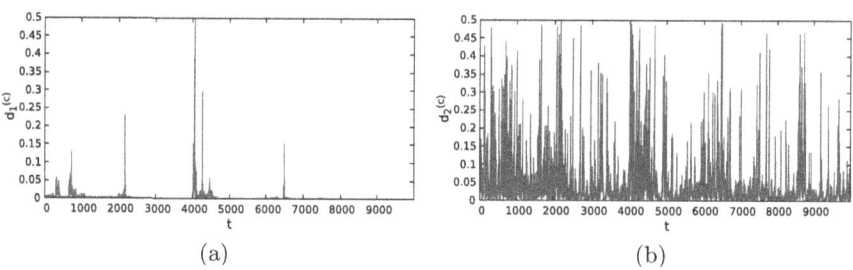

Fig. 7. Time evolution of the distance for $p = 0.499$. (a) Distance $d_1^{(c)}(t)$ between maps 1 and 0; (b) distance $d_2^{(c)}(t)$ between map 2 and 1.

representations of these states coincide. After this, the states remain identical even if there is transversal instability.

The easiest way to check this hypothesis is to repeat the numerical simulations with an increased precision ε, going, e.g., from $\varepsilon = 32$ bit up to $\varepsilon = 512$ bit, and comparing the differences $x_i - y_i$ with the previous simulations, which have obtained with 64 bit precision.

In Fig. 6(a), we present the siblings-distance $s_1 = \langle |x_1 - y_1| \rangle$, computed using different ε. As can be observed, the increase in precision results in a decrease in the width of the synchronized window, which disappears for $\varepsilon > 128$ bit.

In Fig. 6(b), we also show an example of the behavior of the siblings distance $d_1^{(s)}(t) = |x_1(t) - y_1(t)|$ in the first node for $p = 0.180$. Specifically, we display this distance for low ($\varepsilon = 64$) and high ($\varepsilon = 256$) precision. Evidently, higher machine precision allows for a more accurate numerical representation, resulting in a difference in the last digits that leads to a minimum distance between the two nodes on the order of 10^{-20}. For low ε, synchronization is caused by an approximation error: at the given coupling parameter, the distance eventually drops to a value smaller than the machine precision and is numerically zero afterward.

Parent-Child Synchronization: $p \lesssim \pi^c$. We see in Fig. 3 that the distance between parent and children exhibits a sharper transition to zero the larger the layer deep i.

Therefore, we expect that very close to the threshold $p = \pi^c$, the average distance between parent and child grows with the index i, i.e., maps with small i are almost synchronized to map 0, but this synchronization gets lost for large i. One can say that there is a "boundary layer" at small i, which grows to infinity as $p \to \pi^c$.

We note that the time dependence of the distance between parent and child shows an intermittent behavior, implying that its average value is computed over time periods in which this distance is very small alternating with sudden bursts.

This behavior is shown in Fig. 7, where we plot the temporal evolution of distances $d_1^{(c)}(t) = |x_1(t) - x_0(t)|$ and $d_2^{(c)}(t) = |x_2(t) - x_1(t)|$, near the transition

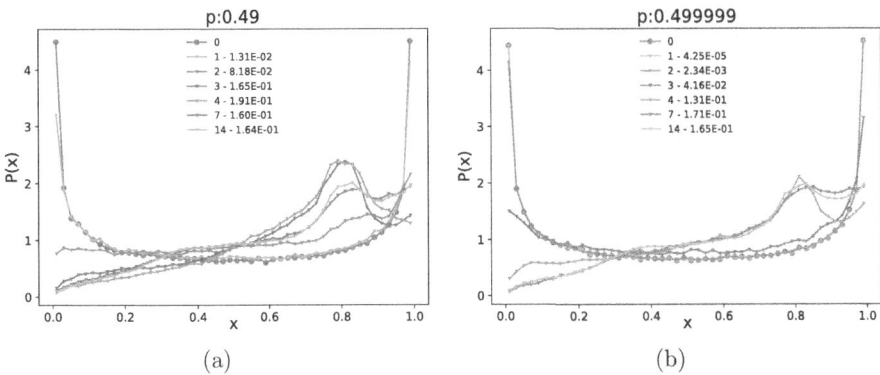

Fig. 8. Probability distribution of the maps in the chain. (a) $p = 0.490000$; (b) $p = 0.499999$. In the legend we also show the Wasserstein distance between the child x_i, $i \geq 1$, and the first node x_0. The distribution of the second layer became similar to the first only when the coupling parameter approaches the critical value (look at the distributions 0 and 1 in panel b).

threshold π^c: going down along the chain the frequency and intensity of bursts amplify.

Remarkably, the Lyapunov exponents $\lambda_i^{(s)}$, Eq (4), are significantly smaller than the exponent $\lambda^{(c)}$, Eq. (2). This can be attributed to the fact that the probability density of maps with $i > 0$ is quite distant from that of the unperturbed logistic map $i = 0$, and more similar to a linear distribution, due to action of the forcing from unit $i-1$. Therefore, the contribution of the highly unstable points at $x = 0$ and $x = 1$ to the Lyapunov exponent is smaller.

This results in the probability distributions at the nodes with large i being very different from that of the original logistic map. One expects that these nodes follow the parent one $i = 0$ only if the parameter difference $\pi^c - p$ decreases exponentially with i. Therefore, a node with large i undergoes a transition to the synchronized state only when the preceding node has a probability distribution sufficiently similar to that of x_0, and this happens extremely close to the critical value π^c.

To quantify this difference between two probability distributions, we compute the Wasserstein distance [17] among the first node x_0 and the other nodes x_i, $i > 0$. The estimated values are reported in the legend of Fig. 8. Very near to the synchronization threshold $\pi^{(c)}$, only the very first nodes approach the probability distribution of x_0. Above $\pi^{(c)}$ all probability distributions coincide, for very large simulation times, with that of x_0, but slightly above $\pi^{(c)}$ this convergence is so slow that one practically observes a shift of the transition threshold for the deeper nodes of the chain ($i > 10$).

Beyond the Synchronization Threshold $p \gtrsim \pi^c$. Let us investigate more in details the divergence of time scales near critical transitions, studying the critical slowing down beyond the synchronization threshold π^c.

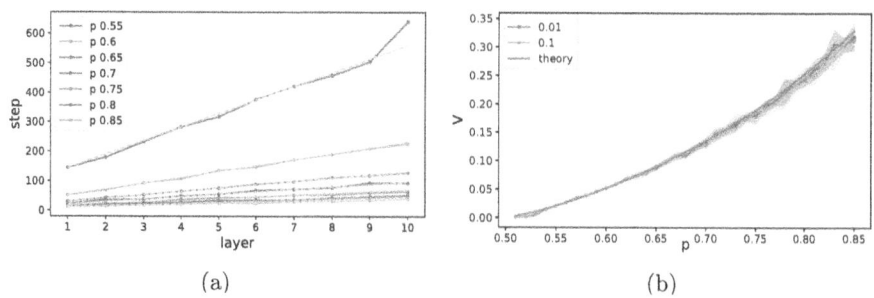

Fig. 9. (a) Number of time step needed to reabsorb a perturbation of amplitude η from a globally synchronized state. In dashed line the linear fit is shown. Here only first ten layers are shown, $T = 10^3$, $\eta = 0.001$. (b) Velocity versus coupling parameter for different initial noise values η (in legend) and theory result (Eq. (9)) for the logistics chain.

Let us assume that we are in the condition of complete synchronization ($p \gtrsim \pi_c$) and initialize the chain in a synchronized state $x_i(0) = x_0(0)$. We add a small perturbation to the initial condition of the first node $x_0(0) \to x_0(0) + \eta$. This perturbation grows and propagates along the chain, but eventually all the nodes return to synchrony. We analyze the time required for individual nodes to return to the synchronized state.

In Fig. 9(a), we show, for different coupling values, the number of time steps t required for the distance $d_i^{(0)}(t) = |x_i(t) - x_0(t)|$ between the first (x_0) and the ith node, to be below a threshold (10^{-8} in the following) in the case where a disturbance of intensity of the order of $\eta = 0.001$. As observed, for layers up to $i \simeq 10$, the trend is approximately linear. The slope estimated from the data allows us to define a propagation velocity of the perturbation, or more accurately, a re-absorption velocity of the disturbance along the chain. This velocity tends to zero at the threshold π^c, implying the divergence of the re-absorption times.

The trend of this velocity as a function of the coupling parameter p can be explained using the definition of the velocity-dependent Lyapunov exponent $\lambda(v)$ [18],

$$\lambda(v) = \lambda_0 + c(y_0, v) + g(y_0, v), \tag{9}$$

where $\lambda_0 = \ln 2$ is the Lyapunov exponent of the unperturbed map, and

$$\begin{aligned} y_0 &= 1 - v, \\ g(y_0, v) &= (1-v)\ln(1-p) + v \ln p, \\ c(y_0, v) &= -(1-v)\ln(1-v) - v \ln v. \end{aligned} \tag{10}$$

The re-absorption velocity V can be estimated imposing $\lambda(V) = 0$. The simulation and the theoretical results are compared in Fig. 9(b), which shows that this approximation is well verified.

4 Conclusions

In summary, in our study we explored the synchronization dynamics of a chain of coupled chaotic maps, focusing on parent-child and siblings synchronization. This configuration captures the essence of a process on a tree structure.

We found that both transitions are influenced by the probability distribution of the node to which they are coupled. Specifically, parent-child synchronization is governed by the probability distribution of the chaotic map defined on each node, while siblings synchronization can be approximated by assuming a flat distribution.

We also analyzed the importance of numerical precision in this type of simulations, showing how synchronization windows can depend on machine precision.

Additionally, we analyzed the reabsorption speed of perturbations starting from a synchronized state. Specifically, we investigated how quickly perturbations introduced at the initial node propagate and are reabsorbed in the synchronized chain.

Our findings contribute to a deeper understanding of synchronization in complex systems, with potential applications in diverse fields. Future research directions may involve investigating synchronization on complex networks and extensions to different map types and continuous dynamical systems.

Acknowledgments. This publication was produced with the co-funding of European Union - Next Generation EU, in the context of The National Recovery and Resilience Plan, Investment 1.5 Ecosystems of Innovation, Project Tuscany Health Ecosystem (THE), CUP: B83C22003920001.

Disclosure of Interests. The authors declare no conflict of interest.

References

1. Kaneko, K.: Overview of coupled map lattices. Chaos: Interdisc. J. Nonlinear Sci. **2**(3), 279–282 (1992). https://doi.org/10.1063/1.165869. issn: 1089-7682
2. Pikovsky, A., Rosenblum, M., Kurths, J.: Synchronization: A Universal Concept in Nonlinear Sciences. Cambridge University Press, Cambridge (2003)
3. Pecora, L.M., Carroll, T.L.: Synchronization in chaotic systems. Phys. Rev. Lett. **64**(8), 821–824 (1990). https://doi.org/10.1103/physrevlett.64.821
4. Pecora, L.M., Carroll, T.L.: Synchronization of chaotic systems. Chaos Interdisc. J. Nonlinear Sci. **25**(9), 097611 (2015). https://doi.org/10.1063/1.4917383
5. Flandoli, F., Gess, B., Scheutzow, M.: Synchronization by noise. Probabil. Theory Related Fields **168**, 511–556 (2017). https://doi.org/10.1007/s00440-016-0716-2. issn: 1432-2064
6. Maritan, A., Banavar, J.R.: Chaos, noise, and synchronization. Phys. Rev. Lett. **72**(10), 1451 (1994). https://doi.org/10.1103/physrevlett.72.1451
7. Rosenblum, M.G., Pikovsky, A.S., Kurths, J.: Phase synchronization of chaotic oscillators. Phys. Rev. Lett. **76**(11), 1804–1807 (1996). https://doi.org/10.1103/physrevlett.76.1804

8. Rulkov, N.F., et al.: Generalized synchronization of chaos in directionally coupled chaotic systems. Phys. Rev. E **51**(2), 980 (1995)
9. Jiang, Y., Parmananda, P.: Synchronization of spatiotemporal chaos in asymmetrically coupled map lattices. Phys. Rev. E **57**(4), 4135–4139 (1998). https://doi.org/10.1103/physreve.57.4135
10. Grassberger, P.: Synchronization of coupled systems with spatiotemporal chaos. Phys. Rev. E **59**(3), R2520–R2522 (1999). https://doi.org/10.1103/physreve.59.r2520
11. Bagnoli, F., Baroni, L., Palmerini, P.: Synchronization and directed percolation in coupled map lattices. Phys. Rev. E **59**(1), 409–416 (1999). https://doi.org/10.1103/physreve.59.409
12. Baroni, L., Livi, R., Torcini, A.: Transition to stochastic synchronization in spatially extended systems. Phys. Rev. E **63**(3), 036226 (2001). https://doi.org/10.1103/physreve.63.036226
13. Ahlers, V., Pikovsky, A.: Critical properties of the synchronization transition in space-time chaos. Phys. Rev. Lett. **88**(25), 254101 (2002)
14. Droz, M., Lipowski, A.: Dynamical properties of the synchronization transition. Phys. Rev. E **67**(5), 056204 (2003). https://doi.org/10.1103/physreve.67.056204
15. Bagnoli, F., Baia, M.: Synchronization, control and data assimilation of the lorenz system. Algorithms **16**(4), 213 (2023). https://doi.org/10.3390/a16040213
16. Pikovsky, A.S.: Comment on "chaos, noise, and synchronization". Phys. Rev. Lett. **73**(21), 2931 (1994)
17. Dobrushin, R.L.: Prescribing a system of random variables by conditional distributions. Theory Probabil. Appl. **15**(3), 458–486 (1970). https://doi.org/10.1137/1115049
18. Pikovsky, A.S.: Local Lyapunov exponents for spatiotemporal chaos. Chaos Interdisc. J. Nonlinear Sci. **3**(2), 225–232 (1993). https://doi.org/10.1063/1.165987. https://pubs.aip.org/aip/cha/article-pdf/3/2/225/18300069/225_1_online.pdf. issn: 1054-1500

Fusing Different Cellular Automata Models for Surface Flows in SCURRI: Viscosity Extension Step

Valeria Lupiano[1], Francesco Chidichimo[2], Paolo Catelan[2], Claudia R. Calidonna[3], and Salvatore Di Gregorio[3,4(✉)]

[1] CNR-IRPI, Research Institute for Geohydrological Protection, Via Cavour 6, 87036 Rende, Italy
valeria.lupiano@cnr.it
[2] Department of Environmental Engineering, University of Calabria, 87036 Rende, Italy
francesco.chidichimo@unical.it
[3] CNR-ISAC, Institute of Atmospheric Sciences and Climate, 88046 Lamezia Terme, Italy
cr.calidonna@isac.cnr.it
[4] Department of Mathematics and Computer Science, University of Calabria, 87036 Rende, Italy
salvatore.digregorio@unical.it

Abstract. Multicomponent or Macroscopic Cellular Automata (MCA) were conceived for modelling and simulating complex "macroscopic" phenomena such as surface flows, forest fires, bioremediation of soils, etc.

Many MCA models were developed for risky surface flows of different typology (lava flows, debris/mud/granular flows, lahars, snow avalanches, pyroclastic flows, rain runoff), these models share many "elementary" processes, but differ in some specificities of the particular phenomenon to be simulated. These specificities could be generalized to a single model (SCURRI: Simulation by Cellular Units of the Rheological RIsks) valid for each surface flow we deal with. The base of SCURRI is given by SCIDDICA, including its derivative MCA models LLUNPIY, VALANCA for simulating single phase surface flows (debris/mud/granular flows, lahars, snow avalanches). We introduce viscosity effects in SCURRI by adopting the approach of the MCA model SCIARA for simulating lava flows: a critical height, beyond which the flows become negligible, is introduced in SCURRI now version 01.

SCURRI-01 was applied to a real event (of course) different from a lava flow: the secondary lahar of February 2005 of Vascún Valley from Tungurahua. This event had been simulated satisfactorily by LLUNPIY. Simulations were also performed by SCURRI-01 with different values of critical height. Higher values of this parameter, produce clear viscosity increasing effects such as speed decrease which can reach so low values that we can no longer talk about lahars, but rather of slow flow landslide.

Keywords: Cellular Automata · Modelling & Simulation · Surface flows · Lahar · Natural Hazard

1 Introduction

This work represents a first, not trivial step to unify Multicomponent or better Macroscopic Cellular Automata (MCA) models [1, 2] for single phase surface flows in a global model SCURRI (Simulation by Cellular Units of the Rheological RIsks).

MCA were initially developed for the simulation of Etnean lava flows (first version of the model SCIARA [3]), and once validated the model would have been used, during emergency situations, for lava flows path forecasting.

Lava flow simulation implies an increase in viscosity which depends on lava cooling. A solution for such a problem was found by defining a "critical height" depending on the lava temperature and introducing an adherence effect: only that part of the lava, which is at a higher level than the critical height above the ground, can flow. The first versions of SCIARA were validated by satisfactorily simulating simple past events, for which viscosity effects emerge correctly [4], furthermore simulations during the 2001 Etnean crisis allowed the forecasting of correct scenarios, which differed by a few meters compared to the real lava flows [5].

The model SCIDDICA for simulating debris/mud/granular flows, that initially followed almost blindly the SCIARA schema [6], was subsequently improved with new sub-states and elementary processes for simulating very rapid debris flows, but the role of the adherence disappeared de facto in these improvements [7–9].

SCIDDICA could be considered as the starting point because it includes its derivative MCA models LLUNPIY [10–13] and VALANCA [14] for simulating lahars and snow avalanches, respectively.

The model SCURRI, here version 01, introduces in SCIDDICA the critical height in order to simulate viscosity effects. A first validation of the model is performed by simulating the February 2005 secondary lahar of Vascún Valley from Tungurahua [10, 13] with different values of critical height.

This choice is dictated by various reasons: this simulation has already been successfully done by LLUNPIY (so we have a comparison); it is possible to approximately consider the viscosity of the lahar constant throughout its entire path apart from the very first initial phase of detachment as the abundant rains have compensated for the erosion of the soil, then the lahar ends up in the Pastaza river in the final phase.

Results are interesting: critical height variation not only generates viscosity increasing effects, but the resulting surface flow may no longer be classified as a lahar, but rather a slow flow landslide.

The next chapter introduces SCURRI version 01 for secondary lahars, the third chapter presents shortly the February 2005 lahar of Vascún Valley from Tungurahua, afterwards a chapter regarding the SCURRI-01 simulations with different values of critical height, at the end discussion and conclusions.

2 The MCA Model SCURRI-01

2.1 The MCA Model SCURRI-01 for M&S Secondary Lahars

Here are reported the implemented and validated version of model SCURRI-01. This SCURRI-01 version is a two-dimensions MCA with a hexagonal tessellation and is defined by the quintuplet $<R, X, P, S, \tau>$ where:

- $R = \{(x, y) : x, y \in \mathbb{N}, 0 \leq x \leq l_x, 0 \leq y \leq l_y\}$ is the set of points with integer co-ordinates, that individuate the regular hexagonal cells; it specifies the region, where the phenomenon evolves, each cell corresponds to a portion of space, whose characteristics are expressed in terms of sub-states.
- $X = \{(0, 0), (1, 0), (0, 1), (-1, 1), (-1, 0), (0, -1), (-1, -1)\}$, the neighbourhood specifies the cells belonging to the neighbourhood: the cell itself (called central cell with index 0) and its adjacent cells with indexes $1 \leq i \leq 6$;
- P is the set of the global physical and empirical parameters, which account for the general frame of the model and the physical characteristics of the phenomenon (Table 1);
- S is the finite set of states of the finite automaton, embedded in the cell; it is equal to the Cartesian product of the sets of the considered sub-states (Table 2);
- $\tau : S^7 \rightarrow S$ is the cell deterministic state transition in R, its elementary processes are summarized in the next sub-section.

Table 1. Physical and empirical parameters

Parameters	Description
p_a, p_t	cell **a**pothem, **t**ime correspondent to a CA step
p_{fc}, p_{crh}	**f**riction **c**oefficient, **cr**itical **h**eight
p_{ed}	**e**rosion **d**issipation of energy
p_{pe}	**p**rogressive **e**rosion
p_{mt}	**m**obilization **t**hreshold
p_{khl}	**k**inetic **h**ead **l**oss (energy dissipation by turbulence)

Table 2. Sub-states

Sub-states	Description
S_A, S_D	cell **A**ltitude, tephra stratum **D**epth
S_{TH}, S_{KH}	the lahar **T**hickness and **K**inetic **H**ead inside the cell
S_X, S_Y	the co-ordinates **X** and **Y** of the lahar barycenter inside the cell
$S_E, S_{EX}, S_{EY}, S_{KHE}$ (6 components)	**E**xternal flow normalized to a thickness, **E**xternal flow co-ordinates **X** and **Y**, **K**inetic **H**ead of **E**xternal flow
$S_I, S_{IX}, S_{IY}, S_{KHI}$ (6 components)	**I**nternal flow normalized to a thickness, **I**nternal flow co-ordinates **X** and **Y**, **K**inetic **H**ead of **I**nternal flow

2.2 SCURRI-01 Implemented Transition Function

In the following, a sketch of the local elementary processes is given, in order to capture the mechanisms of the transition function; the execution of an elementary process updates the sub-states. The values of the cell sub-states are indicated by V plus their subscript; when sub-states need the specification of the neighborhood cell (the component), their index is indicated between round brackets, e.g., $V_I(i)$ means value of ith component of the Internal flow. Δ_Q means variation of the sub-state S_Q value.

Mobilization Effects (Bulking). When the kinetic head value overcomes an opportune threshold ($V_{KH} > p_{mt}$) depending on the soil features and its saturation state then a mobilization of the detrital cover occurs proportionally to the quantity overcoming the threshold, so the detrital cover depth diminishes as the debris thickness increases according to this empirical formula: "$p_{pe} \cdot (V_{KH} - p_{mt}) = \Delta_{TH} = -\Delta_D$"; the corresponding kinetic head loss is specified by: "$-\Delta_{KH} = p_{ed} \cdot (V_{KH} - p_{mt})$".

Turbulence Effect. The effect of the turbulence is modelled empirically by a proportional kinetic head loss at each LLUNPIY step: "$-\Delta_{KH} = p_{td} \cdot V_{KH}$".

Lahar Outflows. Regarding surface flows, all may be normalized to the third dimension, e.g. the lahar volume in the cell is expressed as thickness because the cell surface is constant, kinetic energy may be expressed in terms of kinetic head.

Terms of AMD are the height (h) of the neighborhood cells, whose difference has to be minimized by flows (f), whose sum is equal to the quantity q to be distributed in the neighborhood cells.

Formal specification follows: "$h(0) = V_A(0) + V_{KH}(0) + p_{crh}$"
"$h(i) = V_A(i) + V_{TH}(i), 1 \leq I \leq 6)$"
"$q = V_{TH}(0) - p_{crh} = \sum_{0<i<6} f(i)$"

AMD application minimizes "$\sum_{\{(i,j)|0 \leq i < j \leq 6\}}(|(h(i) + f(i)) - (h(j) + f(j))|)$"

The critical height parameter p_{crh} accounts for the quantity of lahar that cannot leave the cell.

The barycenter coordinates x and y of moving quantities are the same of all the lahar inside the cell and the form is ideally a "cylinder" tangent the next edge of the hexagonal cell. An ideal distance "d" is considered between the central cell lahar barycenter and the center of the adjacent cell i including the slope $\theta(i)$.

The $f(i)$ shift "$s(i)$" $1 \leq i \leq 6$, is computed for lahar flow according to the following simple formula: "$s(i) = v \cdot$", $1 \leq i \leq 6$.

The movement of $f(i)$ $1 \leq i \leq 6$, equates the barycenter movement of a cylindrical body on a constant slope with a constant friction coefficient where "g" is the gravity acceleration, and the initial velocity "$v = \sqrt{2g \cdot V_{KH}(0)}$".

The motion involves three possibilities: (1) only internal flow, i.e., the shifted cylinder is completely inside the central cell; (2) only external flow, all the shifted cylinder is inside the adjacent cell; (3) the shifted cylinder is partially internal to the central cell, partially external to the central cell, the flow is divided between the central and the adjacent cell, forming two cylinders with barycenter corresponding to the barycenter of the internal debris flow and the external debris flow. The kinetic head variation is computed according to the new position of internal and external flows, while the energy dissipation was considered as a turbulence effect in the previous elementary process.

Flows Composition. When debris outflows are computed, the new situation involves that external flows left the cell, internal flows remain in the cell with different co-ordinates and inflows (trivially derived by the values of external flows of neighbor cells) could exist. The new value of V_{TH} is given, considering the balance of inflows and outflows with the remaining debris in the cell. A kinetic energy reduction is considered by loss of flows, while an increase is given by inflows: the new value of the kinetic head is deduced from the computed kinetic energy. The coordinates determination is calculated as the average weight of V_X and V_Y considering the remaining debris in the central cell, the internal flows and the inflows.

3   The February 2005 Vascún Valley Lahar from Tungurahua

The Vascún Valley is located on the NE flank of Tungurahua and leads directly to the Pastaza river; the slopes in the upper 3 km of the valley are steep, being higher than 35−40°, while in the lower 2–3 km the slopes are much gentler ranging from 6−20°. The channel of the Vascún stream is very sinuous in the first 1–2 km, passing through a series of tight 90° bends. The thermal structure 'El Salado Baths' lies along the banks of the river, while the town of Baños is located in the depositional area.

Lahars are one of the most dangerous and destructive natural events for communities located near volcanoes. As the flows of sediment and water travel down the volcano's flanks, they incorporate a large quantity of sediment and rock fragments, making lahars very destructive. Their size and speed constantly change along the path. Moving away from the volcano and with decreasing slopes, the lahar begins to lose its heavy load of sediment and reduces in size.

On February 12, 2005, heavy rainfalls caused the remobilization of ash fall deposits, generating lahars in the Vascún stream. The mean velocities of flows were estimated on the base of records of alert instrumentation. They varied between 7 m/s and 3 m/s, depending on the cross sections [15]. The lahar ran through the valley approximately for 10 km and flooded El Salado Baths during the passage of the flows, then reached the Pastaza river.

This event was chosen to evaluate this first version of SCURRI both because it was simulated with good approximation with LLUNPIY, and because a constant value of viscosity can be assumed, given the considerable precipitation, whose contribution maintained an approximately constant ratio of debris/water during the event, with the sole exception of the detachment phase (6.5 s).

4   Lahar Simulations with Different Critical Heights

SCURRI-01 is implemented in C++ and has run in different Windows PC. Sub-state matrices (1570 × 2345 cells) are memorized in order to save simulation steps and to graphically represent the development of the phenomenon. Of course, a simulation doesn't involve only computation, it could be interrupted for a finer analysis of results: a complete simulation takes one to two days with data saving every 2500 steps.

The simulation of 2005 event was performed on a DEM with 1 m cell size, the same one used in [13, 15] and provided by Prof. Dr M. Sheridan; furthermore, a uniform thickness of 5 m was imposed for detrital cover, because detailed surveys were not available [13]. The simulations did not include the whole lahar extension, in fact, a stretch of about 2.3 km was considered: from an elevation of 2,150 m a.s.l. to the Pastaza river. The considered area, therefore, does not include the lahar source area.

All simulations were run with the parameter values shown in Table 3. Variations in p_{crh} values are made in order to evaluate different viscosity effects:

Table 3. SCURRI-1 adopted parameters

Parameter	Value	Parameter	Value
p_a	5 m	p_{ed}	0.3
p_t	0.065 s	p_{pe}	0.001
p_{fc}	0.08	p_{mt}	0.5 m
p_{khl}	0.00005	p_{crh}	0.001/0.005/0.05/0.25 m

Figure 1 depicts the thickness output of the four simulations with different value of p_{crh} at step 7000, 7' and 30" seconds from triggering. The lahar, in the simulations with $p_{crh} = 0.001$ m and $p_{crh} = 0.005$ m, easily goes beyond the El Salado Bath checkpoint, while with $p_{crh} = 0.05$ m the lahar is about a hundred meters behind, with $p_{crh} = 0.25$ m, then, it even only traveled about 200 m from the starting point. De facto as the critical height increases, the length traveled by the flows decreases, while the arrival times, at the first checkpoint, increase (see also Table 4). The flows simulated with $p_{crh} = 0.001$ m took 63 min to reach Baños checkpoint, 5 more minutes were needed with $p_{crh} = 0.005$ m, while a delay of 24 min, with respect to the first simulation, was observed with $p_{crh} = 0.05$ m (Table 4). The simulation with $p_{crh} = 0.25$ m, at step 250000 (ca 4 h and 30 min from triggering), is still about 225 m far from the Baños checkpoint. This critical height is such that it also affects the flows velocity.

In fact, at step 250000 the maximum velocity is so much less than 1 m/s that the phenomenon can no longer be considered a lahar but, rather, a slow flow landslide.

Moreover, after less than a hundred steps, the flow progression is practically undetectable. When the flows arrive at Pastaza River checkpoint, the simulations were considered ended. The lahars with $p_{crh} = 0.001$ m and $p_{crh} = 0.005$ m flow into the Pastaza River 5 min apart (Table 4, Fig. 2a and 2b), while the one with p_{crh}=0.05m has a delay of 31 min (Table 4, Fig. 2c). As already mentioned the situation with $p_{crh} = 0.25$ m (Fig. 2d) displays a process that stops before reaching Baños.

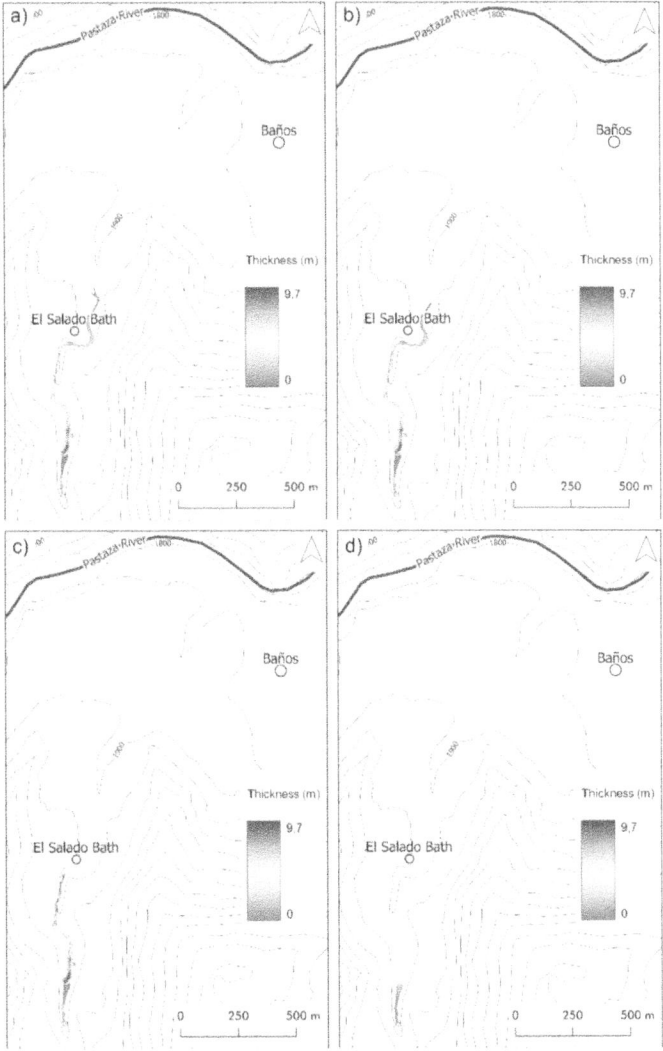

Fig. 1. Thickness output of 2005 simulations at step 7000 with different value of critical height: a) $p_{crh} = 0.001$ m; b) p_{crh}= 0.005 m; c) $p_{crh} = 0.05$ m; d) p_{crh}= 0.25 m.

Table 4. Minutes and simulation steps (ss) elapsed between starting point and checkpoints:

	El Salado Bath	Baños	Pastaza River
estimated time	6'–9'	66'	70'
$p_{crh} = 0.001$ m	6' (5600 ss)	63' (58400 ss)	67' (62000 ss)
$p_{crh} = 0.005$ m	6.30' (6000 ss)	68' (63000 ss)	72' (67000 ss)
$p_{crh} = 0.05$ m	9' (8200 ss)	92' (85400 ss)	98' (90500 ss)
$p_{crh} = 0.25$ m	18' (16600 ss)	-	-

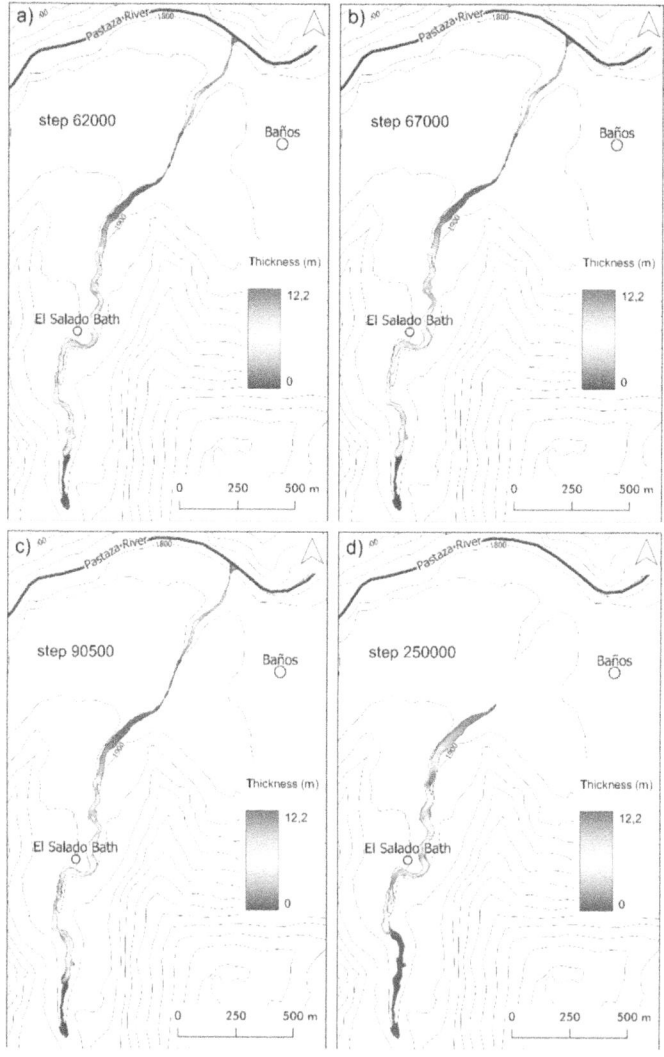

Fig. 2. Thickness output of 2005 simulations at Pastaza River checkpoint, with different values of critical height: a) $p_{crh} = 0.001$ m; b) $p_{crh} = 0.005$ m; c) $p_{crh} = 0.05$ m; d) $p_{crh} = 0.25$ m.

In Table 5 the average velocity in various steps are listed. Higher values correspond to the sections with a greater slope, as the slope decreases the average velocity also decreases. Clearly, as the viscosity increases, from $p_{crh} = 0.001$ m to $p_{crh} = 0.25$ m, the average velocity decreases.

The simulations data are compared with the phenomenon ones: estimated time (however data are limited) taken to reach checkpoints.is the main factor to be considered, the areas covered by the lahar, after the lahar has reached them, do not vary in size over time [13, 15, 16]. The best simulations (Table 4) correspond to p_{crh} minimum values 1 mm, 5 mm and may be positively compared with LLUNPIY simulations [10].

Table 5. Average velocity of SCURRI-1 simulations.

	Step 10000	Step 20000	Step 40000	Step 60000
$p_{crh} = 0.001$ m	2,12 m/s	1,20 m/s	0,67 m/s	0,56 m/s
$p_{crh} = 0.005$ m	2,11 m/s	1,19 m/s	0,67 m/s	0,47 m/s
$p_{crh} = 0.05$ m	1,75 m/s	1,13 m/s	0,64 m/s	0,45 m/s
$p_{crh} = 0.25$ m	0,63 m/s	0,77 m/s	0,58 m/s	0,41 m/s

5 Discussion and Conclusions

MCA models are a semi-empirical and alternative solution to partial differential equation systems. The models (the validated ones) work satisfactorily in well-defined phenomenological contexts; this means that the development of the phenomenon to be simulated can emerge significantly even with opportunely approximated or simplified data.

The MCA models are developed incrementally by adding new sub-states and expanding elementary processes; modeling with MCAs proceeds from "simple" to "complex" in order to capture a more extended phenomenology [17].

The validation of the models based on the simulations of various real events is essential, as well as knowing the scope of application of these models [17]. In SCIARA simulations, the time variation of lava viscosity at different points is not known, but values of the adherence, the reference parameter for the viscosity, are well known. Adherence corresponds in SCURRI-01 to the critical height parameter [3–5]. SCURRI-01 is the fusion of SCIARA and SCIDDICA that includes AVALANCHE and LLUNPIY in their latest versions. Both models were validated in different contexts.

Two simulations with constant critical height of 0.001 m and 0.005 m match the existing partial data regarding the 2005 event as reported in Table 4.

Even if there is uncertainty about the values of the real phenomenon, the travel time data corresponding to the simulated cases $p_{crh} = 0.05$ m and $p_{crh} = 0.25$ m are too far from those estimated; such simulations clearly show a decrease in speed with increasing critical height until the lahars assume a behavior typical of slow single-phase flows.

We thus propose a further verification of these SCURRI-01 results for single-phase flows. An extension of this work to a wider range of cases has already begun.

Acknowledgments. This research took shape in the context of the project DISCURRI-12 (Design and Improvement of Simulations by Cellular automata Units for Rheological Risks Investigation) coordinated for the Italian team by Dr. Francesco Chidichimo.

Disclosure of Interests. The authors have no competing interests to declare that are relevant to the content of this article.

References

1. Di Gregorio, S., Serra, R.: An empirical method for modelling and simulating some complex macroscopic phenomena by cellular automata. FGCS **16**, 259–271 (1999)
2. Avolio, M.V., et al.: An extended notion of cellular automata for surface flows modelling. WSEAS Trans. Comput. **4**(2), 1080–1085 (2003)
3. Barca, D., Crisci, G.M., Di Gregorio, S., Nicoletta, F.: Cellular automata methods for modeling lava flow, simulation of the 1986–1987 eruption, Mount Etna. In: Kilburn, C.R.J., Luongo, G. (eds.) Chapter Twelve of Active Lavas, pp. 291–309. UCL Press London (1993)
4. Barca, D., Crisci, G.M., Di Gregorio, S., Nicoletta, F.: Cellular automata for simulating lava flows, a method and examples of the Etnean eruptions. Transp. Theory Stat. Phys. **23**(1–3), 195–232 (1994)
5. Crisci, G.M., Di Gregorio, S., Rongo, R., Spataro, W.: Application of the cellular automata model SCIARA to the 2001 Mt. Etna Crisis. In: Bonaccorso, A., Calvari, S., Coltelli, M., Del Negro, C., Falsaperla, S. (eds.) Mt. Etna: Volcano Laboratory 2004 Geophysical Monograph Series, vol. 143, pp 343–356. Edited by AGU (2004)
6. Avolio, M.V., et al.: Simulation of the 1992 Tessina landslide by a cellular automata model and future hazard scenarios. Int. J. Appl. Earth Obs. **2**(1), 41–50 (2000)
7. Avolio, M.V., et al.: Debris flows simulation by cellular automata, a short review of the SCIDDICA models. Ital. J. Eng. Geol. Environ. **Special issue V**, 387–397 (2011)
8. Avolio, M.V., Di Gregorio, S., Lupiano, V., Mazzanti, P.: SCIDDICA-SS3, a new version of cellular automata model for simulating fast moving landslides. J. Supercomput. **65**, 682–696 (2013)
9. Lupiano, V., Machado, G.E., Molina, L.P., Crisci, G.M., Di Gregorio, S.: Simulations of flow-like landslides invading urban areas, a cellular automata approach with SCIDDICA. Nat. Comput. **17**, 553–568 (2017)
10. Machado, G., Lupiano, V., Avolio, M.V., Gullace, F., Di Gregorio, S.: A cellular model for secondary lahars and simulation of cases in the Vascún Valley Ecuador. J. Comput. Sci. **11**, 289–299 (2015)
11. Lupiano, V., Machado, G., Crisci, G.M., Di Gregorio, S.: A modelling approach with Macroscopic cellular automata for hazard zonation of debris flows and lahars by computer simulations. Int. J. Geol. **9**, 35–46 (2015)
12. Lupiano, V., et al.: LLUNPIY simulations of the 1877 northward catastrophic lahars of Cotopaxi volcano (Ecuador) for a contribution to forecasting the hazards. Geosciences **11**, 81 (2021)
13. Lupiano, V., et al.: From examination of natural events a proposal for risk mitigation of lahars by a cellular automata methodology, a case study for Vascún Valley, Ecuador. Nat. Hazards Earth Syst. Sci. **20**, 1–20 (2020)
14. Avolio, M.V., Errera, A., Lupiano, V., Mazzanti, P., Di Gregorio, S.: A cellular automata model for snow avalanches. J. Cell. Autom. **12**(5), 309–332 (2017)

15. Williams, R., Stinton, A., Sheridan, M.: Evaluation of the Titan2D two phase flow model using an actual event, case study of the 2005 Vascún Valley lahar. J. Volcanol. Geoth. Res. **177**(4), 760–766 (2008)
16. IGEPN Homepage www.igepn.edu.ec. Reports February 2005. Accessed 10 June 2024
17. Iovine, G., Di Gregorio, S., Sheridan, M.F., Miyamoto, H.: Modelling, computer-assisted simulations, and mapping of dangerous phenomena for hazard assessment. Environ. Model. Softw. **22**(10), 1389–1391 (2007)

Chaos in a Two-Dimensional Magneto-Hydrodynamic System

Franco Bagnoli[1,2](✉) and Raúl Rechtman[3]

[1] Department of Physics and Astronomy and CSDC, University of Florence,
via G. Sansone 1, 50019 Sesto Fiorentino, Italy
franco.bagnoli@unifi.it
[2] INFN, sez. Firenze, Florence, Italy
[3] Instituto de Energías Renovables, Universidad Nacional Autónoma de México,
62580 Temixco, Morelos, Mexico
rrs@ier.unam.mx

Abstract. The flow of an electrolyte in a shallow square horizontal cavity subject to a steady current between opposite sides in the presence of an array of external magnets is simulated using a two dimensional lattice Boltzmann equation method for different values of the Chandrasekhar number. The flow is in a viscous regime for small values of Ch and in an advective one for larger values. In this last regime and in a steady state, a fixed number of pairs of initially close ideal tracer particles are added to the flow. We find that the average distance between each pair grows exponentially in time. Then, an average Lyapunov exponent that grows as a power law of the Chandrasekhar number can be defined.

Keywords: Lattice Boltzmann equation · chaos ·
magnetohydrodynamics · viscous and advective regimes

1 Introduction

Electromagnetic forcing is a practical experimental method used to produce stirring in shallow layers of electrically conducting fluids [1–5].

Oulette and Gollub [4] studied a quasi-2D flow using magnetohydrodynamic forcing in a thin layer of a conducting fluid in a cavity with a square horizontal base above a square lattice of permanent magnets with alternating orientations in the presence of an electric current between two opposite sides of the cavity, see Fig. 1. The alternation of magnets, interacting with the current, introduces a kind of frustration. They showed that the flow can show space time chaos.

This is an interesting and relatively little studied problem, illustrating an example of chaotic behavior in a stationary-forced hydrodynamic system. It is therefore interesting to try to model it using a reasonably simple computational model.

The lattice Boltzmann equation method (LBEM) [6,7] is a compromise between a fully discrete approach, as could be molecular dynamics or lattice gas

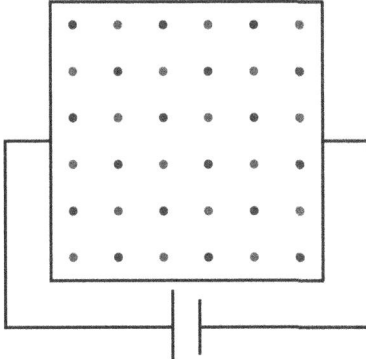

Fig. 1. Scheme of the cavity of sides shown horizontally of sides $6l$ with l being the distance between adjacent magnets located on the external bottom of the cavity with alternating orientations. The distance between the sides of the cavity and the nearest magnets is $l/2$. There is a constant electric current between the opposite sides of the cavity in the y direction.

cellular automata, and a fully continuous one as those used in computational fluid dynamics. It is a deterministic model that can be easily parallelized and distributed on multi-core CPUs or GPUs and can incorporate external forcing terms.

In what follows we simulate a two dimensional electrolyte in a horizontal square cavity in the presence of an external constant electric field and a magnetic field using the LBEM.

The combined effects of magnetic and the current produced by the electric field originates a Lorentz forcing field which may make the flow non-stationary. At low forcing field intensity, the velocity field is stationary. As the magnetic field strength grows, the velocity field becomes time dependent (see Fig. 2).

In this regime, after a reasonable time transient, a certain number of pairs of initially close ideal tracer particles are dropped in the fluid and by following their trajectories during short times (Fig. 3), an average finite time Lyapunov exponent can be defined.

One can see that, despite being the simulated system defined in two dimensions, trajectories can intersect. This is due to the fact that actually the phase space of the system is highly-dimensional, as we shall illustrate in Sect. 2.

The outline of the paper is the following: in Sect. 2 we describe the lattice Boltzmann equation method, followed by a presentation of the electromagnetic model in Sect. 3. In Sect. 4 we illustrate the method for determining trajectories and Lyapunov exponents, using the forced and damped pendulum as a test-bed. In Sect. 5 we present the results on the chaotic behavior of the model, ending with some conclusions in Sect. 6.

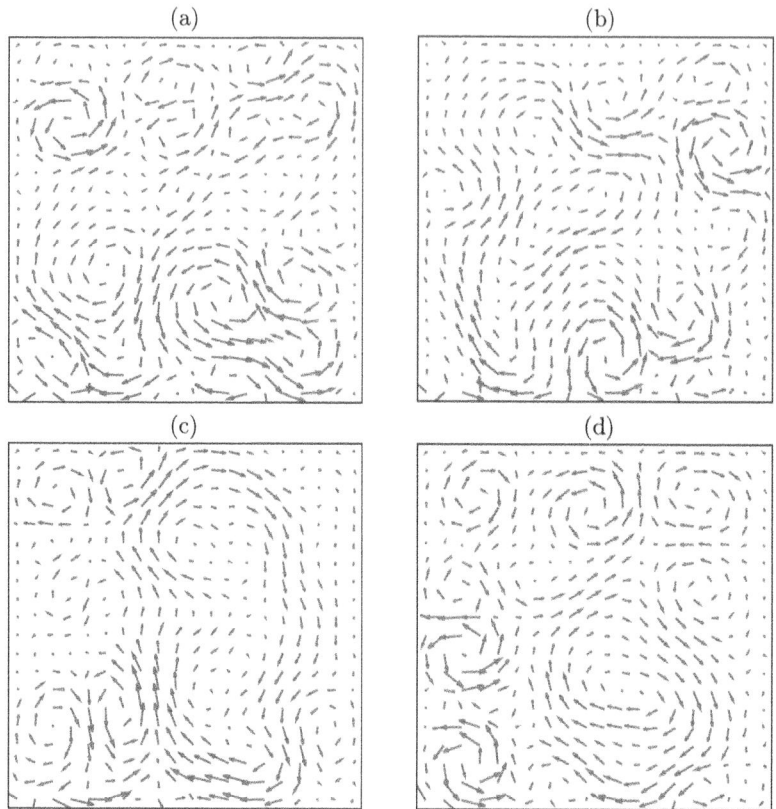

Fig. 2. For $Ch = 6.0 \times 10^3$ and $\delta = 50$, the velocity fields at (a) $t = 0.52$, (b) $t = 0.68$, (c) $t = 0.83$, and (d) $t = 1.1$. In this and the following figures, $l = 196$ and $\delta = 0.255$ in Eq. (7).

2 The Lattice Boltzmann Equation Method

In the lattice Boltzmann equation method (LBEM) [6,7], the system is divided in a $N \times N$ square lattice of cells. In each cell there can be an arbitrary number of particles, represented by a probability distribution of velocities. In our two-dimensional model, we can have rest particles, particles travelling at unit velocity towards the 4 nearest neighboring cells, or particles travelling at velocity $\sqrt{2}$ towards the next-to-nearest neighboring cells, in total nine velocities c_k, $k = 0, \ldots, 9$. The local density, velocity field and energy are given by the appropriate sum of the probability distribution over the possible velocities, as described in the following.

The time evolution of the model consists in an alternation of two phases: streaming and collision (including external forces). In the streaming phase, particles are transferred to neighboring cells according to their velocity. In the collision phase the local velocity field is computed and (possibly) altered by the

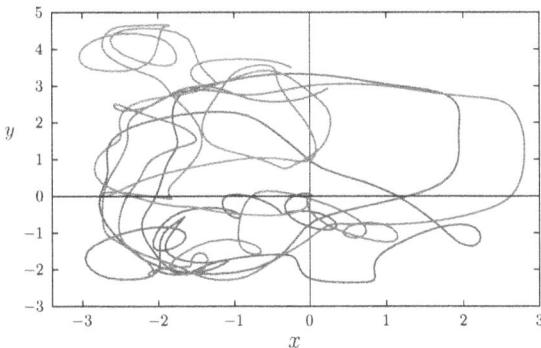

Fig. 3. For $Ch = 6.02 \times 10^3$ and $\delta = 50$, the trajectories of three ideal tracer particles $r - r_0$ for $0.5 \leq t \leq 0.9$ with r_0 the starting position of each tracer at $t_0 = 0.5$.

external forces. In our case the local density is not changed (no sources or wells), and the local energy is computed according to the velocity field and the density. Then the equilibrium distribution is computed according with the new local fields. The updated probability distribution is computed by means of a relaxation to equilibrium procedure [6,8], which determines the viscosity of the fluid.

Let us now introduce some notations. The velocities are

$$c_k = \begin{cases} (0,0) & k = 0, \\ (\cos(\phi_k)c, \sin(\phi_k)c) & k = 1,\ldots,4 \\ (\sqrt{2}\cos(\xi_k)c, \sqrt{2}\sin(\xi_k)c) & k = 5,\ldots,8. \end{cases} \quad (1)$$

In this expression, $\phi_k = (k-1)\pi/2$, $\xi_k = (2k-9)\pi/4$ and $c = \Delta x/\Delta t$ with $\Delta x = \Delta y$ the distance between neighboring cells and Δt the time step. In the following we set $c = 1$, taking $\Delta x = \Delta y = \Delta t = 1$.

By indicating with $f_k(r,t)$ the particle distribution function defined as the average number of particles with velocity c_k at cell r and at time t, the lattice Boltzmann equations in the Bhatnagar-Gross-Krook [6,8] approximation are

$$f_k(r + c_k \Delta t, t + \Delta t) = f_k(r,t) - \frac{\Delta t}{\tau}\left[f_k(r,t) - f_k^{(eq)}(r,t)\right]. \quad (2)$$

In this expression, τ is the relaxation time related to the viscosity ν by

$$\nu = c_s^2(\tau - \tau_0) \quad (3)$$

where $c_s = 1/\sqrt{3}$ is the speed of sound and $\tau_0 = 1/2$.

In Eq. (2), $f_k^{(eq)}$ is the local equilibrium distribution given by

$$f_k^{(eq)}(r,t) = w_k n \left(1 + \frac{c_k \cdot u}{c_s^2} + \frac{(c_k \cdot u)^2}{2c_s^4} - \frac{u^2}{2c_s^2}\right), \quad (4)$$

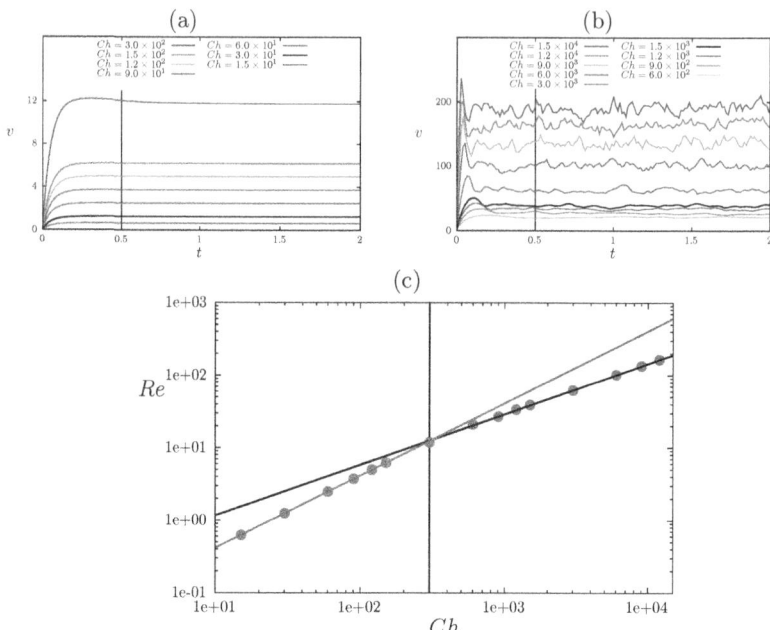

Fig. 4. (a) The characteristic velocity v as a function of time for $\delta = 0.255$ and small values of Ch. (b) The corresponding graph for the larger values of Ch. (c) The Reynolds number Re as a function of Ch. The fit of $Re = aCh^b$ to the data gives $a = 0.0419699$ and $b = 0.995859$ for the values of Ch in (a), curve in red, and $a = 0.234604$ and $b = 0.696974$ for the values of Ch in (b), curve in black. (Color figure online)

where n is the number density and $\boldsymbol{u}$ is the velocity at $(\boldsymbol{r}, t)$,

$$n = \sum_k f_k, \quad \boldsymbol{u} = \sum_k \boldsymbol{c}_k f_k, \tag{5}$$

$\omega_0 = 4/9$, $\omega_k = 1/9$ for $k = 1, \ldots, 4$ and $\omega_k = 1/36$ for $k = 5, \ldots, 8$. [9–13].

The expression in Eq. (4) is a second-order Taylor expansion of the Maxwell-Boltzmann equilibrium distribution, valid for small Mach numbers [14].

The velocity field in a given node is obtained as

$$\boldsymbol{v} = \sum_k \boldsymbol{f}_k \boldsymbol{c}_k. \tag{6}$$

From a dynamical systems point of view, our LBEM scheme corresponds to $9N^2$ coupled maps, so this is the real dimension of the phase space of the system.

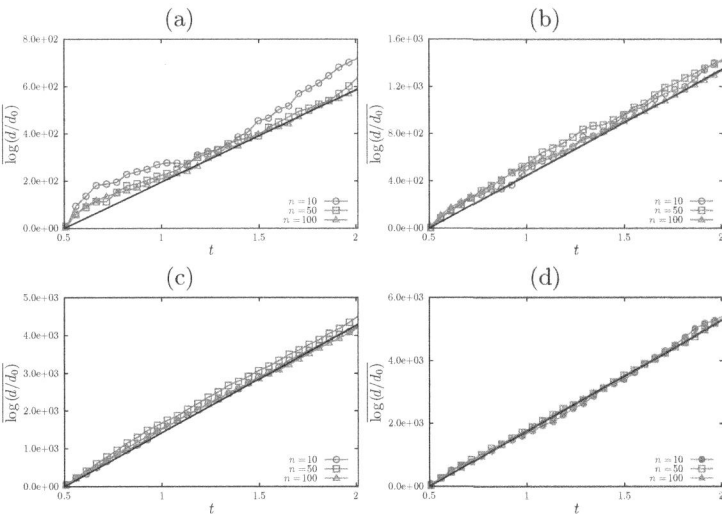

Fig. 5. The average of the logarithm of d_i/d_0, $\overline{\log(d/d_0)}$, where d_i the distance between the i-th of n pairs of ideal tracer particles initially separated by d_0. The black line the best fit of a straight line for $n = 100$, $\overline{\log(d/d_0)} = \lambda(t-t_1)$, $t_1 = 0.5$. (a) $Ch = 6.0 \times 10^2$. (b) $Ch = 1.2 \times 10^3$. (c) $Ch = 6.0 \times 10^3$. (d) $Ch = 9.0 \times 10^3$.

3 The Magnetohydrodynamic Model

The system under study is the flow of an electrolyte inside a horizontal slender cavity of equal sides $L_x = L_y$, and height L_z, $L_z \ll L_x$. In the simulations we used a grid of 1176×1176 cells.

There are M^2 with $M = 6$ equally spaced magnets with alternating orientations placed outside the bottom of the cavity and an electric current produced by electrodes on two opposite sides of the cavity. A scheme of the cavity is shown in Fig. 1.

The magnetic field of a magnet centered at $\bm{r}_{mn} = (l/2+nl, l/2+ml/2)$ with $l = L/6$, the distance between adjacent magnets in the x or y directions and $n, m = 0, \ldots, 5$, points in the z direction and with the applied electric field gives way to a Lorentz force per unit area in the y direction $\bm{F}(\bm{r}) = \sum_{m,n} \bm{F}_{mn}(\bm{r})$ with $\bm{F}(\bm{r}) = (0, F_{mn}(\bm{r}))$ and F_{mn} approximated by [5,15,16]

$$F_{mn}(\bm{r}) = (-1)^{m+n} F_0 \exp\left[\frac{-(\bm{r}-\bm{r}_{mn})^2}{2\delta^2}\right]. \qquad (7)$$

with $\bm{r} = (x, y)$. In the following $L_x = L_y = 6l$ $l = 196$ and $\delta = 50$.

The characteristic length and time are l and l^2/ν and the Chandrasekhar Ch and Reynolds numbers Re are defined by

$$Ch = \frac{F_0 l^3}{\rho \nu^2}, \qquad Re = \frac{vl}{\nu}. \qquad (8)$$

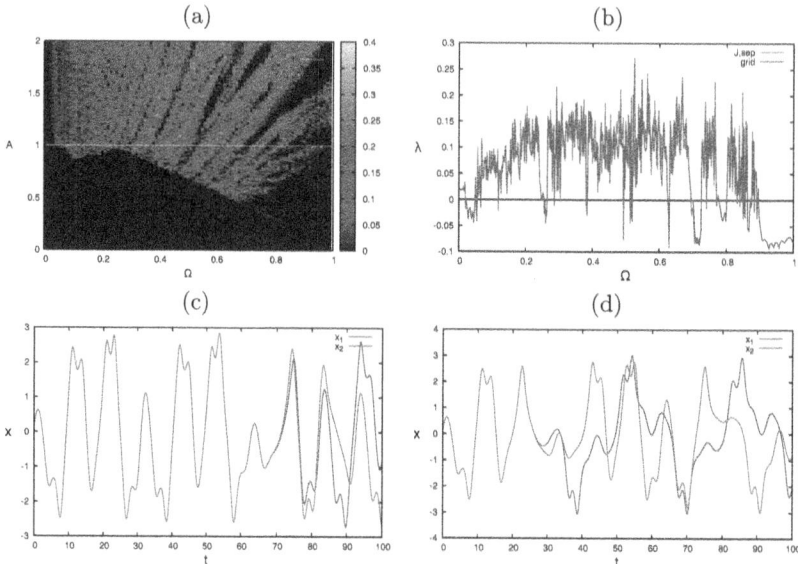

Fig. 6. (a) Color scale of the Lyapunov exponent of the forced and damped pendulum (Eq. (11)) for $\gamma = 0.2$ and several values of A and Ω, random $x(0), y(0)$. The line $A = 1$ is marked. (b) The value of the Lyapunov exponent on the line $A = 1$ computed using method 1 (using the Jacobian, label "J") and 2 (using the separation of trajectories, label "sep"), which are indistinguishable, and using method 3 (interpolating the trajectories and separation using a grid, label "grid"). (c-d) Time plot of one trajectory of the forced and damped pendulum computed using Eq. (11) (x_1) the 256×256 grid interpolation (x_3), for $A = 1$, $\gamma = 0.2$ using a 4th-order Runge-Kutta method with a time interval $\Delta t = 0.01$. (c) $\Omega = 0.61$, $\lambda \simeq 0.096$ and (d) $\Omega = 0.60$, $\lambda \simeq 0.145$.

The Chandrasekhar number is related to the ratio between the Lorentz and viscous forces and v is a characteristic velocity that is the time average after a transient of the average root mean square velocity over all the cells of the cavity [4, 17].

At low forcing field intensity, the velocity field is stationary and this can be defined as the viscous regime. As the magnetic field strength grows, the velocity field becomes time dependent, the advective regime (see Fig. 2). The viscous regime is illustrated in Fig. 4-(a) where the velocity field v is shown as a function time t and small values of Ch with v almost a constant after a time transient $t_t = 0.5$. In Fig. 4-(b) we show v for larger values of Ch, the advective regime [18]. In Fig. 4-(c) we show that Re scales with the Ch as $Re = aCh^b$, with different values of a and b in the two regimes. The values of b compare well with those reported in Ref. [18], where the flow of an electrolyte in the presence of three large rectangular magnets was experimentally studied.

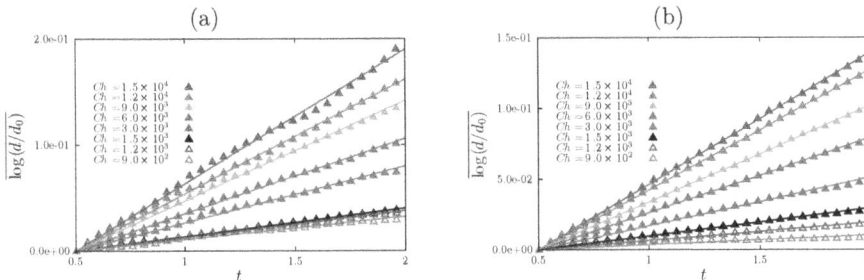

Fig. 7. In the advective regime, the average $\overline{\log(d/d_0)}$ for $n = 100$ pairs of ideal tracer particles and different values of Ch as functions of time t together with the best fit of the straight line $\overline{\log(d/d_0)} = \lambda t$. (a) $\delta = 0.25$, (b) $\delta = 0.5$.

4 Ideal Tracers and Lyapunov Exponents

The LBEM system approximates a continuous flow on a grid of $N \times N$ cells. In each point of the grid we can define a density and a velocity vector as in Eq. (5). By using an interpolation scheme among the nearest grid points, we can obtain the velocity field in any point of system. We use this velocity field to move passive (ideal) tracers (no mass) and obtain their trajectory.

It is possible to obtain an estimate of the Lyapunov exponent related to this trajectory using pairs of initially close tracers. Let us indicate by d_0 their initial distance for each pair.

After a reasonable relaxation time t_1, n pairs of ideal tracer particles are dropped in the fluid. We show in Fig. 3 the trajectories of three tracers.

We iterate all tracers for one LBEM time step, and, for each pair, we measure their mutual distance $d_i(t)$, where i numbers the pairs. After this we move the second tracer of the pair so that their distance returns to be d_0, keeping their mutual orientation. We call this procedure renormalization.

We accumulate the logarithm of $d_i(t)/d_0$,

$$\overline{\log(d(t)/d_0)} = \frac{1}{n}\sum_{i=1}^{n} \log(d_i(t)/d_0). \qquad (9)$$

For n large, this quantity grows almost linearly in time (see Fig. 5). The Lyapunov exponent λ is defined as the best fit of this average distance to a straight line [19, 20],

$$\overline{\log(d(t)/d_0)} \simeq \lambda(t - t_1). \qquad (10)$$

There are other proposals on how to find the Lyapunov exponent of flows [21–23].

We checked this method of computing Lyapunov exponents comparing "synthetic" trajectories from a vector field on a grid applying it to a damped and forced pendulum,

$$\ddot{\theta} = -\gamma\dot{\omega}^2 - \sin(\theta) + A\cos(\Omega t), \qquad (11)$$

which, in phase space, originates a flux.

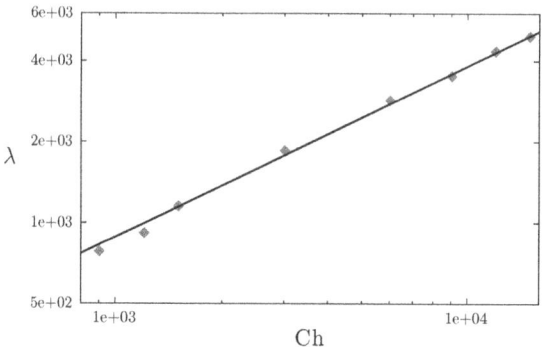

Fig. 8. The Lyapunov exponent λ as a function of the Chandrasekhar number Ch. For $Ch > 8.5 \times 10^2$ the data is fitted by $\lambda = aCh^b$ with $a = 9.313$ and $b = 0.652$.

Taking $x = \theta$, we have a two-dimensional time-dependent flux

$$\begin{aligned} \dot{x} &= y \\ \dot{y} &= -\gamma y - \omega^2 \sin(x) + A\cos(\Omega t). \end{aligned} \tag{12}$$

The corresponding Jacobian is

$$J(x(t), y(t)) = \begin{pmatrix} 0 & 1 \\ -\omega^2 \cos(x(t)) & -\gamma \end{pmatrix} \tag{13}$$

In this way we can compare the Lyapunov exponents λ computed in three ways:

1. Following the trajectory of Eq. (11) (x_1, y_1) and using the Jacobian, Eq. (13), to compute the evolution in tangent space

$$\dot{\boldsymbol{\delta}} = J\boldsymbol{\delta} \quad \Leftrightarrow \quad \begin{pmatrix} \dot{\delta x} \\ \dot{\delta y} \end{pmatrix} = J(x_1(t), y_1(t)) \begin{pmatrix} \delta x \\ \delta y \end{pmatrix},$$

   renormalizing the length of $\boldsymbol{\delta}$ at each time step Δt (see Fig. 6(a));
2. Using a tracer (x_2, y_2) whose trajectory is given by Eq. (11) and another one (x'_2, y'_2) at initial distance d_0, letting them evolve for a time interval Δt using Eq. (11), and then renormalizing their distance [19];
3. Sampling the field of Eq. 12 on a $x-y$ grid, using the interpolation technique described in the previous section to compute the trajectory (x_3, y_3) of one tracer and another (x'_3, y'_3) at initial distance d_0, renormalizing them at each Δt keeping their mutual orientation.

The three results agree within a few percent, as illustrated in Fig. 6-(b). The quality of the grid interpolation can be seen on Fig. 6-(c) and (d), showing the plot of x_1 with respect to x_3. One can see that the two trajectories stay very close for a long time.

In any case, in chaotic systems any initial error leads to the numerical impossibility of exactly following a given trajectory. Since we are interested in measuring the average (over many tracers) Lyapunov exponent, we are not really interested in exact trajectories, but rather in measuring their local separation rate, and therefore the proposed method constitutes a valid approximation.

5 Chaotic Behavior in the Advective Regime

From the results shown in Fig. 5 and similar ones for other values of Ch shown in Fig. 7 we can obtain the Lyapunov exponent λ as a function of Ch, as shown in Fig. 8. We can conclude that λ scales as a power law of Ch or Re in the advective regime, a results consistent with other investigations [20].

6 Conclusions

We have studied the flow of an electrolyte in a shallow square (almost 2D) cavity subject to an external electric field and an array of alternating magnets, by means of a lattice Boltzmann equation method. Above a certain threshold of the magnetic or electric field, the flow becomes non-stationary.

There is a viscous regime for small values of the Chandrasekhar number Ch and an advective one for large values of Ch. In this case, the velocity v changes in time and so does the velocity field.

Using passive tracers, we are able to define trajectories that follow the velocity field and allows that allows the evaluation of the average Lyapunov exponent λ. This method has been validated computing the Lyapunov exponent in several ways for the forced and damped pendulum.

We obtained a reasonable fit of the Lyapunov exponent as a power law of Ch or Re in the advective regime.

References

1. Tabeling, P., Cardoso, O., Perrin, B.: Chaos in a linear array of vortices. J. Fluid Mech. **213**, 511–530 (1990)
2. Paret, J., et al.: Are flows electromagnetically forced in thin stratified layers two dimensional? Phys. Fluids **9**(10), 3102–3104 (1997). https://doi.org/10.1063/1.869419
3. Rothstein, D., Henry, E., Gollub, J.P.: Persistent patterns in transient chaotic fluid mixing. Nature **401**(6755), 770–772 (1999). https://doi.org/10.1038/44529
4. Oulette, N.T., Gollub, J.P.: Curvature fields, topology, and the dynamics of spatiotemporal chaos. Phys. Rev. Lett. **99**, 194502 (2007)
5. Rechtman, R., et al.: Competition of localized thermal buoyancy and Lorentz forces in an electrolyte enclosed in a cavity. Phys. Fluids **33**, 127115 (2021)
6. Succi, S.: The Lattice Boltzmann Equation. Numerical Mathematics and Scientific Computation. Clarendon Press, Oxford (2001)
7. Wolf-Gladrow, D.A.: Lattice-Gas Cellular Automata and Lattice Boltzmann Models. An Introduction. LNM, vol. 1725, pp. 1–13. Springer, Heidelberg (2000)

8. Bhatnagar, P.L., Gross, E.P., Krook, M.: A model for collision processes in gases. I. Small amplitude processes in charged and neutral one-component systems. Phys. Rev. **94**(3), 511– 525 (1954). https://doi.org/10.1103/physrev.94.511
9. Inamuro, T., et al.: A lattice Boltzmann method for a binary miscible fluid mixture and its application to a heat-transfer problem. J. Comput. Phys. **179**, 201–215 (2002)
10. Shan, X.: Simulation of Rayleigh–Bénard convection using a lattice Boltzmann method. Phys. Rev. E **55**(3), 2780–2788 (1997)
11. He, X., Chen, S., Doolen, G.D.: A novel thermal model for the lattice Boltzmann method in incompressible limit. J. Comp. Phys. **146**, 282–300 (1998)
12. Buick, J.M., Greated, C.A.: Lattice Boltzmann modeling of interfacial gravity waves. Phys. Fluids **10**, 1490–1511 (1998)
13. Lallemand, P., Luo, L.S.: Theory of the lattice Boltzmann method: acoustic and thermal properties in two and three dimensions. Phys. Rev. E **68**, 036706 (2003)
14. Chen, S., Doolen, G.D.: Lattice Boltzmann method for fluid flows. Annu. Rev. Fluid Mech. **30**(1), 329–364 (1998). https://doi.org/10.1146/annurev.fluid.30.1.329
15. McCaig, M.: Permanent Magnets in Theory and Practice. Wiley, Hoboken (1952)
16. Cuevas, S., Smolentsev, S., Abdou, M.A.: On the flow past a magnetic obstacle. J. Fluid Mech. **553**, 227–252 (2006)
17. Román, J., Figueroa, A., Cuevas, S.: Wake patterns behind a magnetic obstacle in an electrolyte layer. Magnetohydrodynamics **53**(1), 55–66 (2017)
18. Duran-Matute, M., Trieling, R.R., Van Heijst, G.J.F.: Scaling and asymmetry in an electromagnetically forced dipolar flow structure. Phys. Rev. E **83**(1), 016306 (2011)
19. Wolf, A.: 13. Quantifying chaos with Lyapunov exponents. In: Chaos, pp. 273–290. Princeton University Press (1986). https://doi.org/10.1515/9781400858156.273
20. Fouxon, I., et al.: Reynolds number dependence of Lyapunov exponents of turbulence and fluid particles. Phys. Rev. E **103**(3), 033110 (2021). https://doi.org/10.1103/physreve.103.033110
21. Artale, V., et al.: Dispersion of passive tracers in closed basins: beyond the diffusion coefficient. Phys. Fluids **9**, 3162–3171 (1997)
22. Boffetta, G., et al.: Experimental evidence of chaotic advection in a convective flow. Europhys. Lett. **48**(6), 629 (1999). https://doi.org/10.1209/epl/i1999-00530-3
23. Bian, F., Shi, D., Sun, L., Bao, F.: An efficient approach for computing the finite time Lyapunov exponent in complex three-dimensional flow based on the discrete phase model. Aerosp. Sci. Technol. **133**, 108110 (2023)

Computational Aspects and Applications

Exploring Diverse Configurations of Cellular Automata Based S-Boxes Using Reinforcement Learning

A. Aravind[1]([✉]), Anita John[1,2], and Jimmy Jose[1]

[1] National Institute of Technology Calicut, Kozhikode, India
{aravind_m220264cs,anita_p170007cs,jimmy}@nitc.ac.in
[2] Rajagiri School of Engineering and Technology, Cochin, India

Abstract. This paper presents an approach for constructing efficient substitution boxes (S-boxes) for cryptographic applications by combining cellular automata (CA) and reinforcement learning (RL). Semi-bent Boolean functions derived from CA rules are used to generate the S-box output array with desirable cryptographic properties like high nonlinearity. The selection of optimal CA rules is formulated as a Markov Decision Process (MDP), where a reinforcement learning agent explores the state space of rule combinations to maximize a reward signal based on nonlinearity and differential uniformity. Various configurations for applying the selected CA rules to generate multi-layered S-boxes are explored. The proposed methodology offers advantages such as reduced memory footprint, exploration of a vast solution space, and inherent parallelism suitable for hardware implementations. Experimental results demonstrate that the generated S-boxes outperform previously proposed CA based S-boxes in terms of cryptographic strength.

Keywords: Cellular Automata (CA) · Semi-Bent Boolean Functions · S-boxes · Reinfocement Learning

1 Introduction

Substitution boxes (S-boxes) are critical components in symmetric-key cryptography, introducing nonlinearity to resist cryptanalytic attacks.

CA exhibits complex behavior from simple underlying rules, making them attractive for lightweight cryptographic applications. By employing semi-bent Boolean functions derived from CA rules, we generate the S-box output array with desirable cryptographic properties like high nonlinearity.

To optimize the selection of CA rules, we leverage reinforcement learning, formulating the problem as a Markov Decision Process (MDP). An agent explores the state space of rule combinations, maximizing a reward signal that captures the cryptographic strength of the S-box generated based on nonlinearity and differential uniformity.

This paper presents an approach to construct efficient S-boxes by combining cellular automata (CA) and reinforcement learning, building upon the work of Tarun et al. [1]. Additionally, we investigate multi-layered approaches, where the CA based transformation is applied multiple times, aiming to further enhance the cryptographic properties of the resulting S-boxes.

2 Preliminaries

2.1 Cellular Automata

Cellular Automata (CA) are popular for study due to their low computing cost and parallel processing. CA are suitable for high-bandwidth cryptography applications because of their unique features. CAs are grid-like systems of finite state automata, termed cells. The current cell state and neighbouring cell states determine each cell's state at a particular time step. Localized update rules determine each cell's state transition to the next time step.

Our approach uses binary CA with zero or one as the states. Time and space are discrete, and each CA cell is updated according to its local rule at each time step. Mathematically, we can represent this local rule as a function:

$$f : S^r \to S.$$

Here, r denotes the radius, i.e., the neighbourhood size considered in the update rule. For a one-dimensional CA (1D CA), where cells are arranged in a linear array, a radius of r implies that $2r + 1$ states are considered for the local update rule. There are various boundary conditions for CAs, the most prominent ones being Null Boundary and Periodic Boundary [7].

Null Boundary CA is where the leftmost and rightmost boundary cells assume zero as their left and right neighbours respectively. In *Periodic Boundary CA* the extreme cells are wrapped around as adjacent neighbours [7]. We have used Periodic Boundary CA for our work.

The output for each combination can be concisely represented by a single number, known as the CA's *Wolfram Number* [1]. Another key aspect is that the evolution of the CA is deterministic which facilitates predictability and reproducibility.

Furthermore, each CA rule can be represented in its algebraic normal form (ANF) regarding the states of the considered cells. The ANF of a Boolean function f can be represented as:

$$f(x_1, ..., x_n) = \bigoplus_{I \in 2^{[n]}} a_I x^I, \tag{1}$$

where $x^I = \prod_{i \in I} x_i$, $2^{[n]}$ is the power set of $[n] = \{1, 2, ..., n\}$ and $a_I = 0$ or 1. This is how the rules are written in code. The algebraic degree of the Boolean function f is the cardinality of the largest subset $I \in 2^{[n]}$ in its ANF, such that its coefficient $a_I \neq 0$.

2.2 S-Boxes

Substitution boxes (S-boxes) are an integral part of many encryption schemes. An S-box is a Boolean function that can be represented as:

$$F : \mathbb{F}_2^m \to \mathbb{F}_2^n,$$

where m and n are two positive integers, and $\mathbb{F}_2$ is the Galois Field of two elements. S-boxes can also be referred to as (n, m) functions where n and m represent the number of inputs and outputs respectively.

F can also be represented as a vector $F = (f_1, f_2, ..., f_m)$, where each function f_i is a Boolean function $f_i : \mathbb{F}_2^n \to \mathbb{F}_2$. The functions $f_i, \forall\ i \in \{1, 2, ..., m\}$ are called the coordinate functions of the vector F. Any non-trivial linear combination of the coordinate functions is referred to as the component function of F.

An S-box is said to be secure if it satisfies the following cryptographic properties:

Balancedness. An n variable Boolean function is a mapping $\mathbb{F}_2^n \to \mathbb{F}_2$. The most intuitive way to represent this mapping is through a *truth table*, which is a 2^n bit vector:

$$\Omega_f = (f(0,\ldots,0),\ldots,f(1,\ldots,1)).$$

Put simply, it represents the output of the function f for each possible input vector in lexicographic order. The function f is balanced if and only if Ω_f has an equal number of zeros and ones, which is a basic property of Boolean functions used in cryptography [8].

Nonlinearity. The *Walsh-Hadamard Transform* W_F of (m, n) function F is given by:

$$W_f(u,v) = \sum_{x \in (F)_2^n} (-1)^{v \cdot F(x) \oplus u \cdot x}, v \in \mathbb{F}_2^m, u \in \mathbb{F}_2^n, \quad (2)$$

where $v \cdot F(x)$ represents scalar product between vectors v and $F(x)$ and $u \cdot x$ represent scalar product between vectors u and x [6]. The nonlinearity N_F of a function F is represented as:

$$N_F = 2^{n-1} - \frac{1}{2} \max_{u \in \mathbb{F}_2^n, v \in (\mathbb{F}_2^m)^*} |W_f(u,v)|, \quad (3)$$

where $(\mathbb{F}_2^m)^* = \mathbb{F}_2^m / \{0\}$. A large value for N_F is preferred, as the higher the nonlinearity, the greater the resistance against linear cryptanalytic attacks [6].

Differential Uniformity. For a given (n, m) function F, we can define a *difference distribution table* D_F as:

$$D_F(a, b) = \{x \in \mathbb{F}_2^n : F(x) \oplus F(x \oplus a) = b\}. \tag{4}$$

The differential uniformity of the function δ_F is given by:

$$\delta_F = \max_{a \neq 0, b} \delta(a, b), \tag{5}$$

where $\delta(a, b)$ represent the cardinality of $D_F(a, b)$. Ideally, we want to minimize the differential uniformity of an S-box. A low differential uniformity means that the S-box can withstand differential cryptanalysis. The minimum attainable value of δ_F is 2 and S-boxes that achieve this value are called *almost perfect nonlinear (APN) functions* [8].

2.3 Semi-bent Boolean Functions

Let f denote a Boolean operation. From Eq. 3, we can see that the highest value of N_f is obtained when the max term equals $2^{\frac{n}{2}}$. This yields the following bound: $N_f \leq 2^{n-1} - 2^{\frac{n}{2}-1}$. Such boundary-satisfying functions are referred to as *bent functions*. However, their existence is restricted to even values of n [4]. Due to their unbalanced nature, bent functions are unsuitable for implementation in cryptographic systems.

When n is odd, the quadratic bound is - $N_f \leq 2^{n-1} - 2^{\frac{n+1}{2}-1}$. This bound is achievable by any function of algebraic degree two. In general, when n is an odd number and $n > 7$, this bound is not strict. The precise upper limit on the nonlinearity in that particular case remains an unresolved issue. The equation representing the *Walsh transform* for a Boolean function $f : \mathbb{F}_2^n \rightarrow \mathbb{F}_2$ is as follows:

$$W_f(u) = \sum_{x \in \mathbb{F}_2^m} (-1)^{f(x) \oplus u \cdot x}, \quad \forall u \in \mathbb{F}_2^m, \tag{6}$$

where $u \cdot x = \bigoplus_{i=1}^{n} a_i x_i$ is the scalar product of the vectors u and x. The Walsh transform quantifies the correlation between two Boolean functions, denoted as f and $u \cdot x$. Consequently, it is computed in order to determine whether a Boolean function f is nonlinear. *Semi-bent Boolean functions* are those that have the following definition of the Walsh transform:

$$W_f(u) = \begin{cases} 2^{\frac{n+1}{2}}, & \text{if } n \text{ is odd,} \\ 2^{\frac{n+2}{2}}, & \text{if } n \text{ is even.} \end{cases} \tag{7}$$

The nonlinearity limit of these functions is reached when n is an odd number. Since these functions have the capability of being balanced, we have used them as coordinate functions in our construction of S-boxes [4].

2.4 Reinforcement Learning

Reinforcement Learning (RL) is the third major machine learning paradigm, after supervised and unsupervised learning, and sometimes includes semi-supervised learning. RL studies how agents optimize cumulative reward signals in their environment. The agent makes decisions and acts in its external environment. Modelling the environment as continuous or discrete states allows the agent to choose actions from preset choices, each with a reward signal [1].

The agent aims to maximize reward signal accumulation. This is done by balancing exploration, where the agent tries new things, with exploitation, where it uses its existing knowledge to choose activities that have historically paid off. The exploration-exploitation issue affects most RL problems. The agent must use what it knows to gain rewards and attempt new things to find better ways in the future. In order to find alternate action sequences with higher rewards over time, the agent typically selects less-than-ideal actions even when it knows the best ones [1].

RL problems need states, actions, rewards, policies, reward signals, value functions, and optional environment models. The policy controls agent behavior by mapping states to action probabilities. Reward signals set RL problem goals. The value function provides the agent's expected cumulative reward from a state over time, while state-action values show the expected return from a given state action. Value function emphasizes future reward accumulation, unlike myopic reward signal, which only evaluates next action. When present, the model simulates environmental behavior, allowing dynamical conclusions.

RL problems are commonly expressed as Markov Decision Processes (MDPs). MDPs are mathematical abstractions that serve as an ideal representation of the reinforcement learning problem. They enable us to create mathematical models of theoretical statements. In a finite MDP, the probabilities associated with the states, actions, and rewards are discrete and clearly defined, as indicated by the following equation:

$$\sum_{s' \in S} \sum_{r \in R} p(s', r | s, a) = 1, \quad \forall s \in S, a \in A(s), \tag{8}$$

where S represents all states, $A(s)$ represents all actions in state s, and r represents the reward for migrating from s to s' by taking action a. MDP dynamics are determined by probability p. The probability of each possible state S_t and reward R_t at time t is determined purely by the preceding state S_{t-1} and action A_{t-1}, without any influence from previous states or acts. The RL agent, policy, and state-action pairings actively learn and use positive and negative reinforcements to optimize the reward signal.

3 Literature Survey

S. Wolfram proposed a 1-dimensional CA that generates random sequences. Each site in the CA has a value of zero or one and is updated in parallel according to the rule which is commonly called Rule 30 and is described as $a'_i = a_{i-1} \oplus (a_i \vee a_{i+1})$, where $\oplus$ indicates bitwise XOR, and $\vee$ indicates bitwise OR operations [9].

Meier and Staffelbach critically assess Wolfram's claims in [9] in their research. They found some issues in Wolfram's pseudo-random generator. A new cryptanalytic technique is developed for known plaintext attacks, assuming a unicity distance [5].

Szaban et al. introduced a novel method for generating S-boxes using CA instead of typical set table structures like DES [8]. The authors show that CA based S-boxes can match or exceed DES S-boxes in quality by examining their cryptographic properties. The CA based method allows additional design freedom by allowing the creation of many S-box functions from a single CA by modifying the initial state. These findings indicate a promising approach for creating dynamic, high-quality S-boxes for modern cryptographic algorithms.

Mariot et al. explored the creation of S-boxes for block ciphers using pairs of *Orthogonal Cellular Automata* (OCA) [2,3]. They exhaustively search for nonlinear OCA pairings of various diameters to generate S-boxes of varying sizes. The results show that all S-boxes are linear, limiting their application in block cipher design [3]. However, the authors find an intriguing pattern in these S-boxes' linear components space. The polynomial codes of the S-boxes they found are likewise classified. While OCA cannot be used to create good S-boxes in block ciphers, the findings provide useful insights for theoretically characterizing nonlinear OCA couples.

Picek et al. present a study on evolving S-boxes based on CA with *Genetic Programming* (GP) [6]. Using this approach, they were able to find optimal S-boxes for sizes from 4 × 4 up to 7 × 7. The approach shows great potential and marks the first time that heuristic techniques are able to find optimal S-boxes for sizes larger than 4 × 4 [6]. They also mention that the corpus of obtainable functions is still large enough to give sufficient diversity for future block cipher designs [6].

As a follow-up, Tarun et al. explore the possibility of creating S-boxes using RL [1]. They divided the problem into a three-step process - creating Boolean functions, S-boxes and using reinforcement learning to figure out the optimal rules to create S-boxes. The paper used the well-established MDP as the algorithm used for RL. The results show that the generated S-boxes not only outperform the classical S-boxes in terms of nonlinearity and differential uniformity but are also relatively lightweight in their implementation.

4 Design

As previously mentioned, the S-box design and evaluation resemble the work in [1]. This paper extends the work by focusing on the approach using two semi-bent CA rules. We have used two rules to limit our implementation due to computational constraints. Their research divides the problem into three steps: creating Boolean functions, S-boxes, and reinforcement learning.

4.1 Creation of Boolean Functions

The proposed design utilizes a set of two or more semi-bent Boolean functions to generate the output array. The CA serves as the source of input bits for the substitution boxes (S-boxes). The CA has a length of eight cells, where the state of each cell is determined by the corresponding input bits. Periodic boundary conditions are imposed, such that the neighbourhoods for the cells at the extreme ends wrap around to the opposite ends of the CA.

In this work, we explore the set sizes of two semi-bent Boolean functions to transform the set of input bits into the output array. Each transformation comprises of two operations. The first operation involves applying the CA rule to the set of eight input bits, producing an intermediate array. The second operation entails taking the bitwise XOR of the intermediate array to obtain a single output bit. The size of the intermediate array depends on the neighbourhood size of the employed CA rule (Boolean function). We have fixed the size of the intermediate array as six since we utilize CA rules that have a neighbourhood size of three for both set sizes under consideration.

One bit is output by each intermediate array. The intermediate array is generated eight times in order to obtain the complete 8-bit result. The first cell and the neighbourhood cells do not overlap during each iteration. To produce an output bit, the boundary is fixed to eight cells, extending from the cell indexed at $start$ to the cell indexed at $(start+nbr\_size)\%8$, where nbr_size represents the neighbourhood size. The index $start$ also indicates which bit of the output will be formed by the current iteration of the Boolean function. The initial iteration of the algorithm is illustrated in Fig. 1.

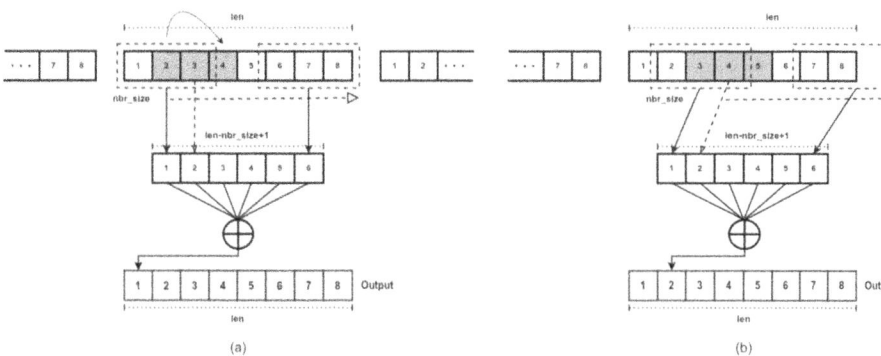

Fig. 1. Iterations of the Boolean functions (a) Generating the first bit for output (b) Generating the second bit for the output

A set of bits of length len is transformed into $len - nbr\_size + 1$ output bits by the first iteration of the algorithm, which employs a CA rule of neighbourhood size nbr_size. These output bits are subsequently XORed to generate a single

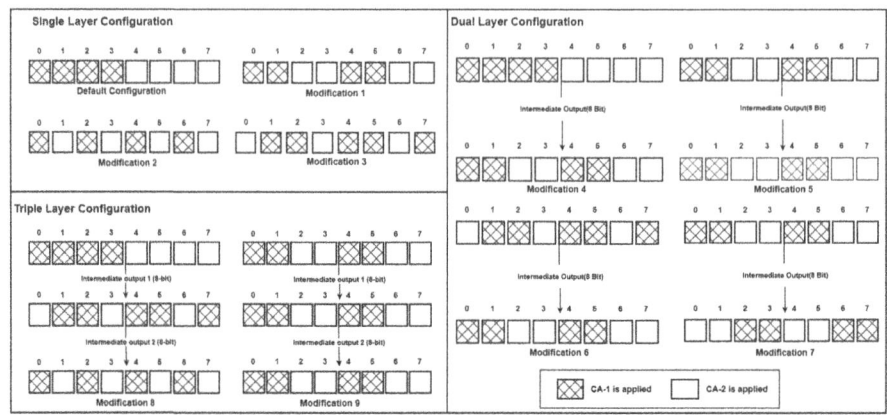

Fig. 2. Different configurations used for the CA based S-boxes

bit. In this context, *len* is equivalent to 8 bits, *nbr_size* is 3 bits, and *start* is 1. This will provide the initial bit of the output.

The same procedure is repeated in the second iteration, with *start* = 2. This will constitute the second bit of the final output.

4.2 S-Box Design

In [1], two CA rules are applied on an 8-bit input to generate an 8-bit output. We refer to these rules as CA-1 and CA-2. The leftmost four bits used CA-1 and the rightmost 4 bits used CA-2.

We explore different ways of using CA-1 and CA-2 among the eight cells, which include spacing CA-1 and CA-2 uniformly and using non-uniform arrangements to investigate which configurations can yield the best results. The selection of these configurations was conducted in a strategic manner. This process involved a thorough examination and analysis of previous configurations to identify patterns and trends that could potentially enhance the effectiveness of the current experiment. Alongside this, we also incorporated a degree of intuitive selection, where we speculated on configurations that might yield significant results based on an informed understanding of the system's dynamics.

We also explore the possibility of implementing multiple layers, where the 8-bit output of the previous iteration of the Boolean function is taken as the input for the next iteration of the Boolean function. We have considered two and three layers for our S-boxes.

In total, 10 configurations were tested including the one used in [1], and nine modified configurations. In Fig. 2, the design of each configuration is illustrated, where the shaded cells indicate the use of CA-1 and the non-shaded cells indicate the use of CA-2 towards the output. The pseudocode is given in Algorithm 1.

Algorithm 1. S-box

Input: An array of 8-bit integers (0, ..., 255)
Output: Array containing 8-bit outputs for each corresponding input
Procedure:
inputs ← [ ]
outputs ← [ ]
$input\_size \leftarrow 8$
$nbr\_size \leftarrow 3$
Generate input combinations:
for $i \in \{0, 1, \ldots, 2^{input\_size} - 1\}$ **do**
    ip ← decimalToBinary(i)
    inputs.append(ip)
end for
Apply rules and generate intermediate outputs:
for ip ∈ inputs **do**
    op ← [ ]
    for start $\in \{0, 1, \ldots, input\_size - 1\}$ **do**
        if start ∈ {Places where CA-1 should be applied} **then**
            op.append(rule_op(rules[0], $nbr\_size, input\_size$, ip, start))
        else
            op.append(rule_op(rules[1], $nbr\_size, input\_size$, ip, start))
        end if
    end for   ▷ After this procedure, we get the final 8-bit output as op
    Select final output:
    outputs.append(op)
end for
Return inputs and outputs:
return inputs, outputs

4.3 Selection of Optimal Configuration Using Reinforcement Learning

Using RL, we determine which configuration is the best fit for each case. Our first step in the Markov Design Process is to identify our states, actions, and rewards. All feasible k-permutations of every semi-bent function comprise our state space.

We thus obtain 3080 states in our state space for our case, with each state differing by at least one semi-bent function.

Because the action is discrete and each swap causes a transition to a different state in the state space, it can be considered as the swapping of one rule for another. Its cryptographic strength, which can be formulated as:

$$strength = (scaled\_NL + (100 - scaled\_DU))/2 \tag{9}$$

where *scaled_NL* and *scaled_DU* are the nonlinearity and differential uniformity values of our S-box scaled with respect to the AES S-box and can be defined as:

$$scaled\_NL = (N_F/112) * 100 \qquad (10)$$

$$scaled\_DU = ((\delta_F - 4)/(128 - 4)) * 100 \qquad (11)$$

where N_F and δ_F are as defined in Eqs. 3 and 5 respectively.

The agent selects a non-greedy action with probability ϵ under the chosen ϵ-greedy policy. This policy dictates that the agent will either take the greedy action, which maximizes the calculated state-action value, or a non-greedy action, with the probability of selecting the non-greedy action being ϵ.

Because this problem deals with a continuous task, the on-policy SARSA (State-Action-Reward-State-Action) algorithm was used as in [1] to calculate the state-action value pairs from the data obtained while exploring the state space. The concept of average reward was employed due to the nature of the problem at hand.

To handle the extensive state space and the nonlinear relationship between the strength of the rules and the states, a value-function approximator method utilizing an Artificial Neural Network (ANN), which approximates a value for a given state, provided the input parameters of that state. For our use case, the input parameters are the CA rules used in our S-box. The ANN is flattened for each rule, producing a unique array for every distinct state. In this design, the input array has a size of $2 * 56 = 112$, allowing the ANN to employ 56 semi-bent rules and two rules per state [4].

Let s1 and s2 be two individual states, i.e., pairs of semi-bent CA rules used for creating S-boxes. The cryptographic strength of the state s2 is what is obtained upon transitioning from state s1 to s2.

The reward obtained at each state is defined as:

$$reward = strength(s2) \qquad (12)$$

This indicates the success of the transition. An alternative would have been to calculate the difference between the strength of s2 and s1, but the aim was to determine which CA rules provide the best result for each configuration. So we only consider the strength of the newly generated S-box since the reward is less if the strength of s2 is less than that of s1.

5 Results

We have analyzed our possible configurations with a set size of two on the various configurations shown in Fig. 2. Each configuration was run five times, with each run traversing 256 unique states. Table 1 shows the best strength, nonlinearity (NL) and differential uniformity (DU) of each arrangement.

Seven of the nine modified configurations improved cryptographic strength compared to [1] by reducing differential uniformity. From these configurations, *Modification-1* has shown comparable strength, nonlinearity, and differential uniformity values to [1], which utilized a set size of three CA rules.

Table 1. Strength, differential uniformity, and nonlinearity for each S-box configuration

S-box Configuration	Strength	DU	NL
Related Work [1]	81.57	32	96
Modification-1	88.02	16	96
Modification-2	67.28	32	64
Modification-3	81.57	32	96
Modification-4	89.54	10	94
Modification-5	**89.63**	**12**	**96**
Modification-6	87.93	14	94
Modification-7	87.04	14	92
Modification-8	87.21	18	96
Modification-9	88.82	14	96

By adding a second layer (*Modification-4 to Modification-7*) we were able to greatly reduce the differential uniformity without drastically affecting nonlinearity, with *Modification-5* achieving the best results overall. Adding a third layer yielded good results, albeit somewhat worse than two layers. So it's not worth the computing complexity.

6 Conclusion

In conclusion, this work has demonstrated an effective approach for constructing cryptographically strong S-boxes by combining CA and reinforcement learning. By leveraging semi-bent Boolean functions derived from CA rules and employing a reinforcement learning agent, we have generated S-boxes outperforming classical designs and previous CA based constructions in terms of nonlinearity and differential uniformity. The exploration of multi-layered CA rule configurations further enhanced the cryptographic strength.

While reinforcement learning proved effective, alternative techniques like evolutionary algorithms, machine learning, and advanced CA models could be explored to further refine and optimize S-box construction. Evolutionary approaches could evolve CA rules and configurations guided by fitness functions capturing desired properties. Machine learning techniques could learn and optimize rules from data. Exploring high-dimensional or hybrid CA models may uncover novel dynamics leading to even more robust S-box designs. The integration of intelligent optimization techniques with CA presents a promising direction for developing efficient and secure cryptographic primitives.

A Appendix

The source code for this experiment can be found at https://github.com/mr-https://github.com/mr-Saamy/RL-based-S-boxes.

References

1. Ayyagari, T., Saji, A., John, A., Jose, J.: Exploring lightweight s-boxes using cellular automata and reinforcement learning. In: Chopard, B., Bandini, S., Dennunzio, A., Arabi Haddad, M. (eds.) ACRI 2022. LNCS, vol. 13402, pp. 17–28. Springer, Cham (2022). https://doi.org/10.1007/978-3-031-14926-9_2
2. Mariot, L., Manzoni, L.: On the linear components space of s-boxes generated by orthogonal cellular automata. In: Chopard, B., Bandini, S., Dennunzio, A., Arabi Haddad, M. (eds.) ACRI 2022. LNCS, vol. 13402, pp. 52–62. Springer, Cham (2022). https://doi.org/10.1007/978-3-031-14926-9_5
3. Mariot, L., Manzoni, L.: A Classification of S-boxes Generated by Orthogonal Cellular Automata, vol. 23, pp. 5–16. Kluwer Academic Publishers, USA (2023). https://doi.org/10.1007/s11047-023-09956-z
4. Mariot, L., Saletta, M., Leporati, A., Manzoni, L.: Exploring semi-bent boolean functions arising from cellular automata. In: Gwizdałła, T.M., Manzoni, L., Sirakoulis, G.C., Bandini, S., Podlaski, K. (eds.) ACRI 2020. LNCS, vol. 12599, pp. 56–66. Springer, Heidelberg (2020). https://doi.org/10.1007/978-3-030-69480-7_7
5. Meier, W., Staffelbach, O.: Analysis of pseudo random sequences generated by cellular automata. In: Davies, D.W. (ed.) EUROCRYPT 1991. LNCS, vol. 547, pp. 186–199. Springer, Heidelberg (1991). https://doi.org/10.1007/3-540-46416-6_17
6. Picek, S., Mariot, L., Leporati, A., Jakobovic, D.: Evolving s-boxes based on cellular automata with genetic programming. In: Proceedings of the Genetic and Evolutionary Computation Conference Companion, GECCO 2017, pp. 251–252. Association for Computing Machinery, New York (2017). https://doi.org/10.1145/3067695.3076084
7. Misra, S., Das, A.K., Chowdhury, D.R., Chaudhuri, P.P.: Cellular automata-theory and applications. IETE J. Res. **36**(3–4), 251–259 (1990). https://doi.org/10.1080/03772063.1990.11436890
8. Szaban, M., Seredynski, F.: Cellular automata-based S-boxes vs. DES S-boxes. In: Malyshkin, V. (ed.) PaCT 2009. LNCS, vol. 5698, pp. 269–283. Springer, Heidelberg (2009). https://doi.org/10.1007/978-3-642-03275-2_27
9. Wolfram, S.: Random sequence generation by cellular automata. Adv. Appl. Math. **7**(2), 123–169 (1986). https://doi.org/10.1016/0196-8858(86)90028-X

Efficient Simulation of Non-uniform Cellular Automata with a Convolutional Neural Network

Michiel Rollier[1(✉)], Aisling J. Daly[1], Odemir M. Bruno[2], and Jan M. Baetens[1]

[1] Department of Data Analysis and Mathematical Modelling, BionamiX, University of Ghent, Coupure Links 653, 9000 Ghent, Belgium
`michiel.rollier@ugent.be`
[2] São Carlos Institute of Physics, University São Paulo, POB 369, BR-13560970, São Carlos, SP, Brazil

Abstract. Cellular automata (CAs) and convolutional neural networks (CNNs) are closely related due to the local nature of information processing. The connection between these topics is beneficial to both related fields, for conceptual as well as practical reasons. Our contribution solidifies this connection in the case of non-uniform CAs (νCAs), simulating a global update in the architecture of the Python package `TensorFlow`. Additionally, we demonstrate how the highly optimised out-of-the-box multiprocessing in `TensorFlow` offers interesting computational benefits, especially when simulating large numbers of νCAs with many cells.

Keywords: Cellular automata · Non-uniform · Convolutional neural networks · `CellPyLib` · `TensorFlow`

1 Introduction

1.1 Elementary and Non-uniform Cellular Automata

Arguably the simplest non-trivial and maximally discrete dynamical system is an elementary cellular automaton (ECA). In this model, a finite or countably infinite number of cells are aligned in one dimension. A cell can be in only two possible states, all cells update their state in discrete time steps based on their own and their direct neighbours's state. Additionally, all cells update their states simultaneously, they do so deterministically, and they all follow the same local update rule (see e.g. [3]). Relaxing any of these conditions results in a CA that belongs to a family of discrete models that typically exhibit a richer behaviour, a more complex mathematical description, and well-defined 'taxonomic' ties to other families. Our forthcoming comprehensive review on this taxonomy [12]) provides an overview.

In particular, allowing certain cells to follow different local update rules results in the family of CAs collectively identified as non-uniform CAs (νCAs).

Our review paper [12] covers non-uniformity in the most general sense, where the 'rule allocation' varies in space and time. However, in the literature [4] νCA rule allocation is typically only spatially non-uniform. For this reason, together with the fact that our proposed implementation is more cumbersome in the general interpretation of a νCA, we will only consider spatially non-uniform CAs in this contribution. Additionally, as we will focus on simulating νCAs, we will only be concerned with finite grids. Figure 1 contains an example of a νCA with $N = 32$ cells and $N_R = 2$ elementary rules.

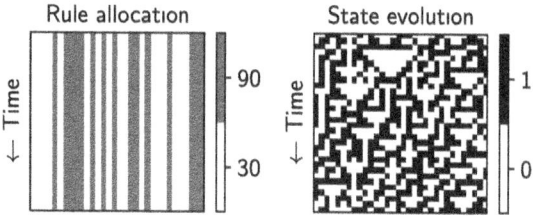

Fig. 1. A non-uniform CA (νCA) is determined by the rules that govern the local update, the allocation of those rules (blue and white), and the initial configuration of the states (black and white). In this example, two rules 30 and 90 are allocated. The allocation (left) is such that the νCA is uniform in time but not in space. In combination with a particular initial configuration, this results in the state evolution shown on the right. (Color figure online)

Clearly, allowing non-uniformity implies that the space of possible CA dynamics increases in size considerably. An ECA with N cells and periodic boundary conditions already has 2^N possible initial configurations, which evolves into different dynamics for each of the 256 elementary rules. A νCA consisting of N cells that each evolve according to one of N_R rules has $2^N \times N_R^N$ such possible initial conditions. This large diversity obstructs mathematical generalisation except in particular cases that are quite remote from applications [6]. An empirical approach to phenomenological classification is therefore imperative, but such a computational task requires an efficient means of simulation.

1.2 CA Classification and Simulation by Means of CNNs

The CA classification problem [5,13] is a challenge at the centre of CA research (see e.g. [14]). Considering the fact that we can interpret the spacetime diagrams of CAs as images, computer vision techniques can be mobilised for their classification, including those researched in the domain of deep learning. Within the spectrum of deep learning, convolutional neural networks (CNNs) are wildly popular, largely due to their undeniable success in image processing and computer vision [7]. We refer to excellent monographs to gain a good understanding of the topic (e.g. [10]), while a good visual introduction is offered by the deep learning series by 3Blue1Brown on YouTube [1].

A lot of diverse data is required in order to effectively train CNNs to identify classes. Fortunately, the local nature of the convolution operation enables not only the identification of CAs, but also their emulation. After all, nodes in a neural network may be identified with CA cells, and a convolutional operation can be interpreted as an update from a local neighbourhood. In fact, as Gilpin [9] shows, the global update mechanism on any kind of CA can be accommodated by the architecture of a CNN. This can be achieved both by a clever choice of weights and biases, or by training the network from a random initialisation.

In the CNN, transforming the input configuration to the neighbourhood encoding is performed by the first 1D convolutional layer, with a kernel of width 3 and fixed weights $(4, 2, 1)$, zero bias, and periodic boundary conditions. The output of this convolution is transformed to a matrix with one-hot vectors as columns, and each of the 8 rows of this matrix corresponds to a channel in the first CNN hidden layer. Next, another convolution layer with a kernel of width 1 essentially sums all channels, where this time the weights are determined by the binary representation of the local update rule. The output is then, by design, the ECA configuration after one global update. With a mere 40 parameters, this is an extremely simple CNN, whose computational complexity scales only with the number of cells N. The subsequent steps required to integrate a global update into a CNN framework are shown schematically in Fig. 2. This concrete example uses ECA rule 54 and a random initial condition, but the required operations are independent of this choice.

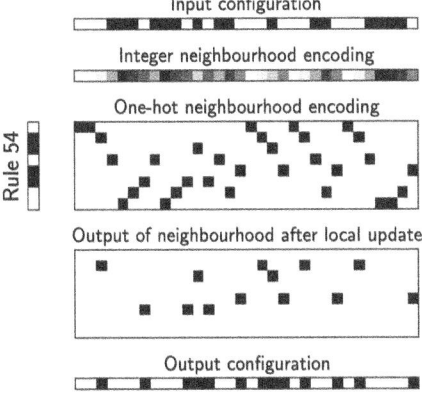

Fig. 2. The subsequent (de)composition steps required for updating an ECA configuration, illustrated for 32 randomly initialised cells, evolved over one time step by rule 54. First, each binary size-3 neighbourhood is translated to an integer from 0 to 7 (shown in grey-scale). This integer is encoded as a size-8 one-hot vector (displayed in columns). Depending on the rule table of the local update rule (displayed on the left-hand side), this columns is kept or removed. As a final step, all columns are summed, resulting in the output configuration.

The parameters within the CNN (weights and biases) can be calculated, but for more general CAs they would typically be trained. In order for the CNN to be in practice (and consistently) trainable, starting from random weights and biases, some additional features are required. We will not focus on the training procedure here, but we may mention that the most important of such additional features would be activation functions [10]. For illustrative purposes we include our convergence towards an optimum in parameter space for a CNN that emulates rule 54 with near perfection in Fig. 3. For details on preferred training procedures for CA emulators, we again refer to [9].

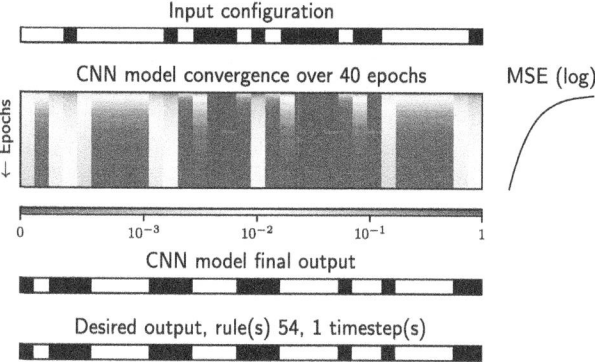

Fig. 3. A CNN can be trained to generate the desired next-timestep configuration of an ECA, illustrated here by showing the evolution of the inference on one of the training samples. Starting from a random choice of weights and biases, the mean square error (MSE) between the generated output and the desired output continues to shrink, while more and more epochs of data are mobilised in the training. This is shown in the qualitative MSE curve on the righthand side, and on a cell-per-cell basis in the log-scaled colour-coded heat map in the centre. For details on similar training procedures, consult [9].

1.3 Scope

The goal of this article is to fill in a gap in the literature, by emulating νCAs by means of CNNs. We can benefit from the extremely streamlined software implementations designed for neural networks, optimised for parallel processing and general performance. That is to say: CNNs can present us with a practical tool for the fast and massive simulation of spacetime diagrams and analyses on these diagrams.

In the next section we will develop a CNN for νCAs, and we will see that this requires only a minimal addition to the architecture outlined above. We discuss some performance characteristics, and conclude with an outlook on the future of the marriage between CAs and CNNs.

2 Methods

In order to assess the performance of a CNN regarding the simulation of νCAs, we first discuss a popular well-established approach, and then introduce two varieties of CNN extensions.

2.1 Existing Approaches

Some programming languages enable very convenient and computationally optimised ways of simulating and analysing CAs. Wolfram Mathematica is an obvious example, which was in fact partially created for this purpose [8]. In Python, the most commonly used package is CellPyLib [2].

It is straightforward to implement a νCA in CellPyLib by defining an array that instructs the evolve method on what rule to apply when to which cell. Adding more or fewer rules (i.e. altering the non-uniformity) should not affect the performance. What does impact the performance, however, is the fact that the non-uniformity of the model no longer allows for caching the states in each step – in CellPyLib this is encapsulated in the memoize option (sic). This means that one cannot make any 'memory shortcuts' which typically speed up the CA simulation considerably. Figure 4 displays an example for an 8-rule νCA simulated in CellPyLib.

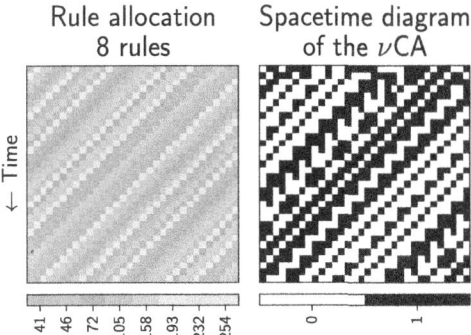

Fig. 4. A νCA is easily simulated using the Python package CellPyLib. One does so in two steps: first by defining the rule allocation (left, dependent on time and cell), and second by parsing this information to the evolve method, which generates the spacetime diagram (right).

2.2 Two Approaches for Non-uniform CAs in TensorFlow

Building on the CNN framework for elementary CAs, below we propose two additions which enable the CNN to emulate νCAs. Both additions involve a change in the CNN architecture between the one-hot neighbourhood encoding and the output layer. For technical details we refer to the code and annotations of the `NucaEmulator` class, available at mrollier/emulating-and-learning-CAs on GitHub, and entirely based on `TensorFlow` modules.

Note that the proposed CNNs emulate a single global update. Generating an entire spacetime diagram of T time steps requires feeding the output of the CNN $T-1$ times back into the input layer, because in the proposed setup, the CNN can only emulate a single global update. Global updates strongly depend on the previous time step, so it is (in general) not possible to distribute the calculation of the CA dynamics into (for example) 'all even time steps' and 'all odd time steps'. This impossibility is related to the so-called computational irreducibility of CAs, and impedes temporal parallellisation of the computation of its dynamics. Note also that if the rule allocation is independent of time (as is conventionally the case for νCAs), the weights and biases of the CNN remain unchanged.

A Locally-Connected Hidden Layer. The first approach includes a locally connected layer, which is essentially a convolution where the kernel weights are allowed to differ for distinct nodes. Like in Fig. 2, all the columns are summed, weighted by the binary representation of the local update rule, but now the weights are not shared. The biases remain zero. The rest of the CNN is identical to that for ECAs. In practice one more intermediate step is added: our model first calculates the entire output configuration as if the CA would be uniform, for each of the N_R rules. Next, the locally connected layer picks out the relevant cells based on which rule was actually allocated to it. While this is less computationally efficient, it does arguably increase interpretability of the model. More relevant in forthcoming research, however, is that this also facilitates flexibility in the training phase. After all, over-parameterisation is one of the key ingredients of deep learning.

Following the required subsequent calculations as explained in Sect. 1.2, this brings the total number of parameters in the model to

$$(3+1) \times 8 + 8 \times N_R + N_R \times N, \tag{1}$$

if we discard the bias parameters that have been set to zero. For the example depicted in Fig. 4, this sums to 352 parameters.

A Sparsely-Populated Dense Layer. A slightly different approach invokes the power of a fully-connected layer, known in the industry as a dense layer. Here, again CA outputs are calculated for each of the rules, but the cell selection now occurs by means of this dense layer. Note that, essentially, a locally connected layer is a dense layer for which all edges have been cut that connect nodes that represent different cells. Whilst this seems superfluous at first, there are two reasons to do so. First, `TensorFlow` is heavily optimised for calculating with large matrices, especially if these are sparse. Second, we again have the consideration of more model power and flexibility in future approaches that also involve training via backpropagation.

Technically, the N_R channels containing size-N outputs of the uniform case are first flattened, i.e. deconstructed into a single vector of length $N_R N$. Next, all elements in this vector (the node values) are connected with the size-N output layer by means of an $N_R N \times N$ weights matrix, where most weights are manually set to 0 or 1.

The total number of parameters in this model is therefore

$$(3+1) \times 8 + 8 \times N_R + N_R \times N^2, \qquad (2)$$

if again we do not count the vanishing bias parameters. For the example in Fig. 4, this now sums to 8288 parameters.

2.3 Comparison of the Three Models in Four Scenarios

It is counter-intuitive that any increase in computing speed is expected at all, considering the clearly large increase in required floating point operations. To be fair, a really well-optimised and parallelised approach tailored to νCA simulation will undoubtedly outperform the proposed over-parameterised CNNs. The main allure, however, is in the combination of ease of use and out-of-the-box high performance of `TensorFlow` (or `PyTorch`, for that matter), as a result of the global scale of its continuous development.

We will briefly examine where the strengths and weaknesses of the CNN approaches lie, compared to the benchmark approach using `CellPyLib`. In particular we will consider four scenarios: the performance when adding more rules, more time steps, more cells, and more samples. Table 1 provides a summary of the parameter values (or ranges) that were found to be appropriate for best illustrating the trends and comparisons: these are the domains in which the overall trend in performance for all approaches is easily discernible. Every sample starts from a random initial condition but an identical rule allocation, such that the CNN needs to be initialised only once. Using the `time` Python package, we simply keep track of how many seconds each model requires for evolving the νCAs, taking the average over ten attempts.

Table 1. Four scenarios that enable the comparison of computational performance ofor νCA simulation by means of `CellPyLib`, and by means of the `TensorFlow` CNNs (locally connected and densely connected).

Scenario	Rules (N_R)	Time steps (T)	Cells (N)	Samples (S)
Alter N_R	[1, 256]	32	256	32
Alter T	4	[10, 100]	64	32
Alter N	4	32	[1, 256]	32
Alter S	4	32	32	[1, 1024]

This small-scale experiment was performed using an Intel Core i7-9850H CPU, 6 cores, at 2.6 GHz. We ran Python 3.11.8 and `TensorFlow` 2.14.0. Note, however, that the numerical value of the timing is secondary to the qualitative comparison.

3 Results

Here we present the computation times of the various scenarios listed in Table 1. We always show the `CellPyLib` data in blue [plus symbols], the data from the locally connected CNN in orange [dots], and data from the fully connected CNN in green [crosses]. In order to value the trends, we always show the mean value and the standard deviation from ten independent computations per unique combination of parameters.

Figure 5 displays the computation times for the all four scenario. First, we show what happens when the number of rules (the 'non-uniformity') is increased by factors of two. Because the number of rules N_R goes up to 256, we also chose $N = 256$, allowing the possibility to allocate each rule at least once. We observe that the computation time is largely independent of N_R for all models, except when a large number of rules is chosen in the densely connected CNN. Rather surprisingly, however, for small values of N_R [the computation time for] this densely connected CNN is significantly smaller than the other two.

For the second scenario, the number of time steps increases linearly between 10 and 100, and the computation time in all models appear to increase linearly as well. Except for small T values, the computation time is similar for all three models.

The results from the third scenario are shown for eight linearly spaced values of N between 32 and 256. As expected, `CellPyLib`'s computation time is proportional to the number of cells. The locally connected CNN also increases more or less linearly – but with a higher 'start-up cost'. The computation time of the densely connected CNN is largely independent of N.

For the fourth and final scenario, we consider 11 logarithmically spaced values of S between 1 and 1024. We again observe that using the larger models requires a certain initial cost, but once we want to simulate a large number of diagrams, they are clearly the least time consuming option.

4 Discussion, Conclusion and Prospects

The highly optimised 'out-of-the-box' multiprocessing of `TensorFlow` is clearly preferred over `CellPyLib` in scenarios where we want to generate many samples of νCAs with many cells. This of course is precisely the condition for obtaining statistically significant results in empirical studies of these discrete dynamical systems, especially when training models for automatic classification.

More surprising, however, is that the densely connected CNN almost always beats the locally connected CNN, despite the fact that mathematically speaking, the latter is a subgraph of the former. This is even the case when simulating more cells, despite, as Eq. (2) shows, the quadratic growth of the number of parameters. This precisely demonstrates the point: `TensorFlow` is so cleverly optimised, that more complex models can outperform the simpler ones. This is arguably the reason why the `LocallyConnected1D` layer is discontinued in more recent versions of `TensorFlow`.

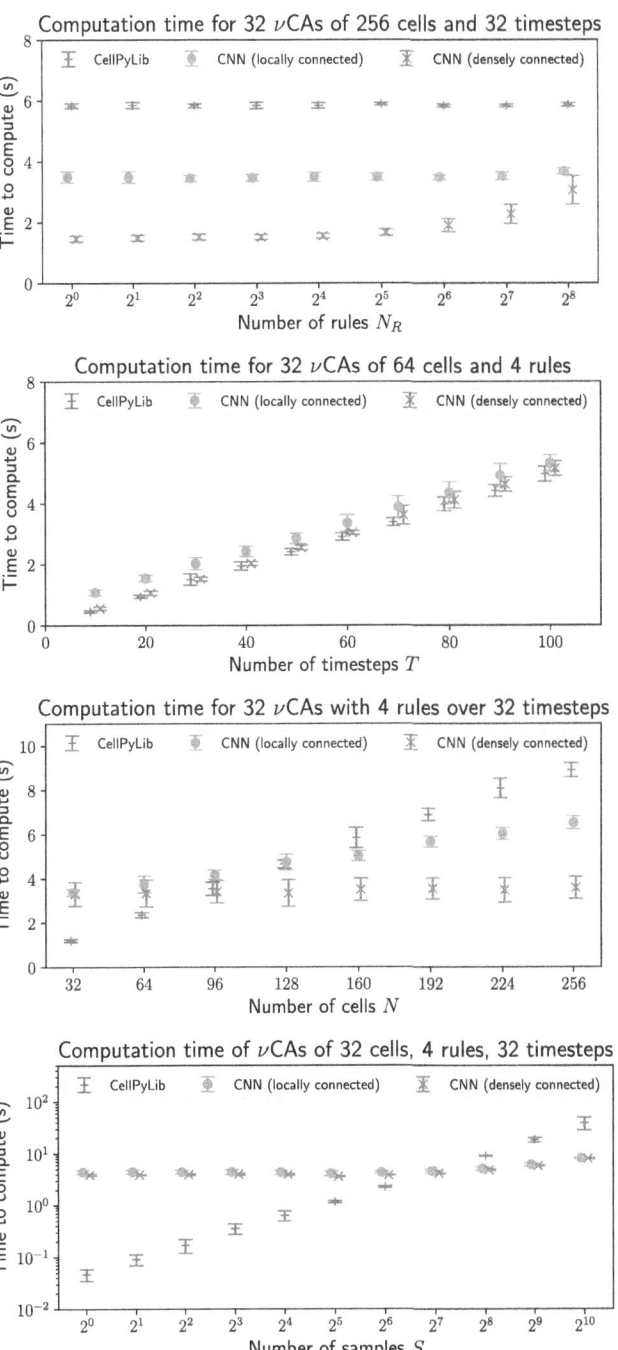

Fig. 5. Results for all four scenario as listed in Table 1, comparing three different simulation methods.

CNNs are a great tool for efficient simulation, which enables a more thorough exploration of the computational landscape of CAs. Similar approaches will enable the simulation of other types of CAs. As we show in forthcoming work, for example, graph CNNs are quite straightforward to mobilise in the simulation of network automata as well. While we do not claim that `TensorFlow` is the computationally optimal solution for CA simulation, it does present the CA researcher with an educational, ergonomic and flexible engine for efficient simulation.

CNNs are more than a tool for studying CAs, however. Arguably the most promising possibilities are created when, inversely, CAs serve the theoretical study and practical applications of CNNs. The training of CAs in the CNN framework may be used to better understand the information flow and learning process of CNNs [9]. Another exciting avenue is the mobilisation of CAs for generative neural networks, as was elegantly illustrated in [11]. In any case, increased efforts in joining discrete dynamical modelling and deep learning, such as the one shared in this work, offer interesting benefits for both research domains.

Disclosure of Interests. The authors have no competing interests to declare that are relevant to the content of this article.

References

1. 3Blue1Brown: Deep learning (2017). https://www.youtube.com/watch?v=aircAruvnKk
2. Antunes, L.M.: Cellpylib: a python library for working with cellular automata. J. Open Source Softw. **6**(67), 3608 (2021). https://doi.org/10.21105/joss.03608
3. Bhattacharjee, K., Naskar, N., Roy, S., Das, S.: A survey of cellular automata: types, dynamics, non-uniformity and applications. Nat. Comput. **19**(2), 433–461 (2020). https://doi.org/10.1007/s11047-018-9696-8
4. Cattaneo, G., Dennunzio, A., Formenti, E., Provillard, J.: Non-uniform cellular automata. In: Dediu, A.H., Ionescu, A.M., Martín-Vide, C. (eds.) LATA 2009. LNCS, vol. 5457, pp. 302–313. Springer, Heidelberg (2009). https://doi.org/10.1007/978-3-642-00982-2_26
5. Comelli, T., Pinel, F., Bouvry, P.: Comparing elementary cellular automata classifications with a convolutional neural network. In: ICAART (2), pp. 467–474 (2021)
6. Dennunzio, A., Formenti, E., Provillard, J.: Non-uniform cellular automata: classes, dynamics, and decidability. Inf. Comput. **215**, 32–46 (2012). https://doi.org/10.1016/j.ic.2012.02.008
7. Dhillon, A., Verma, G.: Convolutional neural network: a review of models, methodologies and applications to object detection. Prog. Artif. Intell. **9**(2), 85–112 (2020). https://doi.org/10.1007/s13748-019-00203-0
8. Gaylord, R.J., Nishidate, K.: Modeling Nature: Cellular Automata Simulations with Mathematica®. Springer, Heidelberg (1996)
9. Gilpin, W.: Cellular automata as convolutional neural networks. Phys. Rev. E **100**(3), 032402 (2019). https://doi.org/10.1103/PhysRevE.100.032402
10. Goodfellow, I., Bengio, Y., Courville, A.: Deep Learning. MIT Press, Cambridge (2016)

11. Mordvintsev, A., Randazzo, E., Niklasson, E., Levin, M.: Growing Neural Cellular Automata. Distill (2020). https://doi.org/10.23915/distill.00023
12. Rollier, M., Zielinski, K.M.C., Daly, A.J., Bruno, O.M., Baetens, J.M.: A comprehensive taxonomy of cellular automata (2024)
13. Silverman, E.: Convolutional Neural Networks for Cellular Automata Classification (2019). https://doi.org/10.1162/isal_a_00175
14. Vispoel, M., Daly, A., Baetens, J.: Progress, gaps and obstacles in the classification of cellular automata. Physica D Nonlinear Phenomena **432**, 30 (2022). http://hdl.handle.net/1854/LU-8740550

A Scheme for Symmetric Cryptosystem Using Large Cycle Reversible Cellular Automata

Tarun Lywait, Kiran Srinivasan, Krishnadas Nair,
and Kamalika Bhattacharjee[✉]

National Institute of Technology, Tiruchirappalli, Tiruchirappalli 620015, India
kamalika@nitt.edu

Abstract. In this paper, we expand upon Cipher Block Chaining (CBC), and explore large cycle reversible non-uniform cellular automata for secure encryption and decryption. Both our encryption and decryption algorithms have an $O(n)$ time complexity. We have conducted statistical tests like NIST, *dieharder*, and *smallCrush* to prove its robustness, along with brute force attacks and strict avalanche criterion (SAC) to further test its security. Analysis for known plaintext attack and chosen plaintext attack shows that neither of the attacks are feasible. Lastly, a visual test is conducted to further show the aparant randomness of the ciphertexts.

Keywords: Reversible Cellular Automata · Encryption · Decryption · Non-Uniform · Large Cycle · Cipher Blcok Chaining (CBC)

1 Introduction

In this day and age, the amount of data being transferred at any given instance is reaching a staggering amount, and ensuring the integrity and security of data is of major concern. Efficient cryptosystems are required which do not compromise security and still offer fast encryption and decryption times. A good candidate that can serve as a basis for a performant cryptosystem both when speed and security is concerned is cellular automaton (CA).

Cellular automata (CAs), also known as cellular spaces or tessellation automata, are discrete models of computation explored in automata theory. With applications in physics, theoretical biology, and microstructure modeling, they exhibit versatile behavior. Due to the inherent complexity through emergent behaviors from simple rules, parallelism for efficient computation, and resilience to attacks like differential and linear cryptanalysis, cellular automata have been a popular choice for cryptosystems since long [1–4, 7, 9, 10].

T. Lywait, K. Srinivasan and K. Nair—These authors contributed equally to this work. This work is partially supported by Start-up Research Grant (File number: SRG/2022/002098), SERB, Department of Science & Technology, Government of India.

Elementary cellular automata (ECAs) are a specific class of CA with simple rules. Each cell evolves to one of two states 0 or 1 based on its state and that of its immediate neighbors. This research leverages one-dimensional n-cell non-uniform reversible ECAs with large cycles under null boundary condition for the implementation of a robust cryptosystem. Each CA is represented by a *rule vector* $\mathcal{R} = \langle \mathcal{R}_0, \mathcal{R}_1, \cdots, \mathcal{R}_{n-1} \rangle$ where $\mathcal{R}_i$ is an ECA. The rules take as argument the combination of the present states of the left neighbor, the cell itself and the right neighbor; we call this combination as *Rule Min Term* or RMT. Unless explicitly mentioned, by a 'CA', we shall mean such a CA only. If all cells use the same rule, it is *uniform*.

In case of a reversible cellular automata, its global evolution rule is reversible, allowing the unique determination of the previous configuration based on the current one. The main goal of this research is to explore the potential of reversible non-linear non-uniform cellular automata with large cycle lengths as the basis for building a symmetric-key cryptosystem, capitalizing on their inherent security and complexity. The base of the theory for this work is the paper by Mukherjee et al., 2021 [8] which presents a stochastic method for synthesizing non-linear non-uniform reversible CAs with large cycle lengths. In this paper, it has been argued that, most of the CAs synthesized by this method have a cycle at least as large as 2^{n-1} [8]. We utilize this algorithm to generate our keys for the cryptosystem with exponentially large cycles having many non-linear rules. The reason for focusing on non-linear non-uniform CAs is that their non-linear dynamics and very large rule space enhance the security of the designed cryptosystem.

The paper is divided into 4 sections. Section 2 talks about the architecture and the working of our cryptosystem. It also describes our function, *prevConf* that provides the previous configuration of an input configuration based on a given CA rule vector and is instrumental in linear time decryption. Section 3 discusses about analysis and the tests done on the proposed cryptosystem to determine its security and robustness. Finally, Sect. 4 concludes the paper.

2 The Proposed Scheme for Symmetric Cryptosystem

Our proposed symmetric key cryptosystem utilizes the Cipher Block Chaining (CBC) mode of operation. CBC mode encrypts plaintext by XORing each block with the previous ciphertext block, using an Initialization Vector (IV) for the first block, enhancing security through a chaining effect. The initialization vector is a block-sized vector of randomly generated bits. For populating the bits of IV, we use Mersenne Twister (MT19937) [13], a widely used pseudo-random number generator (PRNG) algorithm with a period of $2^{19937} - 1$, known for its speed and good statistical properties and used in many programming languages and applications. Any other good PRNG will also serve the same purpose.

The plain text is divided into a number equal sized blocks of size 128 bits treated as the initial configurations. To enable parallel processing inside CBC mode, we define BLOCKS_PER_GROUP as the number of plain text blocks assigned to each thread. All blocks are clumped into groups of size BLOCKS_PER_GROUP. Each group operates in CBC mode with its own IV and concurrent thread.

The key used by the cryptosystem is a large cycle non-linear reversible CA synthesized using the stochastic algorithm on a Gaussian distribution [8]. The same key is used in every block. As the key is a randomly synthesized CA, the performance of the system depends on the generated key.

Encryption is done in 2 layers: the first layer is a non-linear CA followed by a linear CA layer. Initially, the block is advanced 64 time steps forward through the key using the *nextConf* method. This is followed by 128 steps forward through a linear uniform CA with ECA rule 153. Decryption retraces the above steps, first by advancing 128 steps through the linear layer followed by 64 steps forward through the reverse of the key. This is equivalent to moving 64 steps backward by the original key, achieved using *prevConf* algorithm. As the same key is used for both encryption and decryption, this scheme is a symmetric-key cryptosystem. Figures 2 provides an overall view of our cryptosystem. Next we describe the major components in our scheme: *encrypt, blockEncrypt, nextConf, prevConf, blockDecrypt* and *decrypt*.

2.1 Encryption

The plaintext is divided into blocks, and a set number of these blocks are combined to form groups, each with an initialized IV that is shared across all blocks within the group. Each block is encrypted using the *blockEncrypt* (Fig. 2a) module to produce a ciphertext block. The current ciphertext block is generated by XORing the current plaintext block with the previous ciphertext block, this implements the CBC mode of operation (Fig. 1). Encryption for each group is managed by a single thread instance, enabling multi-threading. Algorithm 1 describes the overall encryption process. Note that the function CBC_Enc() used in Algorithm 1 takes initialization vector, list of plaintext blocks and the block-wise encryption function as parameters and produces the corresponding list of ciphertext blocks (including the IV) as output.

Algorithm 1. encryption function

---

  procedure ENCRYPT(PTblocks) ▷ PTblocks is a list of 128-bit plaintext blocks
   $threadCount \leftarrow len(PTblocks)\ /\ \text{BLOCKS\_PER\_GROUP}$
   $CTblocks \leftarrow$ empty list
   for $tIdx \leftarrow 0$ to $threadCount - 1$ **do**
    New thread is generated for the current group
    $tBlocks \leftarrow$ PTBlocks[tIdx*BLOCKS_PER_GROUP...(tIdx+1)*BLOCKS_PER_GROUP-1]
    $IV_{tIdx} \leftarrow$ random 128-bit number using MT19937
    CTblocks.append(CBC_Enc(IV_{tIdx}, tBlocks, blockEncrypt))
   return CTblocks

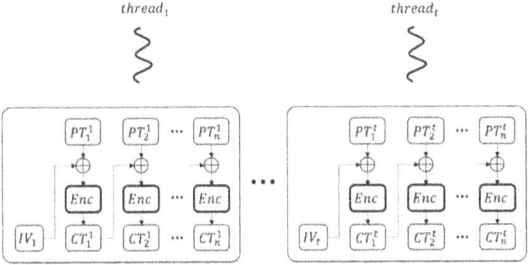

Fig. 1. Block diagram of encrypt Module

2.2 blockEncrypt

This function, when passed a plaintext block, is tasked with encrypting the given block and then returning the corresponding ciphertext block. The 128 bit block is first passed through a layer of non-linear CA provided as a single key. The block is evolved by using the *nextConf* function for 64 steps in accordance to the rule vector given in the key. The output block thus obtained is further passed through a linear layer containing a uniform CA of rule 153. In this layer, the block advances 128 time steps. The ciphertext generated this way is then returned. A schematic of this module is given in Fig. 2a.

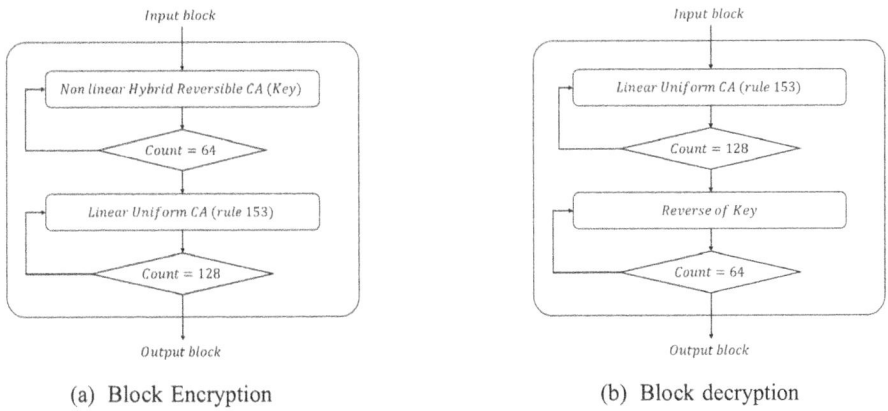

(a) Block Encryption (b) Block decryption

Fig. 2. Encryption and Decryption

2.3 nextConf

This function generates the next configuration of a block from a binary sequence representing current configuration, using the rule vector chosen as the key.

2.4 prevConf

The decryption process in this scheme involves tracing a number of steps backwards in time along a large cycle CA. In order to go backwards towards the original plaintext block, one naive approach might be to start from the ciphertext block and store all the intermediate configurations until the same configuration is repeated completing the cycle. Then using this stored data, one can simply reverse this order, and then go forward 64 steps from the ciphertext block to reach the plaintext block. See for example, Fig. 3 for a CA with small cycle. However, this approach is inefficient and impractical because we are interested in choosing our key having cycles of exponential size to n.

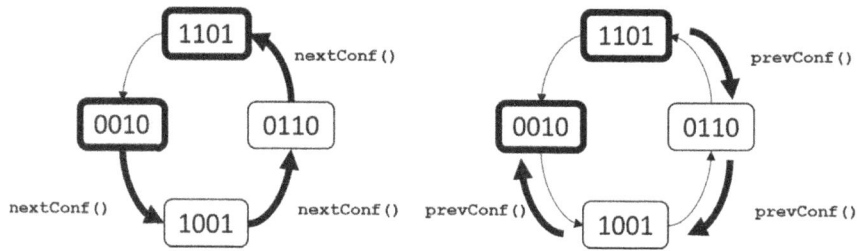

Fig. 3. Representative functionality of *nextConf* and *prevConf* to move forwards and backwards in time respectively for rule vector $\langle 51, 153, 51, 153 \rangle$

To address this issue efficiently, this paper proposes a linear time algorithm that can find the previous configuration of a given configuration using the *prevConf* function. We apply this algorithm 64 times, starting from the ciphertext block, to get the original plaintext block. Using this strategy, we need not explore the entire cycle but rather a very small subset of it saving time and space.

This function is crucial for the purpose of decryption. It operates through a series of steps, involving both left-to-right and right-to-left passes over the bit positions (cells). The details of the algorithm to find the previous configuration *prev* of a given configuration *conf* is as follows:

1. For every cell (position) i in *conf* from left to right, collect all possible neighborhood configurations (RMTs) that give output conf[i] provided the RMTs satisfy the following conditions:
   (a) RMTs should consider null boundary at left (0xx) if $i = 0$
   (b) RMTs should consider null boundary at right (xx0) if i is at rightmost position
   (c) Any bits already known in *prev* must be present in the RMTs at the right position
   (d) Already known equality/inequality relationships between adjacent bits must be obeyed by the RMTs

2. Among the RMTs thus narrowed down, check if the bit in the left neighbor, right neigbor or the center cell is the same. If so, then for that corresponding position in *prev*, that bit is resolved. For instance, if narrowed down RMTs are {100, 010}, we can resolve the rightmost bit to be 0.
3. Among the RMTs, check if any equality or inequality relationship exists between left neighbor and right neighbor, left neighbor and center cell, or right neighbor and center cell. If any relation exists, keep a note of the relation between the two corresponding positions in *prev*. For instance, we can see an inequality relationship between left and middle bits in {100, 010}.
4. After traversing every position in *conf* from left to right, traverse *prev* from right to left to address any unresolved bits. At this stage, each unresolved bit will have a defined relationship with another position in *prev*. During the backward traversal, if we encounter an unresolved bit, we identify its relation to an already resolved position. By leveraging this relationship, the value of the unresolved bit can be determined. Upon reaching the beginning of *prev*, the previous configuration of *conf* is completely resolved.

Table 1. RMTs of a sample rule vector $\langle 5, 165, 169, 80 \rangle$

Rule	RMT							
	111	110	101	100	011	010	001	000
5	0	0	0	0	0	1	0	1
165	1	0	1	0	0	1	0	1
169	1	0	1	0	1	0	0	1
80	0	1	0	1	0	0	0	0

Let us take an example to understand how *prevConf* works. Consider Table 1 that contains the RMTs for the rule vector $\langle 5, 165, 169, 80 \rangle$ taken as the key and find the previous configuration of 1001. During the initial pass we identify candidate RMTs for each position from left to right. For cell 0, we observe that the candidate RMTs are of the form 0x0 because the only two RMTs that can produce next state as 1 while also maintaining null boundary condition for left are 010 and 000. With this information we infer that cell 1 in previous configuration has to be 0. Next, the candidate RMTs for cell 1 are observed to be 001 and 100, giving the form $x0\bar{x}$. No new resolutions are obtained. It is noted that the left and right don't care bits in the candidate RMTs are related by inequality relation. Moving on, we find out that the candidate RMTs for cell 2 are 010 and 001, which is in the form $0x\bar{x}$. Once again there are no new resolutions obtained, but the center and right don't care bits are observed to be related by inequality relation. Now, we go to cell 3, where the candidate RMTs are 110 and 100, which is of form 1x0. This leads to the resolution of the bit at cell 2 to 1.

Currently, only values of cells 1 and 2 have been determined, with values 0 and 1, respectively. Starting at cell 3, we use the inequality relation between cells 2 and 3. Since cell 2 is 1, cell 3 is set to 0. Likewise, given the inequality relation between cells 0 and 2, and knowing that cell 2 is 1, cell 0 is set to 0. Therefore, the resulting previous configuration is determined to be 0010.

2.5 blockDecrypt

As shown in Fig. 2b, the evolution path is retraced, first by passing through the linear layer with uniform ECA rule 153 for 128 time steps. Following this, the block advances 64 time steps on the non-linear layer containing the reversed key. Instead of computing the reverse of the key, we perform the equivalent operation of advancing *backward* on the original key itself using our novel algorithm *prevConf*.

2.6 Decryption

The ciphertext is segmented into blocks, and analogous to the encryption process, a predetermined number of blocks are grouped together to form groups. Each block within these sets is decrypted using the *blockDecrypt* module. A dedicated thread handles the decryption process for each set, ensuring efficient parallel processing. Figure 4 showcases the block diagram which depicts the CBC mode of operation that is being implemented, whereas, the steps for Decryption are shown in Algorithm 2. Note that the function CBC_Dec() used in Algorithm 2 takes list of ciphertext blocks and the block-wise decryption function as parameters and produces the corresponding list of ciphertext blocks as output.

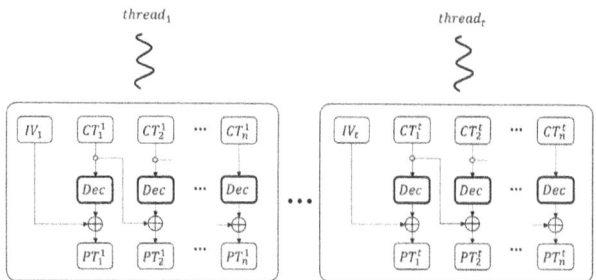

Fig. 4. Block diagram of decrypt Module

Algorithm 2. decryption function

procedure DECRYPT(CTblocks)   ▷ CTblocks is a list of 128-bit ciphertext blocks
   CT_BLOCKS_PER_GROUP ← BLOCKS_PER_GROUP + 1   ▷ +1 because of IV
   $threadCount$ ← $len(CTblocks)$ / CT_BLOCKS_PER_GROUP
   $PTblocks$ ← empty list
   for t ← 0 to $threadCount - 1$ **do**
     New thread is generated for the current group
     $tBlocks$ ← PTblocks[t*CT_BLOCKS_PER_GROUP...(t+1)*CT_BLOCKS_PER_GROUP-1]
     CTblocks.append(CBC_Dec(tBlocks, blockDecrypt))
   return PTblocks

3 Analysis

Now we analyze the implementation details along with the security of the scheme.

3.1 Design Considerations

Choice of Time Steps: It is known that the length of cycle for an n-cell uniform CA containing ECA rule 153 is $l = 2^{\lfloor \log n \rfloor + 1}, n \geq 2$ [6]. Hence, for 128-cell CA, the cycle length is 256. By advancing 128 time steps forward both during encryption and decryption, a total of 256 time steps are taken, which brings the configuration back to the original input of the linear layer. The is the rationale for advancing 128 time steps forward in the linear layer.

In the case of the non-linear layer containing a large cycle reversible cellular automaton as a key, the input is advanced by 64 steps during encryption and reversed by the same number of steps (using prevConf) during decryption. This decision was made to achieve a balance between the security of the cryptosystem, which improves with a greater number of time steps, and the time required to perform encryption/decryption, which worsens with more time steps.

Ordering of Layers. The encryption module contains a non-linear layer provided by the reversible key followed by a uniform linear layer of ECA rule 153. The decision to place this linear layer after the non-linear layer is to ensure that, even if in the worst case, the selected key puts the plaintext into a very small cycle, the linear layer of ECA rule 153 ensures that all configurations are present in some fixed 256-length cycle. Furthermore, ECA 153 is an equal-length cycle CA which guarantees that there are a huge number of such cycles available for 128 cell size where the encrypted text from the non-linear layer will be put which gives an extra layer of security.

Number of Keys. The cryptosystem uses a single 128-cell CA as the symmetric key. This key is used for encrypting and decrypting all blocks in every thread. A known plaintext attack designed to expose the underlying cyclic structure of the largest cycle in the CA would involve the encryption of around 2^{127} 128-bit blocks [8], which is computationally infeasible.

Mode of Encryption. The mode of encryption for this cryptosystem is Cipher Block Chaining (CBC). This mode was chosen instead of the more common Substitution Permutation Network (SPN) mode in order to protect against linear and differential cryptanalysis attacks that arise as a consequence of having static S-boxes.

3.2 Choice of Keys

The proposed cryptosystem is designed to use a single 128-cell non-uniform non-linear reversible CA as the symmetric key for all blocks. Consequently, the robustness of the cryptosystem is largely influenced by the strength of the key, for which one of the key factors is the length of the cycles generated by that CA for 128 cells for any initial configuration (that corresponds to the plaintexts here). To empirically ensure that the cycles are always exponentially large, an experimental evaluation is conducted on a set of randomly generated keys derived from the key synthesis algorithm outlined in Ref. [8]. This evaluation mimics a cycle discovery attack, where an initial configuration of 128 bits is created and subsequent configurations are computed iteratively until returning to the original configuration. The test concludes upon rediscovering the initial configuration or reaching one hour of elapsed time, whichever comes first. Listed below are some of the randomly selected keys synthesized using the algorithm which have passed this cycle discovery attack for one hour run time over different arbitrary seeds.

Key 1: <10, 90, 90, 150, 156, 169, 105, 105, 149, 150, 150, 106, 90, 90, 150,105, 166, 90, 90, 150, 90, 105, 90, 150, 165, 165, 105, 105, 90, 90, 90, 90, 90, 165, 105, 105, 150, 105, 150, 90, 150, 150, 150, 105, 165, 150, 90, 150, 165, 150, 106, 165, 106, 165, 149, 105, 165, 105, 150, 150, 165, 90, 105, 150, 90, 165, 150, 150, 165, 150, 165, 105, 150, 150, 150, 150, 90, 165, 90, 90, 150, 150, 150, 150, 105, 150, 105, 105, 150, 165, 150, 150, 90, 90, 150, 90, 90, 101, 150, 105, 166, 150, 150, 86, 150, 105, 101, 150, 90, 150, 106, 105, 165, 165, 150, 150, 90, 166, 105, 105, 89, 165, 86, 165, 150, 165, 90, 5>

Key 2: <9, 142, 90, 90, 165, 90, 105, 166, 165, 106, 165, 166, 90, 165, 105,150, 150, 165, 150, 105, 90, 105, 150, 165, 90, 90, 165, 150, 90, 150,89, 150, 165, 90, 105, 105, 105, 165, 150, 150, 90, 150, 150, 105, 90, 105, 150, 105, 165, 90, 105, 105, 165, 105, 165, 165, 90, 165, 90, 90, 90, 165, 150, 105, 150, 165, 105, 150, 165, 105, 165, 169, 150, 150, 90, 150, 150, 89, 105, 90, 90, 165, 165, 150, 150, 165, 105, 90, 90, 105, 105, 150, 105, 105, 150, 165, 169, 165, 105, 150, 165, 90, 165, 165, 105, 166, 90, 166, 105, 165, 150, 149, 105, 90, 90, 90, 165, 105, 106, 90, 105, 150, 150, 106, 165, 105, 105, 65>

Key 3: <6, 90, 105, 90, 166, 90, 90, 165, 169, 90, 150, 165, 101, 105, 105, 165, 150, 150, 165, 105, 165, 150, 169, 150, 150, 150, 165, 165, 90, 154, 105, 105, 165, 150, 150, 90, 150, 165, 90, 150, 89, 90, 169, 150, 150, 150, 105, 90, 90, 105, 150, 165, 165, 150, 150, 105, 105, 150, 154, 105, 165, 90, 105, 106, 150, 150, 150, 165, 150, 150, 165, 150, 105, 105, 90, 105, 169, 90, 106, 165, 150, 150, 165, 105, 90, 150, 165, 90, 150, 150, 150, 105, 150, 105, 150, 166, 165, 166, 150, 150, 90,

150, 150, 165, 150, 165, 90, 90, 105, 150, 105, 150, 90, 105, 90, 105, 90, 90, 150, 150, 150, 105, 165, 86, 150, 105, 90, 80>

Key 4: <9, 113, 165, 90, 150, 154, 165, 150, 165, 89, 90, 169, 105, 150, 105, 90, 105, 165, 106, 150, 105, 105, 105, 165, 150, 86, 150, 105, 149, 150, 90, 105, 90, 105, 150, 105, 90, 150, 105, 90, 90, 90, 150, 165, 105, 150, 105, 90, 150, 150, 165, 150, 105, 90, 89, 105, 105, 105, 90, 105, 105, 90, 105, 90, 165, 105, 150, 165, 90, 165, 90, 150, 149, 165, 90, 165, 105, 150, 150, 90, 105, 165, 150, 150, 165, 165, 165, 101, 165, 150, 105, 105, 169, 90, 166, 150, 165, 90, 90, 90, 165, 165, 150, 154, 165, 105, 165, 90, 165, 149, 90, 105, 90, 105, 90, 90, 165, 89, 165, 89, 105, 165, 150, 165, 90, 165, 105, 20>

Key 5: <10, 150, 90, 75, 150, 149, 105, 105, 166, 90, 90, 165, 149, 165, 165, 90, 150, 90, 166, 150, 150, 106, 150, 105, 90, 150, 90, 165, 165, 90, 105, 101, 90, 105, 105, 105, 165, 90, 90, 90, 150, 165, 90, 90, 101, 150, 90, 150, 150, 90, 169, 90, 90, 105, 90, 150, 106, 165, 165, 105, 90, 105, 169, 150, 150, 101, 90, 101, 90, 165, 105, 105, 90, 150, 90, 105, 105, 105, 150, 89, 150, 150, 149, 150, 165, 90, 105, 165, 90, 101, 105, 150, 150, 90, 150, 165, 150, 150, 105, 86, 165, 165, 90, 106, 150, 90, 165, 105, 150, 105, 105, 105, 165, 150, 150, 165, 90, 90, 105, 165, 165, 165, 165, 105, 105, 105, 150, 5>

Key 6: <9, 178, 165, 90, 150, 106, 150, 105, 165, 105, 165, 90, 105, 150, 150, 105, 89, 150, 105, 165, 165, 154, 150, 105, 150, 105, 150, 90, 165, 165, 165, 90, 150, 165, 105, 150, 165, 90, 150, 165, 165, 90, 150, 105, 105, 150, 169, 90, 90, 105, 105, 165, 150, 90, 150, 149, 105, 165, 165, 90, 90, 165, 150, 105, 90, 169, 105, 165, 150, 150, 105, 150, 165, 90, 90, 90, 89, 105, 150, 89, 90, 101, 105, 165, 90, 165, 150, 150, 105, 105, 105, 105, 105, 106, 90, 165, 165, 166, 90, 169, 165, 90, 165, 105, 165, 105, 90, 86, 105, 105, 166, 165, 89, 90, 105, 105, 90, 105, 165, 105, 150, 90, 90, 106, 150, 165, 90, 65>

We observe that, in all tested scenarios, none of the initial configurations were retrieved within the one-hour test period, suggesting the existence of exceedingly large cycles. A more comprehensive result of the test is present on GitHub [15].

3.3 Strict Avalanche Criterion Test (SAC)

The Strict Avalanche Criterion (SAC) is a property used to evaluate the avalanche effect in cryptographic functions. It measures how much the output of a function changes when a single input bit is flipped. A cryptographic function satisfies the SAC if, for each output bit, changing any single input bit causes the output bit to change with a probability of approximately 50%. In simpler terms, the SAC requires that flipping any single input bit should, on average, change approximately half of the output bits [5].

For conducting the SAC test, a random 128-bit plaintext p and a random set of keys are chosen. We iterate over the plaintext, and flip each bit at a time, which generates a new input p'. Here there is a one bit difference between p and p'. Both of the inputs are encrypted using the cryptosystem to give ciphertexts c and c'. We count the number of bit difference between c and c', to get the SAC score. This process is repeated a certain amount of times (1000, in this case) and the final SAC score is calculated as the average of the individual SAC scores. In our case, because a 128-bit input produces a 256-bit output, the ideal SAC score should be 128. We ran the test accordingly and Table 2 displays the result. As the mean is approximately 128, we can conclude that the cryptosystem passes the SAC test.

Table 2. Results of SAC test

Metric Used	Metric Value
Expected Mean	128
Observed Mean	127.992
Variance	63.89
Std. Deviation	7.99312
Coeff. of Variance	0.0624499

3.4 Resistance Against Brute Force Attack

The reversibility of CAs, coupled with high entropy and diffusion propagated with the help of the IV creates a challenging environment for most statistical or analytical attacks due to the absence of linearity or patterns. This resistance often leaves brute force as the primary attack vector, which is computationally infeasible with sufficiently long key lengths as explained below.

In the scenario of a brute force attack, two possible techniques can be employed. The first technique involves the attacker attempting to deduce the 128-cell key by systematically generating keys using Ref. [8]. However, this approach is computationally impractical and would require at least $\Omega(2^n)$ time to complete for an n-rule key.

Another technique entails the attacker constructing the transition diagram to identify the cyclic structure of the key where each evolution of the ciphertext is situated. Empirical evidence indicates that this method is similarly impractical, taking $O(2^{n-1})$ time for an n-rule key if chosen properly, due to the extensive cycle lengths of the utilized CAs.

3.5 Resistance Against Known Plaintext Attack

In known plaintext attack, a set of random plaintexts and their corresponding ciphertexts are known to the adversary. By analyzing multiple pairs of known plaintext and ciphertext, the attacker aims to deduce the encryption key or reveal weaknesses in the encryption algorithm. Chosen ciphertext attack is same as known plaintext attack but the difference is that, the adversary can choose the set of plaintexts to be encrypted.

However, in the proposed cryptosystem, due to the high amount of non-linearity and very large cycle length, there is no clear relation between the plaintext and ciphertext that can be exploited. However, if a plaintext block is chosen in such a way that it occurs in a small cycle in the CA (the corresponding intermediate ciphertext block also exists in this cycle), then the adversary simply needs to explore the small cycle, that may potentially reveal certain information about the key. But, because the CA used, is not known to the attacker, the only way to find such a plaintext would be through brute force. The attacker would need to go through 2^n states (where n is the size of the rule vector of the CA) in the worst case scenario, which is not feasible. Thus, Chosen plaintext attack, and by extension, Known plaintext attack will fail on our cryptosystem.

3.6 Resistance Against Cycle Attack

In a cycle attack, an attacker would try to find a cycle in the CA's state transitions that leads back to the initial state after a certain number of steps. This could potentially reveal information about the key or the internal state of the CA, compromising the security of the cryptosystem. However, as large cycle CAs are being generated as keys [8], where it has been empirically proven that the large cycle length would be at least 2^{n-1}, we can safely say that, if the block size is sufficiently long (128-bit long, in our case), the attacker would be unable to detect the cycle in polynomial time. The probability for the plaintext and ciphertext to be in a small cycle is negligibly low. Thus, the cryptosystem is resistant against cycle attacks.

3.7 Statistical Tests

Statistical tests performed on our cryptosystem involve three rigorous test suites namely, dieharder [11], NIST [12] and *smallCrush* [14] which tests for detecting patterns/non-randomness. In case of dieharder and NIST, a binary file of size 1GB is created which consisted of encrypted form of the same plaintext repeated over and over again. Due to the design of our system, the same plaintext with the same set of keys will not necessarily yield the same ciphertext every time. For generating the ciphertext, our cryptosystem uses 12 threads with 16 blocks per thread. In case of smallCrush, whenever the test requested for a random number, a specific plaintext (128-bit long) was encrypted that yielded a 256-bit output. This output was divided into 8 32-bit sub-blocks, each of which is provided as output consecutively. When 256 bits are exhausted, the program generates a new ciphertest from the same plain test. It is observed that, in case of Dieharder, all 114/114 tests have passed successfully out of which only 6 tests resulted as Weak. Similarly for NIST, 187/188 tests have successfully passed and for smallCrush, 13/15 tests have passed successfully. The weakness results are presented in Table 3, 4 and 5.

Table 3. Results of dieharder test

S. No.	Test Name	No. of tests	Passes	Weaks	Fails
1	dieharder_dna	1	0	1	0
2	dieharder_sums	1	0	1	0
3	sts_serial	16	15	1	0
4	rgb_lagged_sum	33	30	3	0

Table 4. Results of NIST test

S. No.	Test Name	No. of tests	Passes	Fails
1	Universal_lot	1	0	1

Table 5. Results of SmallCrush test

S. No.	Test Name	Result
1	BirthdaySpacings	Fail
2	MaxOft AD	Fail

3.8 Visual Test

Lastly, we apply visual test similar to space-time diagram in order to see if there is any pattern in the encrypted texts. Here we consider as plaintext the space-time diagram of a cellular automaton with high fractal characteristics. In this case, we chose Rule 90 for 128 cells generated for 100 time steps. Each row in the space-time serves as plain text whose ciphertext is presented in the corresponding row of another diagram. From the results displayed in Fig. 5, it is observed that the predictable pattern of the plain text space-time diagram is completely randomized upon encryption. Figure 6 depicts the ciphertext space-time diagrams for 12 more distinct keys using the same plaintext shown in Fig. 5a, suggesting that all keys are capable of producing this randomization effect.

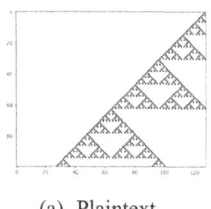
(a) Plaintext

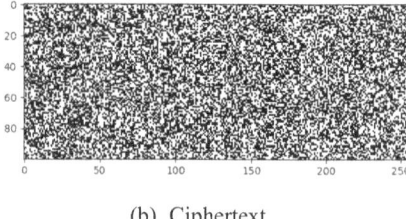
(b) Ciphertext

Fig. 5. Visual Test

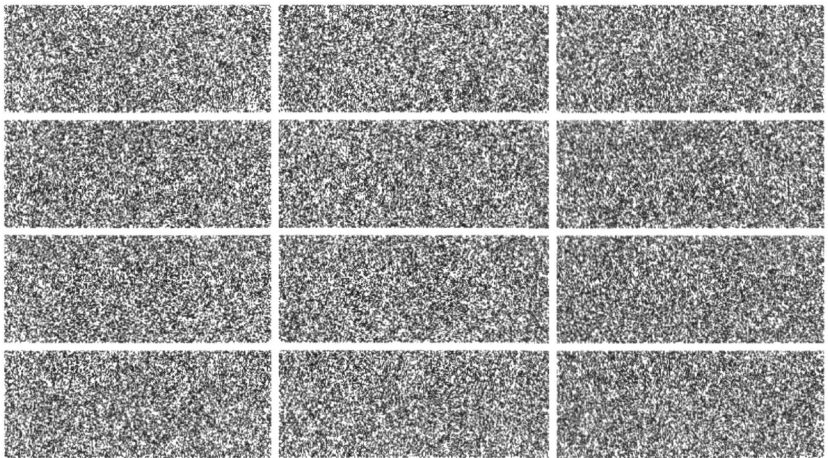

Fig. 6. Ciphertext for 12 more keys over plaintext of Fig. 5a

4 Discussion

In conclusion, the combination of a reversible cellular automaton with large cycle properties, CBC mode of encryption, and IV generation with a good PRNG like Mersenne Twister [13] presents a promising avenue for developing robust and efficient cryptosystems. The implementation of the scheme described in this paper is available on GitHub [15]. However, the research considers that a good key has been generated by the stochastic algorithm of Reference [8]. As it is not theoretically possible to find the actual cycle length of such a CA, we have empirically tested some randomly synthesized CAs for cycle discovery attack and reported such sample good keys. Furthermore, the concept of isomorphism can be realized to find the a larger set of good keys from a given good key as part of the key generation algorithm. These are the tasks of the future.

References

1. Wolfram, S.: Cryptography with cellular automata. In: Williams, H.C. (ed.) CRYPTO 1985. LNCS, vol. 218, pp. 429–432. Springer, Heidelberg (1986). https://doi.org/10.1007/3-540-39799-X_32
2. Chaudhuri, P.P., Chowdhury, D.R., Nandi, S., Chattopadhyay, S.: Additive Cellular Automata: Theory and Applications, Volume 1 (Vol. 43). Wiley, Hoboken (1997)
3. Nandi, S., Kar, B., Pal, P.: Theory and applications of cellular automata. IEEE Trans. Comput. **43**(1346), 1357 (1995). https://doi.org/10.1109/12.338094
4. Tomassini, M., Perrenoud, M.: Cryptography with cellular automata. Appl. Soft Comput. **1**(2), 151–160 (2001)
5. Banerjee, T., Roy Chowdhury, D. (2021). EnCash: an authenticated encryption scheme using cellular automata. In: Gwizdałła, T.M., Manzoni, L., Sirakoulis, G.C., Bandini, S., Podlaski, K. (eds.) ACRI 2020. LNCS, vol. 12599, pp. 67–79. Springer, Cham (2020). https://doi.org/10.1007/978-3-030-69480-7_8

6. Mukhopadhyay, D., RoyChowdhury, D.: Cellular automata: an ideal candidate for a block cipher. In: Ghosh, R.K., Mohanty, H. (eds.) ICDCIT 2004. LNCS, vol. 3347, pp. 452–457. Springer, Heidelberg (2004). https://doi.org/10.1007/978-3-540-30555-2_52
7. Das, S., RoyChowdhury, D.: CAR30: a new scalable stream cipher with rule 30. Cryptogr. Commun. **5**, 137–162 (2013)
8. Mukherjee, S., Adak, S., Bhattacharjee, K., Das, S.: Non-uniform non-linear cellular automata with large cycles and their application in pseudo-random number generation. Int. J. Mod. Phys. C **32** (2021). https://doi.org/10.1142/S0129183121500911.
9. Mariot, L., Leporati, A.: A cryptographic and coding-theoretic perspective on the global rules of cellular automata. Nat. Comput. **17**, 487–498 (2018)
10. Lira, E.R., de Macêdo, H.B., Lima, D.A., Alt, L., Oliveira, G.M.: A reversible system based on hybrid toggle radius-4 cellular automata and its application as a block cipher. Nat. Comput. 1–17 (2023)
11. Brown, R.G., Eddelbuettel, D., Bauer, D.: Dieharder: A Random Number Test Suite. https://webhome.phy.duke.edu/~rgb/General/dieharder.php
12. Rukhin, A., et al.: Statistical test suite for random and pseudorandom number generators for cryptographic applications, NIST special publication. revision 1a, volume 800-22. National Institute of Standards and Technology, Technology Administration, U.S. Department of Commerce (2010)
13. Matsumoto, M., Nishimura, T.: Mersenne twister: a 623-dimensionally equidistributed uniform pseudo-random number generator. ACM Trans. Model. Comput. Simul. 8(1), 3–30 (1998). https://doi.org/10.1145/272991.272995
14. L'Ecuyer, P., Simard, R.: TestU01: a C library for empirical testing of random number generators. ACM Trans. Math. Softw. **33**(4), 22:1–22:40 (2007)
15. https://github.com/ghadilion/large-cycle-rca-symmetric-key-cryptosystem/tree/main

Reversible Decimal First Degree Cellular Automata for Data Classification

C. J. Baby$^{(\boxtimes)}$ and Kamalika Bhattacharjee

Department of Computer Science and Engineering, National Institute of Technology, Tiruchirappalli, Tiruchirappalli 620015, Tamil Nadu, India
babycj1120@gmail.com, kamalika@nitt.edu

Abstract. The classification problem predicts labels for input data based on the training dataset. In this paper, the cyclic spaces of first degree reversible cellular automata are used to solve the classification problem. Every dataset of classification problem contains different classes. Based on the class labels in each configuration of cycles of the first degree reversible cellular automaton, they are grouped into different classes. The main advantage of this proposed model is that we can use real-world numerical data directly without doing any type of encoding. When selecting a reversible CA for classification, it is essential to maintain a minimum distance property within the same cycle's configurations, while ensuring a significant distance between configurations from different cycles. This study identifies (linear) CAs that satisfy this criterion. Our model performs well in comparison to the existing machine learning models.

Keywords: First Degree Cellular Automata (FDCAs) · Decimal CA · Reversibility · Cyclic Space · Classification

1 Introduction

Reversible cellular automata are cellular automata (CAs) in which the global transition function is bijective. That means, for each of the configurations in the reversible cellular automaton (CA), there is a unique predecessor. The configurations of a reversible cellular automaton belong to a set of cycles. Let $S_c = \{cl_1, cl_2, cl_3, \cdots cl_i, \cdots cl_{m-1}, cl_m\}$ be the set of cycles formed by all configurations of a reversible cellular automaton. In this paper, these cycles are used to solve the classification problem.

In a classification problem, every dataset contains a number of classes. Based on the labels of the classes in the training data, the label of the new input data is predicted. In this paper, the cycles of the reversible cellular automata are grouped into different sets based on the labels of the objects in the cycle. The label of the new input object can be predicted based on this cycle. Suppose

---

This work is carried out as a project in the Summer School on Cellular Automata Technology 2023.

© The Author(s), under exclusive license to Springer Nature Switzerland AG 2024
F. Bagnoli et al. (Eds.): ACRI 2024, LNCS 14978, pp. 147–162, 2024.
https://doi.org/10.1007/978-3-031-71552-5_13

a reversible CA has two cycles, where $Cycle_1$ is associated with *class A* and $Cycle_2$ with *class B* according to the training dataset. If a new input object is located within $Cycle_1$, we can predict the label of the new object as *class A*.

In machine learning problems, we need to analyze complex data. The majority of data includes decimal numbers. Most of the existing work done for classification using cellular automata is by using multiple-attractor cellular automaton (MACA) [8]. The pattern classification can also be done by asynchronous cellular automata [12] using Rule 219 and fixed-point attractors [11]. In all these existing works, elementary cellular automata have been used for classification, so, some encoding method is required to convert the decimal data or any data into binary form suitable for the binary cellular automata. However, this encoding often results in loss of the properties of the individual features which causes poorer performance in comparison to the other machine learning models.

If we can use a CA that can handle decimal numbers directly, then we can avoid this data encoding step. Therefore, to address this, the natural choice for a cellular automata based classifier for classifying data elements is using decimal cellular automata where each cell can take any of the states 0 to 9. However, for a decimal CA with m-neighborhood dependency, there are 10^{10^m} number of rules possible which is gigantic for any characterization. So, in this work, we concentrate over a small class of decimal CAs with only the nearest neighborhood dependency, namely the *first degree CAs* which are very simple CAs represented by only eight parameters. Hence, the number of possible rules becomes 10^8. Using these first degree cellular automata, we can utilize the decimal numbers directly as input without any encoding step. This is one of the main advantage of using decimal first degree cellular automata for classification in comparison to other existing classification algorithms using cellular automata.

In traditional pattern classification using cellular automata, the central concept is an attractor state. However, in this work, we introduce the cycle as the primary concept for grouping data classes. When an appropriate CA is selected for a dataset, closely related data elements tend to cluster within the same cycle. Utilizing a cycle allows us to identify a group of data such that the classification depends on all configurations within the cycle. Whereas relying solely on a single attractor means the entire class depends on this distant attractor. Thus, in this work, we utilize cycles as instruments for classification.

2 First Degree Cellular Automata

Introduced in Ref. [3], first degree cellular automaton is a very effective representation for a subset of CAs where number of states per cell can be large but neighborhood is restricted to only three. Formally, a first degree rule is defined as the following:

Definition 1. *A cellular automaton rule* $R : S^3 \to S$ *is of first degree, if the rule* $R : S^3 \to S$ *can be represented in the following form:*

$$R(x, y, z) = c_0 xyz + c_1 xy + c_2 xz + c_3 yz + c_4 x + c_5 y + c_6 z + c_7 \pmod{d}$$

Here S is the finite state set, d is the number of states of CA, and x, y, z are the states of the three neighbors where $x, y, z \in S$. The constants $c_i \in \{0, 1, 2, 3, \cdots d - 1\}$ [3,4].

As each of the variables in the above equation is of degree one, such cellular automata rules are called *rules of first degree*, and a CA with such a rule as simply a *first degree CA (FDCA)* [3]. Any rule following the above equation can be represented only by the values of the constants $\langle c_0, c_1, c_2, c_3, c_4, c_5, c_6, c_7 \rangle$; these values are the *parameters* of the first degree CA. Note that, if $c_0 = c_1 = c_2 = c_3 = 0$, the CA is a linear CA. Therefore, the set of *linear* CAs is a subset of the set of first-degree CAs. Our rule space is effectively reduced by selecting only CAs represented by these parameters [4]. In this work, we consider $d = 10$, that is, a decimal first-degree CA under the null boundary condition. Hereafter, if not explicitly mentioned, by a CA, we shall mean only finite decimal first-degree cellular automaton under the null boundary condition.

The combination of neighbors in the rule of a CA is also called a *Rule Min Term (RMT)*. Therefore in the equation of Definition 1, each possible value of the combination xyz is an RMT. Obviously, for a decimal CA, there are $10^3 = 1000$ such RMTs. These RMTs can be grouped based on some relation:

Definition 2. *For an m-neighborhood CA, a set of RMTs is called k-equivalent or $\mathcal{E}^k$, if the values of all neighbors of them are invariant except the kth neighbor, where $0 \leq k \leq m - 1$. Mathematically, $\mathcal{E}_i^k = \{a_1 a_2 \cdots a_{k-1} \mathrm{x} a_{k+1} \cdots a_m \in S^m \mid \mathrm{x} \in S\}$. Here, i is the decimal equivalent of $a_1 a_2 \cdots a_{k-1} a_{k+1} \cdots a_m$, S is set of states of the CA and m is the number of neighbors.* [10].

An example of the possible k-equivalent sets for our 3-neighborhood decimal FDCAs where $0 \leq k < 3$ is shown in Table 1.

Table 1. The grouping of RMTs based on neighborhood equivalence

$\mathcal{E}^2$		$\mathcal{E}^1$		$\mathcal{E}^0$	
$\mathcal{E}_0^2$	{000,001,002,...,009}	$\mathcal{E}_0^1$	{000,010,020,...,090}	$\mathcal{E}_0^0$	{000,100,200,...,900}
...	...	...	...	...	...
$\mathcal{E}_{99}^2$	{990,991,992,...,999}	$\mathcal{E}_{99}^1$	{909,919,929,...,999}	$\mathcal{E}_{99}^0$	{099,199,299,...,999}

Let G_n be the global transition function of the n-cell CA and C_n be the set of all n-cell configurations. Then the configurations can be represented by the configuration transition diagram, or simply, a transition diagram. For example, consider the FDCA with rule $R = \langle 0, 0, 0, 0, 2, 1, 6, 3 \rangle$ for cell length 4, that is, $c_0 = 0, c_1 = 0, c_2 = 0, c_3 = 0, c_4 = 2, c_5 = 1, c_6 = 6, c_7 = 3$. If the initial configuration is 1540, then by applying rule R, the next configuration will be 4471. Figure 1 shows only two cycles from the complete configuration transition diagram of this CA. In this transition diagram, the configuration 1540 means cell 1 has value 1, cell 2 has value 5, cell 3 has value 4, and cell 4 contains value 0. When we apply the rule, the next configuration is 4471 and so on. We call, configuration 4471 is *reachable* from 1540.

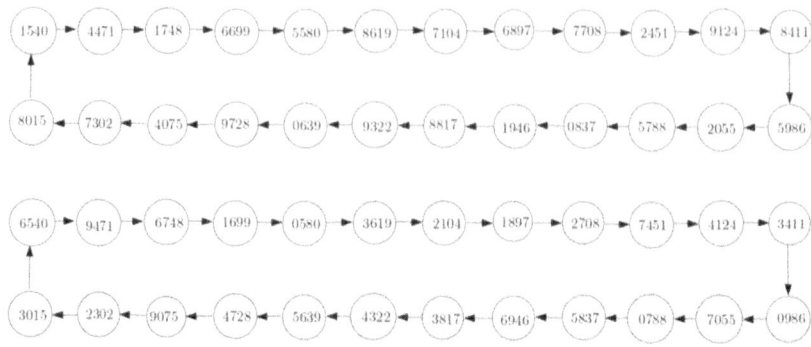

Fig. 1. Two cycles of the 4-cell reversible first degree CA $\langle 0, 0, 0, 0, 2, 1, 6, 3 \rangle$

Definition 3. *Let $x, y \in C_n$ be two configurations of a CA $G_n : C_n \to C_n$. The configuration y is called reachable from x if there exists $k \in \mathbb{N}$ such that $G_n^k(x) = y$; otherwise it is not reachable from x [9].*

Definition 4. *A finite CA of size n with global transition function $G_n : C_n \to C_n$ is reversible, if for each $y \in C_n$, where C_n is the set of all configurations of length n, there exists exactly one $x \in C_n$, such that, $G_n(x) = y$. That is, the CA is reversible if and only if G_n is a bijection. Otherwise, the CA is irreversible for that n [2].*

The configuration space of a reversible CA is a collection of *cyclic spaces* and there is no non-reachable configuration. For example, in Fig. 1, the configuration 8015 is reachable from 1540, but the configuration 0580 is not reachable from 1540. However, there is no non-reachable configuration, because every configuration has a unique predecessor. Therefore, $\langle 0, 0, 0, 0, 2, 1, 6, 3 \rangle$ is a reversible first degree CA for cell length 4.

3 Classification Using Reversible Cellular Automata

Here first degree decimal reversible cellular automata with $d = 10$ under null boundary condition are used. Since there are only 10 states for the cell, we call these CAs as decimal cellular automata (Decimal CAs). The main advantage of using a decimal CA is that we can utilize the numerical dataset for classification without doing any encoding. We simply concatenate the values of each feature corresponding to an object to create a decimal string which is equivalent to the object. This decimal string is treated as a configuration of the decimal CA. That means, using decimal CA, we can reduce the execution delay for encoding.

In a classification problem, we group data into two or more sets based on the label of the data points in the dataset. For this grouping, we can use the cyclic spaces of reversible cellular automata. Each dataset is divided into training and testing datasets. Based on the labels of the training dataset, the cycles of the

CA are grouped into different classes. During the testing phase, when new input data is given, the label can be predicted based on the cycle information received during the training phase.

Suppose the ith cycle of a CA is $cl_i = \{o_1, o_2, o_3, \cdots o_j, \cdots o_l\}$ where o_j is a configuration in cl_i and $T_s = \{d_1, d_2, d_3, \cdots d_i, \cdots d_t\}$ where T_s is the training dataset with data elements d_i in form of decimal strings. Let there be some common elements between the cycle and the training dataset, then, based on the label of those common elements, the cycle cl_i will be assigned a label. That is, for a dataset with k classes,

$$D = cl_i \cap T_s \neq \emptyset \implies cl_i \in class\ A_1 | class\ A_2 | \cdots | class\ A_k$$

Based on the label of the data points in D, we can decide the class of cycle cl_i. Ideally,

$$class(cl_i) = \begin{cases} class\ A_1, & \text{if } \forall x \in D, label = class\ A_1 \\ class\ A_2, & \text{if } \forall x \in D, label = class\ A_2 \\ \cdots \\ class\ A_k, & \text{if } \forall x \in D, label = class\ A_k \end{cases}$$

For example, suppose $cl_1 = \{o_1, o_2, o_3, o_4\}$ and $T_s = \{o_1, o_2\}$. During the training phase, if the labels of $o_1 \& o_2$ are $class\ A$, then we can take $class(cl_1) = class\ A$. If a new input data o_3 comes in the testing phase, then the label for this o_3 can be predicted as $class\ A$.

Table 2. Hypothetical Sample dataset

Data ID	Attribute 1	Attribute 2	Attribute 3	Class	Target configuration
1	97	2	8	A	9728
2	40	7	5	A	4075
3	57	8	8	A	5788
4	36	1	9	B	3619
5	21	0	4	B	2104
6	74	5	1	B	7451
7	66	9	9	A	6699
8	65	4	0	B	6540

Let us consider a hypothetical dataset of Table 2 for classification. In this dataset, attribute 1 is of length 2, attribute 2 of length 1, and attribute 3 of length is 1. So we need a 4-cell CA since the total length of any data element is 2+1+1 = 4. Suppose we use the cyclic space of first degree CA $\langle 0, 0, 0, 0, 2, 1, 6, 3 \rangle$ for 4 cells as shown in Fig. 1. We can divide the dataset of Table 2 into the training and testing phases. Suppose $Data\ ID = 1, \cdots, 6$ belong to the training dataset. In this dataset, 9728 (Data 1), 4075 (Data 2), and 5788 (Data 3) belong to the same class (*class A*). In Fig. 1, these three data points belong to the same cycle, so, this cycle (say, $Cycle_1$) gets the label as *class A*. Similarly, the other cycle of Fig. 1 is mapped to *class B*. In order to predict the label of any new input data, we can use this cyclic space. When a new input data 6699 is given during the testing phase, then as in Fig. 1, 6699 belongs to $Cycle_1$, from that cycle information, we can predict the label of 6699 as *class A*. The same procedure can be followed for *class B* as well.

4 Selection of Proper Rules

Section 3 gives a basic idea of classification using a reversible CA. However, any classification is good if the distance between the elements of the same class is very small in comparison to that for the elements in another class. So, while selecting a reversible CA for the classification, we need to satisfy that, for the chosen CA, the configurations that belong to the same cycle maintain this minimum distance property while the distance between a pair of configurations from two different cycles is large. This section describes our method of looking for the CAs that hold this property and gives the final list of candidate CAs suitable for classifying any dataset (see Fig. 2 for an overview).

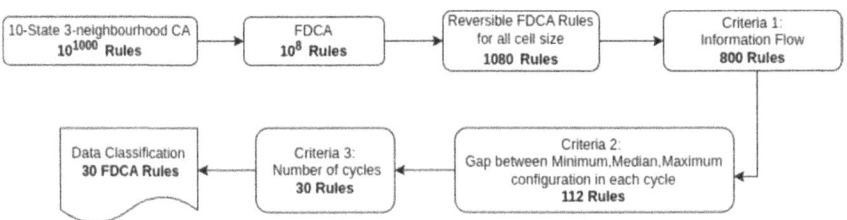

Fig. 2. Criteria for selecting proper FDCA rules for Data classification

4.1 Filtering Based on Information Propagation

For 3-neighborhood 10-state CAs, there are a total of 10^{1000} possible rules. However, with FDCA, the rule space is effectively reduced to only 10^8. Among these 10^8 FDCA rules, 1080 rules are reversible for each $n \in \mathbb{N}$ using the algorithm of [9]. Interestingly, for all these rules, the first four parameter values are zero, indicating them as linear rules. Among these 1080 rules, our initial criterion is to filter out rules where there is minimal change in the states of the cells while moving from one configuration to the next. This occurs when there is low information propagation from one configuration to the next. Additionally, the rule must exhibit a high rate of self-replication. That means that the rule has a tendency to maintain the state of a cell unchanged during evolution. As a result, configurations that are similar belong to the same cycle.

Now, this information propagation in the CA is calculated with respect to every neighbor [7,10]. Formally, for each neighbor k, $0 \leq k \leq 2$, of our decimal

FDCA, information propagation to the kth neighbor due to change in the current cell is $\Lambda_i^k = \frac{1}{10^2} \sum_{i=0}^{10^2-1} \lambda_i^k$ where $\lambda_i^k = \frac{1}{90} \sum_{r,s \in \mathcal{E}_i^k, r \neq s} \delta_i^k(r,s)$ and

$$\delta_i^k(r,s) = \begin{cases} 1 & \text{if } R[r] \neq R[s] \text{ where } r,s \in \mathcal{E}_i^k, \ r \neq s \\ 0 & \text{otherwise} \end{cases}$$

The information propagation rate considering the cell itself as its neighbor, that is, for $k=1$, is the self-replication rate which is the information propagation due to the change in the cell itself. The maximum information propagation rate for any neighbor is 100% and the minimum is 0%. To choose the rules, we set our first filtering criterion as:

Criterion 1: *Select the CA rules such that the rate of information propagation for each neighbor is $\leq 90\%$ and the rate of self-replication is $\geq 63\%$.*

There are 800 reversible FDCA rules out of the 1080 reversible FDCA rules for all cell lengths that satisfy Criterion 1. We need further filtering to reduce this rule-set.

Table 3. Potential candidate CAs are based on Criterion 1 (Information Propagation) and Criterion 2 (Median). Here, all rules have an information propagation rate of 0.64444 for $k=1$.

FDCA Rule	Information Propagation Rate	FDCA Rule	Information Propagation Rate	FDCA Rule	Information Propagation Rate	FDCA Rule	Information Propagation Rate
Decimal CA	k=0,2	Decimal CA	k=0,2	Decimal CA	k=0,2	Decimal CA	k=0,2
0,0,0,0,0,1,0,0	0.00	0,0,0,0,0,1,0,5	0.00	0,0,0,0,0,1,5,0	0.56	0,0,0,0,0,1,5,5	0.56
0,0,0,0,0,3,0,1	0.00	0,0,0,0,0,7,0,3	0.00	0,0,0,0,0,9,0,1	0.00	0,0,0,0,0,9,0,2	0.00
0,0,0,0,0,9,0,3	0.00	0,0,0,0,0,9,0,4	0.00	0,0,0,0,0,9,0,5	0.00	0,0,0,0,0,9,0,6	0.00
0,0,0,0,0,9,0,7	0.00	0,0,0,0,0,9,5,1	0.56	0,0,0,0,0,9,5,2	0.56	0,0,0,0,0,9,5,6	0.56
0,0,0,0,0,9,5,7	0.56	0,0,0,0,1,1,0,0	0.00	0,0,0,0,1,1,0,5	0.00	0,0,0,0,1,3,0,1	0.00
0,0,0,0,1,7,0,3	0.00	0,0,0,0,2,1,0,0	0.89	0,0,0,0,2,1,0,5	0.89	0,0,0,0,2,1,5,0	0.89
0,0,0,0,2,1,5,5	0.89	0,0,0,0,2,3,0,1	0.89	0,0,0,0,2,7,0,3	0.89	0,0,0,0,3,1,0,0	0.00
0,0,0,0,3,1,0,5	0.00	0,0,0,0,3,3,0,1	0.00	0,0,0,0,3,9,0,1	0.00	0,0,0,0,3,9,0,2	0.00
0,0,0,0,3,9,0,3	0.00	0,0,0,0,3,9,0,4	0.00	0,0,0,0,3,9,0,5	0.00	0,0,0,0,3,9,0,6	0.00
0,0,0,0,3,9,0,7	0.00	0,0,0,0,4,1,0,0	0.89	0,0,0,0,4,1,0,5	0.89	0,0,0,0,4,1,5,0	0.89
0,0,0,0,4,1,5,5	0.89	0,0,0,0,4,7,0,3	0.89	0,0,0,0,4,9,0,1	0.89	0,0,0,0,4,9,0,2	0.89
0,0,0,0,4,9,0,3	0.89	0,0,0,0,4,9,0,4	0.89	0,0,0,0,4,9,0,5	0.89	0,0,0,0,4,9,0,6	0.89
0,0,0,0,4,9,0,7	0.89	0,0,0,0,4,9,5,1	0.89	0,0,0,0,4,9,5,2	0.89	0,0,0,0,4,9,5,6	0.89
0,0,0,0,4,9,5,7	0.89	0,0,0,0,5,1,0,0	0.56	0,0,0,0,5,1,0,5	0.56	0,0,0,0,5,3,0,1	0.56
0,0,0,0,5,7,0,3	0.56	0,0,0,0,5,9,0,1	0.56	0,0,0,0,5,9,0,2	0.56	0,0,0,0,5,9,0,3	0.56
0,0,0,0,5,9,0,4	0.56	0,0,0,0,5,9,0,5	0.56	0,0,0,0,5,9,0,6	0.56	0,0,0,0,5,9,0,7	0.56
0,0,0,0,6,1,0,0	0.89	0,0,0,0,6,1,0,5	0.89	0,0,0,0,6,1,5,0	0.89	0,0,0,0,6,1,5,5	0.89
0,0,0,0,6,3,0,1	0.89	0,0,0,0,6,9,0,1	0.89	0,0,0,0,6,9,0,2	0.89	0,0,0,0,6,9,0,3	0.89
0,0,0,0,6,9,0,4	0.89	0,0,0,0,6,9,0,5	0.89	0,0,0,0,6,9,0,6	0.89	0,0,0,0,6,9,0,7	0.89
0,0,0,0,6,9,5,1	0.89	0,0,0,0,6,9,5,2	0.89	0,0,0,0,6,9,5,6	0.89	0,0,0,0,6,9,5,7	0.89
0,0,0,0,7,1,0,0	0.00	0,0,0,0,7,1,0,5	0.00	0,0,0,0,7,7,0,3	0.00	0,0,0,0,7,9,0,1	0.00
0,0,0,0,7,9,0,2	0.00	0,0,0,0,7,9,0,3	0.00	0,0,0,0,7,9,0,4	0.00	0,0,0,0,7,9,0,5	0.00
0,0,0,0,7,9,0,6	0.00	0,0,0,0,7,9,0,7	0.00	0,0,0,0,8,3,0,1	0.89	0,0,0,0,8,7,0,3	0.89
0,0,0,0,8,9,0,1	0.89	0,0,0,0,8,9,0,2	0.89	0,0,0,0,8,9,0,3	0.89	0,0,0,0,8,9,0,4	0.89
0,0,0,0,8,9,0,5	0.89	0,0,0,0,8,9,0,6	0.89	0,0,0,0,8,9,0,7	0.89	0,0,0,0,8,9,5,1	0.89
0,0,0,0,8,9,5,2	0.89	0,0,0,0,8,9,5,6	0.89	0,0,0,0,8,9,5,7	0.89	0,0,0,0,9,3,0,1	0.00
0,0,0,0,9,7,0,3	0.00	0,0,0,0,9,9,0,1	0.00	0,0,0,0,9,9,0,2	0.00	0,0,0,0,9,9,0,3	0.00
0,0,0,0,9,9,0,4	0.00	0,0,0,0,9,9,0,5	0.00	0,0,0,0,9,9,0,6	0.00	0,0,0,0,9,9,0,7	0.00

4.2 Filtering Based on the Minimum, Median, and Maximum of Configurations in the Cycle

To classify a cycle as a group representing a data class, the configurations within the cycle are expected to fall within a specific range (considering the decimal strings as decimal numbers). A wide range of minimum and maximum values within the configurations of the cycle indicates a weaker intra-relationship among the elements in the cycle. Taking this into consideration, we filter rules from the 800 rules based on the criterion that the difference between the median of elements in the cycle ($\mathcal{C}_{med}$) and the minimum element ($\mathcal{C}_{min}$) (resp. maximum element $\mathcal{C}_{max}$) in the cycle should be less than l_1 percent of the maximum configuration possible for that CA. For this, we calculate the gap between elements ($\mathcal{G}$) as follows:

$$\mathcal{G} = (\mathcal{C}_{med} - \mathcal{C}_{min}) + (\mathcal{C}_{max} - \mathcal{C}_{med})$$

To select appropriate rules for classification, we prioritize rules based on ($\mathcal{G}$) with respect to the maximum value of configuration in the CA. Hence, our second filtering criterion is:

Criterion 2: *For any n, choose those CA rules such that in each of the cycles, the gap $\mathcal{G}$ is $\leq l_1$ percent of the maximum configuration of the CA.*

This l_1 can be set according to the user's requirements. For example, when $n = 5$ the maximum configuration is $10^5 - 1$. If l_1 is 80, then the maximum possible gap is 80% percent of $10^5 - 1$. For any cell length, when considering $l_1 = 80$ over Criterion 1, we get a list of 112 CA rules as shown in Table 3.

Table 4. List of FDCA parameters satisfying Criteria 1, 2 and 3

0,0,0,0,1,1,0,5	0,0,0,0,1,3,0,1	0,0,0,0,1,7,0,3	0,0,0,0,2,1,5,5	0,0,0,0,3,1,0,5
0,0,0,0,3,3,0,1	0,0,0,0,3,9,0,1	0,0,0,0,3,9,0,3	0,0,0,0,3,9,0,5	0,0,0,0,3,9,0,7
0,0,0,0,4,1,5,5	0,0,0,0,4,9,5,1	0,0,0,0,4,9,5,7	0,0,0,0,6,1,5,5	0,0,0,0,6,9,5,1
0,0,0,0,6,9,5,7	0,0,0,0,7,1,0,5	0,0,0,0,7,7,0,3	0,0,0,0,7,9,0,1	0,0,0,0,7,9,0,3
0,0,0,0,7,9,0,5	0,0,0,0,7,9,0,7	0,0,0,0,8,9,5,1	0,0,0,0,8,9,5,7	0,0,0,0,9,3,0,1
0,0,0,0,9,7,0,3	0,0,0,0,9,9,0,1	0,0,0,0,9,9,0,3	0,0,0,0,9,9,0,5	0,0,0,0,9,9,0,7

4.3 Filtering Based on Cycle Structure

A successful classification involves grouping related data points together. In the context of CAs, elements (configurations) inside each cycle play a crucial role in determining which data points belong to the same class. Therefore, it is essential to ensure that in the chosen CA, each cycle contains at least a minimum number of configurations, that is, at least a minimum number of cycles exist in the cyclic space. Nevertheless, for any n, there is always a CA rule for which every configuration is forming an independent cycle. Therefore, the maximum number of cycles possible is equal to the number of configurations $= 10^n$. For our purpose, we select CA rules from 112 rules such that for any n, its configuration space contains l_2 percentage of the maximum possible number of cycles. Hence our third criterion for selecting rule is:

Criterion 3: *For any n, choose those CA rules out of Table 3 for which number of cycles in the CA for any n is l_2 percentage of the total number of configurations.*

The maximum allowed number of cycles l_2 can be selected based on user requirements. For instance, here, we fix $l_2 = 25$ and obtain 30 CA rules following Criterion 3. These rules are presented in Table 4. In this paper, we work with only these rules.

5 The Proposed Classification Algorithm

Classification in machine learning involves categorizing data into different classes based on their features. In our proposed approach, we perform multi-label classification utilizing the cyclic space of decimal FDCAs.

For instance, in binary classification, there are typically two data classes, such as *class A* and *class B*. We map the cyclic space of reversible first degree CAs to the training data points, determined by the intersection of training data points and configurations in the cyclic space.

As an example, in Fig. 3, three cycles containing green data points represent *class A*, while cycles with yellow data points represent *class B*. As indicated by their respective mappings, we label the cycles with class information. However, certain cycles were missed due to the lack of training data, resulting in ambiguous cycles. To address this, we assign labels to these ambiguous cycles based on the training data point closest to the median of all elements within the cycle. This ensures that all cycles are appropriately mapped to class labels, even in the absence of explicit training data. Algorithm 1 shows the steps for the training phase. Algorithm 2 shows the steps for the testing phase. An example with a real dataset explains the applicability of the algorithm.

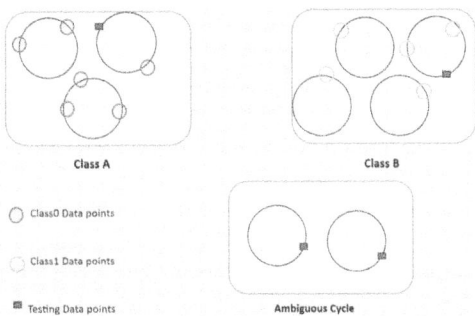

Fig. 3. Classification based on the cycles formed by training data

Algorithm 1: Training Algorithm using Decimal Reversible FDCA

Input: Training Dataset $T_r = \{d_1, d_2, d_3, \cdots d_i, \cdots d_t\}$
Output: Class Information of each cycle of the reversible FDCA
1: **for** Each cycle cl_i of reversible FDCA R **do**
2:    **if** $cl_i \cap T_r \neq \emptyset$ **then**
3:       Count the number of instances of each class within the cycle.
4:       $Class_X \leftarrow$ Label of the majority class within the cycle cl_i
5:       Label all configurations in cycle cl_i as $Class_X$
6:    **end if**
7: **end for**
8: **for** Each cycle cl_i of reversible FDCA R not labeled any class **do**
9:    Find the median M of all elements in cycle cl_i
10:   **if** Training data T_r is nearest to median M **then**
11:      Label all configurations in cycle cl_i as Class of Training data T_r
12:   **end if**
13: **end for**

Algorithm 2: Testing Algorithm using Decimal Reversible FDCA

Input: Testing Dataset $T_s = \{t_1, t_2, t_3, \cdots t_i, \cdots t_t\}$, Labeled cycles of reversible FDCA
Output: Predicted classes of Testing data set
1: **for** Each $t_i \in T_s$ **do**
2:    **if** $t_i \in$ Cycle cl_i **then**
3:       $Class(t_i) \leftarrow$ class of cycle cl_i
4:    **end if**
5: **end for**
6: Calculate the accuracy using predicted and actual data

Example: Classification of Haberman's Survival Dataset

Haberman's Survival Dataset contains cases from a study conducted on the survival of patients who had undergone surgery for breast cancer [6]. This dataset contains four attributes: age of patient at time of operation (numerical), patient's year of operation (numerical), number of positive auxiliary nodes detected (numerical), and survival status (class attribute). Each attribute contains a two-digit value. For example, consider two data points $(54, 63, 19)$ and $(50, 58, 01)$ of this dataset. These two data points have class label 1. To classify this dataset, we take a first degree reversible decimal CA of cell length 6 with constants $\langle 0, 0, 0, 0, 4, 7, 7, 4 \rangle$: $(c_0 = 0, c_1 = 0, c_2 = 0, c_3 = 0, c_4 = 4, c_5 = 7, c_6 = 7, c_7 = 4)$. One partial cycle from the transition diagram of this CA is shown in Fig. 4a. Here, the configurations 546319 and 505801 belong to the same cycle. Based on the label of data points and configurations in the cycle, we can group cycles into different sets corresponding to each class in the dataset. In Fig. 4b, all the configurations of the cycle belong to *class* 1 since the majority of labels for this

(a) Two training objects (b) Attached class information

Fig. 4. One partial cycle from the transition diagram of CA $\langle 0,0,0,0,4,7,7,4\rangle$ and corresponding label from the Haberman's Survival Dataset

cycle belong to *class* 1. During the testing phase, if any input data belongs to this cycle, then we can predict the label as 1.

6 Implementation and Performance Analysis

Our implementation of the proposed algorithm treats the original dataset as decimal strings suitable for reversible cellular automata and considers only the best rule set (Table 4). We have transformed our code into a Python module [1] for convenient access, usage, and replication. Our package is adaptable to different datasets. Figure 5 depicts the different stages comprising our proposed algorithm, which are outlined next.

Step 1: Input a numerical or categorical dataset, then perform preprocessing steps including concatenating data and mapping it to the target configuration, as outlined in Sect. 3

Step 2: Generate all cycles corresponding to the selected FDCA rule from the rule list.

Step 3: Create the trained model by assigning class labels to all cycles using Algorithm 1.

Step 4: Calculate the performance of the model by predicting the class label of the new data and evaluating its accuracy.

---

[1] GitHub repository: https://github.com/kamalikaB/Classification

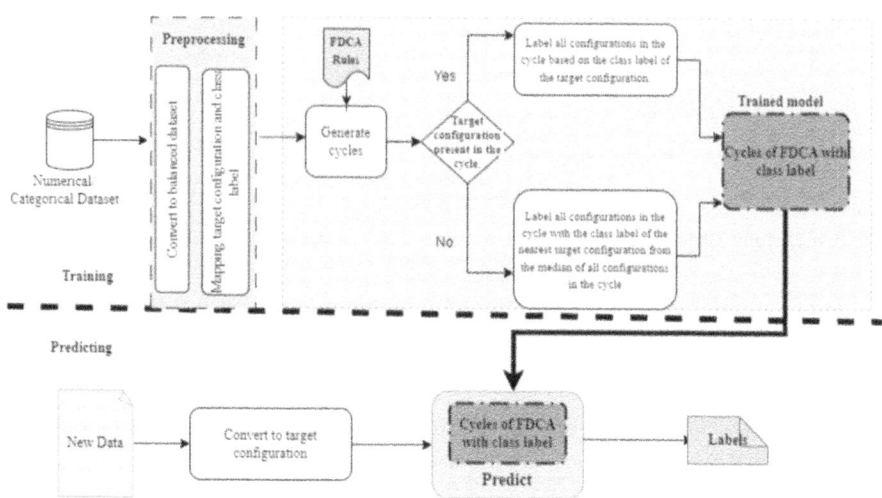

Fig. 5. Classifying numerical datasets using FDRCA (First Degree Reversible Cellular Automata)

We analyse the performance of our proposed algorithm on standard benchmark machine learning metrics based on some real standard datasets.

6.1 Datasets and Preprocessing

If the dataset is imbalanced, the minority class needs to be upsampled to make the majority and minority classes equal. If the dataset contains non integer values, it has to be pre-processed to suit the cell length of the CA. In some datasets, attribute data may be of different sizes. So we convert it into the same size by padding zero. If both single and double-valued data are present for an attribute, then single-valued data needs to be converted to two digits by padding zero in front of attribute values. For example, in the Haberman's Survival Dataset, the attribute *auxiliary nodes detected* have both single-digit and double-digit values. Consider one data point $(30, 64, 1)$; it has to be pre-processed and converted to $(30, 64, 01)$ in order to suit the 6-cell cellular automata.

In this work, due to computational constraints, we restrict our chosen datasets to be such that each object can be fitted within a CA size 1 to 9. This limits the number of real datasets available for testing the efficiency of our algorithm. The description of training and testing data considered in this paper is given in Table 5. Here, n denotes the cell length of each CA used. The unbalanced dataset is upsampled using the predefined python package *sklearn.utils.resample*, ensuring that the two classes have an equal number of instances. The initial data sample count and the upsampled data count are also shown in the Table 5.

Table 5. Description about datasets used

Dataset	Samples (Upsampled)	Attributes	Training Dataset	Test Dataset	Cell length (n)
Haber-man cancer	306(450)	4	360	90	6
balloons	16(18)	5	14	4	4
Monk 1	432(432)	7	345	87	6
Monk 2	432(580)	7	464	116	6
Monk 3	S432(456)	7	364	92	6
Insurance	27(28)	2	22	6	2
Gender	66(66)	5	54	12	4

6.2 Performance Analysis Measures

Performance metrics in classification are fundamental in assessing the quality of learning methods and learned models. However, many different measures have been defined in the literature with the aim of making better choices in general or for a specific application area [5]. In this work, we use four performance analysis measures, such as accuracy, precision, recall and F-score for comparison with other machine learning models. The formula for these performance analysis measures are given below:

$$\text{Accuracy} = \frac{\text{Number of correct predictions}}{\text{Total predictions}}$$

$$\text{Recall} = \frac{\text{Correctly classified positives}}{\text{Total positives}}$$

$$\text{Precision} = \frac{\text{Correctly classified positives}}{\text{Total predicted as positives}}$$

$$\text{F-measure} = \frac{2 \cdot \text{recall} \cdot \text{precision}}{\text{recall} + \text{precision}}$$

6.3 Result on Real Dataset

In order to calculate the performance of the classification algorithm, each data set is divided into two sets: 20% of the dataset is considered for the testing phase, and remaining is considered for the training phase. Our proposed algorithm is verified with seven different datasets taken from Ref. [6,13] and the accuracy is calculated for all these datasets. Rule $\langle 0,0,0,0,0,1,0,0 \rangle$ is a special case that yields good results for the majority of datasets as it creates cycle with each individual element. However, our selection criteria are not met by this rule. The accuracy for these different datasets and the corresponding decimal CA constants are given in Table 6.

Table 6. Performance of Reversible Decimal FDCA for Classification

Dataset	Constants c0 c1 c2 c3 c4 c5 c6 c7	Accuracy
Haber-man cancer	0,0,0,0,1,7,0,3	0.96
balloons	0,0,0,0,1,1,0,5	1
Monk1	0,0,0,0,0,1,0,0	1
Monk2	0,0,0,0,8,7,5,3	0.93
Monk 3	0,0,0,0,0,1,0,0	0.68
Insurance	0,0,0,0,1,1,0,5	0.833
Gender	0,0,0,0,6,1,5,5	0.71

Table 7. Performance of different model in classification

Dataset	Measure	Decimal CA	SVM	NB	DT	LR	KNN	MLP
Haber-man cancer	Accuracy	**0.966**	0.578	0.633	0.822	0.811	0.811	0.711
	Precision	1.000	0.528	0.567	0.825	0.879	0.879	0.629
	Recall	0.929	0.905	0.905	0.786	0.690	0.690	0.929
	F1-Score	0.963	0.667	0.697	0.805	0.773	0.773	0.750
balloons	Accuracy	**1.000**	0.250	0.250	0.750	1.000	1.000	0.500
	Precision	1.000	0.250	0.000	0.500	1.000	1.000	0.333
	Recall	1.000	1.000	0.000	1.000	1.000	1.000	1.000
	F1-Score	1.000	0.40	0.000	0.667	1.000	1.000	0.500
Monk 1	Accuracy	**1.000**	0.644	0.667	0.897	0.805	0.897	0.908
	Precision	0.915	0.684	0.674	0.929	0.741	0.891	0.911
	Recall	0.956	0.578	0.698	0.867	0.956	0.911	0.911
	F1-Score	0.935	0.627	0.681	0.897	0.835	0.901	0.911
Monk 2	Accuracy	**0.9224**	0.621	0.534	0.957	0.644	0.638	0.681
	Precision	0.982	0.573	0.512	0.932	0.579	0.592	0.644
	Recall	1.000	0.839	0.534	0.982	0.982	0.804	0.750
	F1-Score	0.991	0.681	0.603	0.957	0.728	0.682	0.694
Monk 3	Accuracy	0.688	0.826	0.793	1.000	1.000	1.000	0.989
	Precision	0.955	0.735	0.67	1.000	1.000	1.000	1.000
	Recall	1.000	0.786	0.857	1.000	1.000	1.000	0.976
	F1-Score	0.977	0.805	0.791	1.000	1.000	1.000	0.988
Insurance	Accuracy	0.833	1.000	0.167	1.000	1.000	1.000	1.000
	Precision	0.833	1.000	0.000	1.000	1.000	1.000	1.000
	Recall	1.000	1.000	0.000	1.000	1.000	1.000	1.000
	F1-Score	0.909	1.000	0.000	1.000	1.000	1.000	1.000
Gender	Accuracy	**0.714**	0.429	0.500	0.571	0.857	0.857	0.643
	Precision	1.000	0.500	0.667	0.750	1.000	1.000	0.800
	Recall	0.500	0.250	0.250	0.375	0.750	0.750	0.500
	F1-Score	0.667	0.333	0.364	0.500	0.857	0.857	0.615

6.4 Performance Comparison with Existing ML Models

We compare the accuracy, precision, recall, and F-score or F-measure of first-degree reversible CA-based classification algorithm with some standard machine learning (ML) algorithms [1] such as Decision Trees (DT), Naive Bayes (NB), Logistic regression (LR), K-Nearest Neighbours (KNN), Multi-Layer Perceptron (MLP) and Support Vector Machine (SVM). The same training and testing dataset is used for computing the performance of different models. In ML models, the original dataset is used. In decimal CA, some preprocessing steps are done in order to match the input set of decimal CA. The detailed result analysis is given in the Table 7. We can see that, the decimal CA-based Classifier performs well compared to other ML models.

7 Conclusion

This work has reported a CA-based classification technique where we have used the bijective functions of first degree reversible decimal CAs. In a reversible CA, the configuration space is divided into a number of cyclic spaces, where the configurations inside a cyclic space are reachable from each other. In this work, we use these cyclic spaces of decimal first degree reversible CAs for classification. Since we have used decimal CA, we can use any decimal data as a dataset without any encoding. It is shown that, the performance of our classification algorithm is better than that of the existing machine learning classification algorithms for small real-life datasets.

One of the limitations of using decimal CA is that when the dimension of the dataset increases, the cyclic space of first degree CA becomes exponentially large. To limit our computation, in this work, the maximum cell length of the first degree CA is chosen as 9. However, in a real dataset, we need to handle high-dimensional data. So the future scope of this work includes updating this algorithm to handle high-dimensional dataset. Furthermore, as CAs are homogeneous structures, there is no bias in feature positions or more significant digit values. In future, work may be directed to explore whether this aspect holds any significance in the classification.

References

1. Alauthman, M., et al.: Enhancing small medical dataset classification performance using GAN. In: Informatics, vol. 10, p. 28. MDPI (2023)
2. Bhattacharjee, K.: Cellular automata: reversibility, semi-reversibility and randomness. Ph.D thesis, IIEST, Shibpur (2019). http://arxiv.org/abs/1911.03609
3. Bhattacharjee, K.: First degree cellular automata as pseudo-random number generators. In: Das, S., Martinez, G.J. (eds.) Asian Symposium on Cellular Automata Technology, pp. 123–137. Springer, Cham (2022). https://doi.org/10.1007/978-981-19-0542-1_10
4. Bhattacharjee, K., Vicky, V.: Study of first degree cellular automata for randomness. J. Cellular Automata **17** (2023)

5. Ferri, C., Hernández-Orallo, J., Modroiu, R.: An experimental comparison of performance measures for classification. Pattern Recogn. Lett. **30**(1), 27–38 (2009)
6. Haberman, S.: Haberman's Survival. UCI Machine Learning Repository (1999). https://doi.org/10.24432/C5XK51
7. Kamilya, S., Das, S.: A study of chaos in cellular automata. Int. J. Bifurcat. Chaos **28**(03), 1830008 (2018)
8. Maji, P., Shaw, C., Ganguly, N., Sikdar, B.K., Chaudhuri, P.P.: Theory and application of cellular automata for pattern classification. Fund. Inform. **58**(3–4), 321–354 (2003)
9. Mukherjee, S., Bhattacharjee, K., Das, S.: Reversible cellular automata: a natural clustering technique. J. Cell. Automata **16** (2021)
10. Paul, S., Bhattacharjee, K.: Modeling spread of contagious disease by temporally stochastic cellular automata. In: Das, S., Martinez, G.J. (eds.) ASCAT 2023. AISC, vol. 1443, pp. 161–175. Springer, Singapore (2023). https://doi.org/10.1007/978-981-99-0688-8_13
11. Sethi, B., Das, S.: Modeling of asynchronous cellular automata with fixed-point attractors for pattern classification. In: 2013 International Conference on High Performance Computing & Simulation (HPCS), pp. 311–317. IEEE (2013)
12. Sethi, B., Roy, S., Das, S.: Asynchronous cellular automata and pattern classification. Complexity **21**(S1), 370–386 (2016)
13. Wnek, J.: MONK's Problems. UCI Machine Learning Repository (1992). https://doi.org/10.24432/C5R30R

Sentiment Analysis for Code-Mixed Data Using Cellular Automata with Deep Learning Models

M. J. Elizabeth[✉][iD], Avinash Krishna Kommineni, and Raju Hazari[iD]

Department of Computer Science and Engineering,
National Institute of Technology Calicut, Kerala, India
libiyajose@gmail.com

Abstract. The proposed work presents a cellular automata-based approach to sentiment analysis in code-mixed data. Our method demonstrates promising results in effectively analyzing sentiment across multilingual tweets or sentiments. By leveraging the dynamic properties of cellular automata, our model navigates the complexities of code-mixed data, where multiple languages or dialects are intertwined within the same text. Through extensive experimentation and evaluation, we showcase the robustness and efficacy of our approach in accurately identifying sentiment-bearing components in diverse linguistic contexts. The research contributes to advancing sentiment analysis techniques in the realm of code-mixed data, offering valuable insights for understanding user sentiment in multilingual communities and enhancing communication strategies in linguistically diverse environments. Our supervised classification approach produces 89% of the F1_score for Bi-LSTM without using any kind of pre-trained word embeddings or language models.

Keywords: Cellular Automata (CA) · Code-mixed data · Text Classification · Machine Learning · Sentiment Anlaysis · Deep Learning · Bi-LSTM

1 Introduction

Analyzing sentiment in code-mixed data presents unique challenges due to the mixing of multiple languages or dialects within the same text. However, advancements in natural language processing techniques have enabled the development of models capable of handling such complexities. By leveraging machine learning algorithms trained on diverse datasets, researchers have made significant strides in accurately determining sentiment in code-mixed text. These models typically employ techniques such as tokenization, language identification, and sentiment lexicons, to extract and analyze sentiment-bearing components across languages. Despite the inherent complexity, sentiment analysis in code-mixed data is vital for understanding user sentiment in multilingual communities and facilitating more effective communication strategies in diverse linguistic contexts.

Code-mixed data refers to linguistic contents of multiple languages that are combined within the same communication instance, whether it's spoken or written. This mixing can occur at various linguistic levels [2], including lexical, syntactic, or discourse. Code-mixing encompasses various forms, including intra-sentential, intra-word, and inter-sentential mixing [12]. Intra-sentential code-mixing denotes the blending of languages within a single sentence, while intra-word code-mixing occurs when languages are mixed within a single word. On the other hand, inter-sentential code-mixing involves the mixture of languages across different sentences. Most of the applications of the code-mixed data are presented briefly in the survey provided by Thara et al. in 2018 [23]. Handling code-mixing in social media text presents a significant challenge for natural language processing due to differences in spelling, grammar, and language conventions across languages. Various machine learning (ML) methods such as Naïve Bayes (NB), Support Vector Machine (SVM), Decision Tree (DT), and Logistic Regression (LR) for analyzing code-mixed data have seen significant advancement in recent years [24]. A lot of research has been done on deep learning models, recurrent neural networks (RNNs), convolutional neural networks (CNNs) [20], and transformer-based architectures like BERT and its variants [3]. These methods leverage the power of neural networks to capture complex patterns and relationships present in code-mixed text, enabling effective sentiment analysis, language identification, and named entity recognition. Additionally, ensemble learning techniques, such as stacking and boosting, have been employed to enhance the performance of the machine learning models on code-mixed data by combining predictions of multiple ML-based models. With the increasing availability of annotated datasets and the continuous evolution of ML algorithms, the field of code-mixed data analysis is poised for further advancements in understanding and processing multilingual text.

Our approach focused on a new method using cellular automata (CA) without depending on linguistic features. CA generates cycle length (CL) values by utilizing rule vectors derived from the Unicode Standards. Those CL values are employed as inputs for deep learning models to predict class labels. The Bi-LSTM algorithm showed higher accuracy and an F1_score of 89% on the SentMix-3L dataset. Our model operates entirely independently of linguistic features. We analyzed the performance of the model using four code-mixed datasets, and the results, along with a comparative analysis, indicate the superiority of the proposed approach.

The rest of the paper is structured as follows: In Sect. 2, we discussed the related work in the area of code-mixed data. The preliminary study of CA is explained in Sect. 3. Section 4 presents the proposed methodology using CA. The results and performance analysis of the model are done in Sect. 5. The conclusion and future work are given in Sect. 6.

2 Related Work

We have found a few survey papers on sentiment analysis of code-mixed datasets. Recently, a comprehensive survey was presented by Perara et al. [16] for feature extraction, is a crucial step in the sentiment analysis process. In 2020, Tho et al. [24] presented a systematic literature survey on code-mixed sentiment analysis. They have explored various previous studies of code-mixed data in detail. For example, they explained a range of language pairs such as English-Punjabi, English-Hindi-Marathi, Hindi-English [6,14,21], Indonesian-English, Malayalam-English [19], English-Tamil-Telugu-Hindi-Bengali, English-German-French, English-Bengali, and Arabic-English in detail. Srinivasan et al. [22] contributed a lexicon dictionary for a code-mixed corpus. They experimented to see how well resampling methods like SMOTE and ADASYN worked with code-mixed data to fix the Tamil-English dataset's class imbalance problem. They used Levenshtein distance as a preprocessing technique to improve the performance. Uthpala et al. [25] introduced a sentiment-labeled corpus for Sinhala-English code-mixed language, and they have completed the annotation process of that corpus. The logistic regression algorithm performs better than all other ML algorithms in their experiment with 72% accuracy. The GPT 3.5 Turbo model generates 62% of the weighted F1_score for the dataset SentMix-3L. For Bangla-English-Hindi code-mixed sentiment analysis, Raihan et al. [18] introduced SentMix-3L dataset in 2023 to evaluate zero-shot prompting with GPT-3.5, which outperforms all existing transformer-based models.

To perform the sentiment analysis process effectively, we need to identify the language of the sentiment in the dataset. In 2022, Hidayatullah et al. [12] systematically reviewed the language identification (LID) task of code-mixed text and its challenges. LID is a fundamental task, particularly in sentiment analysis and machine translation. Gokul et al. [8] demonstrated the effectiveness of a conditional random field-based approach for identifying languages at the word level of the code-mixed data. Recently, Dey et al. [9] presented a study on language identification for Indian code mix languages, and their framework produces 92.67% accuracy. While Language Identification (LID) has shown promising results, sentiment analysis performance remains subpar. Recognizing this gap, we propose a language-independent model tailored for conducting sentiment analysis tasks on code-mixed data. The proposed approach aims to address the challenges associated with varying linguistic compositions in mixed-language text, ultimately improving the accuracy and efficacy of sentiment analysis in such contexts.

3 Cellular Automata Preliminaries

John von Neumann [15] began the voyage of cellular automata by modeling a biological self-reproduction mechanism. A discrete, spatially extended dynamical system that has been thoroughly investigated as a physical system model is called a cellular automaton (CA). A CA is made up of a lattice of cells, each of which has a discrete variable representing the cell's current state. At time 't', a cell's current state and the current states of its neighbors influence its next state at the time '$t+1$'. Each cell uses a set of rules and the states of the nearby cells to evolve with a finite set of values in a defined amount of time. In our approach, we made use of Wolfram's Elementary Cellular Automata (ECA) [26], which is the most basic type of CA in which every CA cell stores a binary state at a time 't'. The next state is determined by its current state as well as the current states of its two closest neighbours, which can be determined as follows:

$$S_i^{t+1} = f(S_{i-1}^t, S_i^t, S_{i+1}^t) \tag{1}$$

Table 1. Look-up table for rules 35 and 118

Present State:	111	110	101	100	011	010	001	000	Rule
	(7)	(6)	(5)	(4)	(3)	(2)	(1)	(0)	
(i) Next State:	0	0	1	0	0	0	1	1	35
(ii) Next State:	0	1	1	1	0	1	1	0	118

where S_{i-1}^t, S_i^t, and S_{i+1}^t denote the current states of the left, self, and right neighbors of the ith cell at time 't', and 'f' is the next state function. A look-up table can be used to describe the function $f : \{0,1\}^3 \mapsto \{0,1\}$ (see Table 1). "Rule" is the decimal representation of the eight outputs [26]. $2^8 (= 256)$ ECA rules in total. Table 1 displays two of these rules, 35 and 118.

Typically, if each of the CA cells adheres to the same next-state function then such type of CA is *uniform* CA. Conversely, if the CA cells are permitted to adhere to distinct next-state functions then it becomes a *non-uniform* CA. In this study, we used non-uniform ECA with periodic boundary conditions.

4 Proposed Weight-Based Approach Using CA

In our study, we leveraged cellular automata theory to design a language-independent ML model tailored for sentiment analysis tasks. The basic description of the model was presented in the paper [11]. The process of CA evolution entails updating each cell's state concurrently with the rules, which are typically based on its state and the states of its neighboring cells. In this context, cellular automata are employed as a computational framework to handle the complexities of mixed-language text. This approach allows for the extraction of meaningful features and context from mixed-language text, making it more accurate in multilingual environments. The design of our model is illustrated in Fig. 1. Initially, every word and text entry in the dataset undergoes conversion into rule vectors using rule tables. These rule vectors are then fed into the cellular automata (CA) for processing, generating cycle-length (CL) values. Subsequently, these

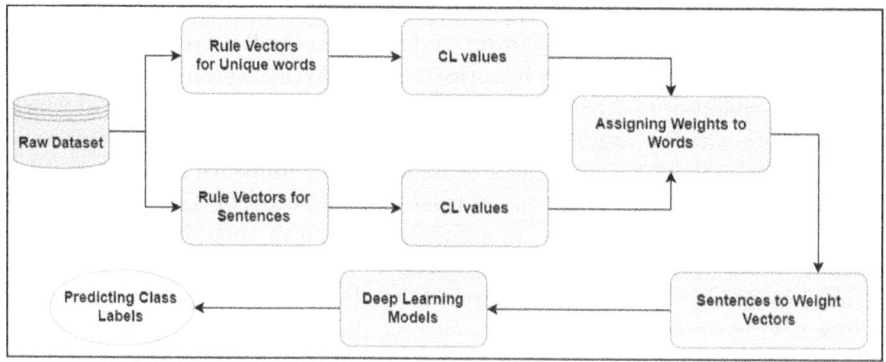

Fig. 1. Design of our Proposed Model

CL values serve as inputs for deep learning models for predicting the positive and negative classes. The entire process is structured into three distinct modules: initial conversion into rule vectors, processing by the CA, and classification using deep learning models.

4.1 Text to Rule Vector Conversion

In our approach, we utilize a universal character encoding system to generate rule vectors. The Unicode standard [1] is utilized for generating rule vectors of texts of any language. This strategy ensures consistency and compatibility across different languages, allowing for effective rule vector generation irrespective of the language being processed. Following are the steps executed for the generation of the CA rule vectors for any text:

- Identify the Unicode position for each character from the Unicode charts.
- If the Unicode position is represented in the format: 0x (p1 p2 p3 p4), where p1, p2, p3, p4 are hexa decimal characters, then the decimal value derived for the character is calculated with the expression: $256^1(16p1 + p2) + 256^0(16p3 + p4)$. Subsequently, the corresponding CA rule can be calculated using the expression: $<(16p1 + p2), (16p3 + p4)>$.
- If p1 and p2 both are '0' then the corresponding CA rule calculated using the expression is: $<(16p3 + p4)>$

For example, the text of the English-Malayalam dataset is:

Kidu song അതിലെ scenes എനിക്ക് ഒരുപാട് ഇഷ്ടപ്പെട്ടു

In the above sentence, there are three words in English and four words in Malayalam. Here, each word is separated with a space. So the words 'Kidu' and 'song' are converted to [75, 105, 100, 117] and [115, 111, 110, 103] respectively, using Unicode values. In converting Malayalam words into rule vectors using Unicode values, each letter is mapped to a specific set of numerical values based on its Unicode representation. For instance, the first letter " അ " is transformed into the rule vector [13, 5], while the second letter " തി " is transformed into [13, 36], and so forth. This transformation is applied to every letter in the word, resulting in a vector representation known as the "Rule vector" for that particular text. So rule vectors for the above text are: [75, 105, 100, 117, 32, 115, 111, 110, 103, 32, 13, 5, 13, 36, 13, 63, 13, 50, 13, 71, 32, 115, 99, 101, 110, 101, 115, 32, 13, 14, 13, 40, 13, 63, 13, 21, 13, 77, 13, 21, 13, 77, 32, 13, 18, 13, 48, 13, 65, 13, 42, 13, 62, 13, 31, 13, 77, 32, 13, 7, 13, 55, 13, 77, 13, 31, 13, 42, 13, 77, 13, 42, 13, 70, 13, 31, 13, 77, 13, 31, 13, 65].

In our study, all the rule vectors are stored in a dedicated table known as the Rule Vector Table. This table serves as a comprehensive repository, containing the letters of each language and their corresponding values as rules. Specifically, we have utilized four code-mixed datasets, namely English-Malayalam, English-Kannada, Sinhala-English, and Bangla-English-Hindi. Each entry in the rule-vector table comprises letters from these languages and their associated values, facilitating efficient rule-based processing for each language pair within our model.

4.2 Processing of CA and Generation of Cycle Length Values

In our experiment, we initialized the Cellular Automata (CA) by assigning 1 s and 0 s alternatively. Then the CA evolved over 1,000 iterations. The iteration value for this evolution was determined based on the length of the text, specifically set to half of the maximum text length. For a dataset with a maximum text length of approximately 2,000 characters, we observed that after 1,000 iterations, there were no further changes in the cycle length (CL) values. The outcome remains consistent even with different initial sequences, as the cell updation depends on the rules applied to each cell and its neighbours. Thus, the same initial condition is used for each case in the dataset. The evolution of the CA is depicted in Fig. 2. During the CA evolution, patterns on a given cell start to replicate themselves throughout the CA evolution process after a specific number of iterations. The cycle length value during CA evolution indicates the length of the repeating pattern. If the same pattern appears at least 32 times, we consider this number as the CL value for that specific CA cell. We can use any value instead of 32 (A threshold value), which will not have an impact on performance because CL values are the length of the repeated pattern. The cell's CL value is set to '−5' (or any negative number) if no repeated patterns are found in the CA cell. The bold lines in Fig. 2 indicate that certain patterns, which may vary in length, eventually repeat themselves during the CA evolution. The details are explained in Paper [10]. For example, the CL sequence of the above English-Malayalam text is: [8, 8, 8, 4, 4, 2, 6, 3, 3, 3, 3, 1, 1, 1, 1, 1, 1, 4, 4, 4, 4, 2, 6, 6, 6, 6, 3, 3, 3, 3, 3, 1, 1, 1, 1, 1, 1, 1, 1, 1, 1, 1, 1, 1, 1, 1, 2, 2, 1, 1, 1, 1, 1, 1, 1, 1, 1, 3, 3, 3, 6, 2, 2, 1, 1, 1, 1, 3, 3, 1, 1, 3, 3, 1, 1, 1, 1, 1, 1, 1, 1].

....	1	0	1	0	1	0	1	0	....
....	0	1	0	1	0	1	1	0	....
....	1	0	1	0	1	1	1	0	....
....	1	0	0	1	0	0	1	1	....
....	0	0	1	0	0	1	1	1	....
....	1	1	0	1	1	0	1	0	....
....	1	1	0	0	1	0	0	0	....
....	0	0	0	1	1	0	0	1	....
....	1	1	0	0	0	1	1	0	....
....	1	0	0	1	1	0	0	0	....
....	0	1	1	0	0	0	0	0	....
....	1	0	0	1	0	0	0	0	....
....	1	0	0	0	0	1	0	1	....
....	⋮	⋮	⋮	⋮	⋮	⋮	⋮	⋮	....

Fig. 2. CA Evolution: The periodic boundary turns around to the opposing edge during the evolution process.

4.3 CL Values to Weight Matrix Transformation

The proposed model functions by extracting every unique word from the dataset and generating CL values for each word. These CL values, both for individual words and entire sentences, are then utilized to calculate the CL weight of each specific word. If the word 'α' is present in n number of sentences then the weight of 'α' can be calculated using the Eq. 2.

$$\frac{\sum_{i=1}^{n} X_i}{Y_\alpha} \qquad (2)$$

where 'X' denotes the sum of the CL values of the sentence and 'Y' denotes the sum of the CL values of the word 'α'. For example, if we consider two texts, then the weight of the words in the sentences can be calculated as follows:

Text-1:
കണ്ടിട്ട് കിടുക്കും എന്ന് തോന്നുന്നുണ്ട് Waiting for vidyaji magic

The CL values of Text-1: [3, 1, 1, 3, 3, 1, 1, 1, 1, 2, 2, 2, 2, 1, 1, 1, 1, 1, 1, 1, 1, 1, 1, 1, 1, 2, 1, 1, 1, 1, 1, 1, 2, 2, 1, 1, 1, 1, 3, 3, 1, 1, 1, 1, 1, 1, 1, 3, 3, 1, 1, 1, 1, 1, 1, 1, 1, 1, 1, 1, 1, 1, 1, 1, 1, 1, 3, 3, 3, 1, 3, 3, 1, 1, 1, 3, 3, 3, 3, 1, 1, 1, 5, 5, 5, 5, 5, 15, 15, 3, 3, 3] and the sum of those CL values (Sentence CL-Sum) is 196.

Text-2: "Vidyasagar bgm vidyaji fans like here"
CL values of Text-2: [1, 1, 1, 2, 1, 2, 6, 3, 3, 3, 6, 2, 2, 1, 2, 1, 1, 1, 4, 2, 4, 4, 1, 4, 2, 4, 4, 1, 1, 1, 1, 1, 1, 4, 4, 4] and the sum of those values is 87.
To calculate the weight of the word 'vidyaji', which appears in both sentences, first, we find out the CL sequence of that word using CA, which is [6, 2, 6, 3, 6, 6, 6] and then the sum of the CL values which is 35. So weight (vidyaji) = (196 + 87)/35 = 8.08. The weight of each word is assigned to words in the sentence to form a weight matrix for each text. That can be used as input for deep learning models.

4.4 Prediction Process

We used two deep learning (DL) models, Long short-term memory (LSTM) and Bi-directional LSTM (Bi-LSTM) for the prediction Process. The weight matrix of each text is used as input for the DL models. LSTMs can capture dependencies in sequences over long distances [13], which is crucial for many real-world applications. The basic architecture of a cell in LSTM is shown in Fig. 3. The cell state C_t acts as a memory that carries information across different time steps in the sequence. There are three gates such as input, output and forget gates (i_t, o_t, f_t) to control the flow of information and also help to capture long-term dependencies in sequential data. During inference, the trained LSTM model takes a sequence of input data and produces a sequence of output predictions. At each time step, the model updates its hidden state and cell state based on the current input and the previous states and produces an output prediction.

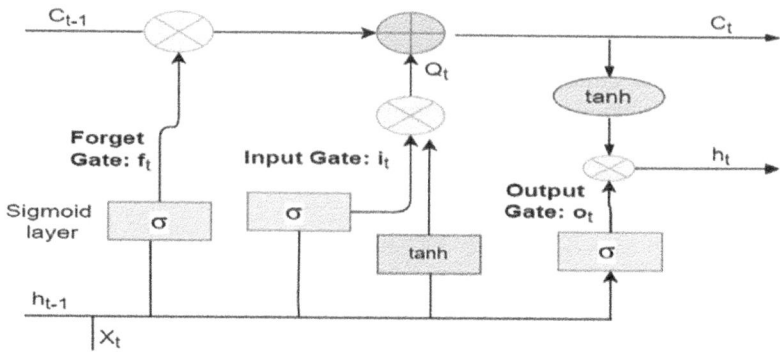

Fig. 3. The basic architecture of an LSTM cell

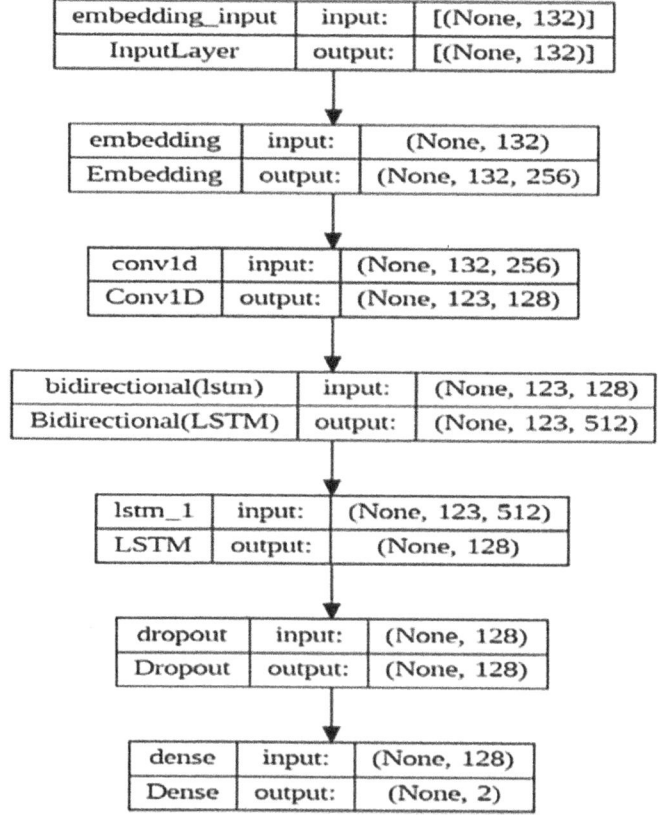

Fig. 4. The details of the LSTM model used for processing two-class problems of various code-mixed datasets

Table 2. Hyperparameter details of the Proposed Approach

Hyperparamets used	Values	Hyperparamets used	Values
Number of neurons in the LSTM layer	64	Learning rate	0.0005
Dropout Rate	0.2	Recurrent Dropout	0.5
Spatial Dropout	0.25	No. of epochs	10
Number of neurons in the Dense layer	5	Optimizer	Adam
Batch size	64	Activation Function	Softmax
patience	5	Loss function	Binary_crossentropy
Embedding Dimension	128		

The LSTM and Bi-LSTM deep-learning models underwent fine-tuning, where various aspects were meticulously adjusted to optimize their performance. These adjustments encompassed a spectrum of parameters, including the learning rate, batch size, choice of optimizer, regularization techniques, and configurations of network architecture. This holistic approach aimed to finely calibrate the models to achieve optimal accuracy and efficiency in handling the given task. The hyperparameters used with LSTM are shown in Table 2 and the details of the LSTM layers used are shown in Fig. 4. For Bi-LSTM, we used the 'ReLU' activation function with an embedding dimension of 256. The processing steps of the proposed approach are given in Algorithm 1.

Algorithm 1. Processing steps of the proposed approach

Require: $words \geq 1$
Ensure: $class\_labels \geq 2$
  $Clean\_Text \leftarrow clean\_dataset()$
  $unique\_words \leftarrow Find\_unique\_words(Clean\_Text)$
  $Word\_rule \leftarrow Unicode\_rule(unique\_words)$
  $Word\_CL \leftarrow CA\_Processing(Word\_rule)$
  $Sentence\_rules \leftarrow Unicode\_rule(Clean\_Text)$
  $Sentence\_CL \leftarrow CA\_Processing(Sentence\_rules)$
  $Word\_cl\_Sum \leftarrow Sum\_CL(Word\_CL)$
  $Sentence\_cl\_Sum \leftarrow Sum\_CL(Sentence\_CL)$
  for each word in unique_words **do**
    $weight \leftarrow 0$
    for each sentence in Clean_Text **do**
      if word in sentence **then**
        $weight \leftarrow weight + \text{Sentence\_cl\_Sum}(sentence)$
      end if
    end for
    $\text{word\_weight}[word] \leftarrow \frac{weight}{\text{Word\_cl\_Sum}(word)}$
  end for
  $weight\_vectors \leftarrow Convert\_Sen\_to\_weight(Clean\_Text, word\_weight)$
  $class\_labels \leftarrow LSTM(weight\_vectors)$

5 Results and Performance Analysis

In this section, we have given the dataset descriptions and the results obtained for the LSTM and Bi-LSTM models. A detailed comparative analysis is also conducted to highlight the strengths and weaknesses of the models.

Details of Datasets Used for Code-Mixed Sentiment Analysis: We have used four datasets of code-mixed languages with our proposed approach. The descriptions are given below:

- SentMix-3L (Bangla-English-Hindi) dataset [18] contains 1007 texts of positive, negative, and neutral classes.
- Kannada-English dataset [7] contains 7,671 comments of five different classes, such as positive, negative, neutral, mixed feelings, and unknown state.
- Malayalam-English dataset [17] contains 19,616 comments of the same five different classes.
- Sinhala-English dataset [25] contains 7832 texts of positive, negative, and neutral classes.

The proposed approach is a new strategy specifically tailored for code-mixed datasets. So we focused solely on 2-class and 3-class problems. The performance of the proposed approach is given in Table 3. The LSTM model augmented with a CA-based weight matrix, showcases superior performance compared to numerous existing models. Notably, it achieves the highest accuracy on the SentMix-3L dataset.

Table 3. Accuracy, Precision, Recall and F1_score of our proposed approach with LSTM model

Dataset	No. of Classes	Accuracy	Precision	Recall	F1_Score
Bangla-English-Hindi [18]	2	0.88	0.89	0.88	0.88
Bangla-English-Hindi [18]	3	0.89	0.91	0.89	0.89
Kannada-English [7]	2	0.78	0.79	0.78	0.78
Sinhala-English [25]	2	0.88	0.87	0.88	0.87
Malayalam-English [17]	2	0.87	0.87	0.87	0.86
Malayalam-English [17]	3	0.71	0.71	0.71	0.71

The accuracy, precision, recall, and F1_score of Bi-LSTM are given in Table 4. The Bi-LSTM model consistently achieves over 80% accuracy across all datasets, except the Sinhala-English dataset. Figure 6 displays the ROC curve of the model on the SentMix-3L dataset, illustrating its performance across various classification thresholds. The training and validation accuracy graphs are plotted in Fig. 5. The area under the ROC curve of 0.88 provides valuable insights into the model's ability to discriminate between different classes and its overall classification performance.

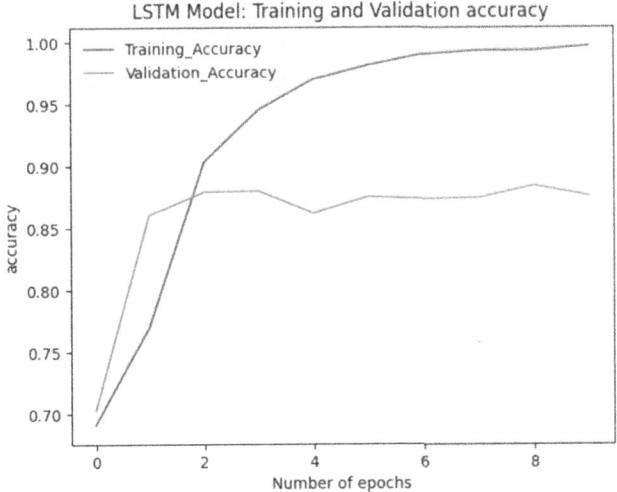

Fig. 5. The training and validation accuracy of LSTM model

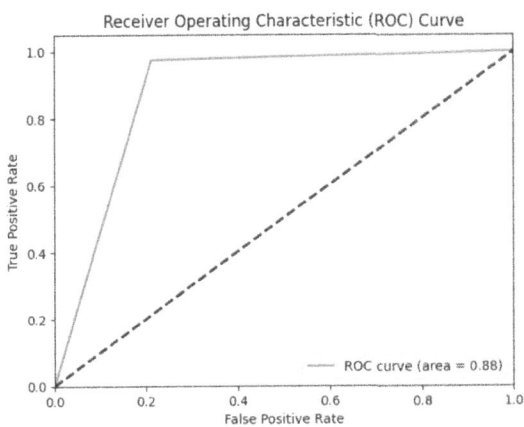

Fig. 6. ROC curve of the proposed approach on SentMix-3L dataset

Table 4. Accuracy, Precision, Recall and F1_score of our proposed approach with Bi-LSTM model

Dataset	No. of Classes	Accuracy	Precision	Recall	F1_Score
Bangla-English-Hindi [18]	2	0.88	0.89	0.88	0.88
Bangla-English-Hindi [18]	3	0.75	0.78	0.75	0.75
Kannada-English [7]	2	0.80	0.80	0.80	0.80
Sinhala-English [25]	2	0.738	0.72	0.74	0.70
Malayalam-English [17]	2	0.85	0.84	0.85	0.84
Malayalam-English [17]	3	0.70	0.69	0.70	0.69

The comparative Analysis of the proposed model is shown in Table 5. We conducted both 2-class and 3-class analyses on the Bangla-English-Hindi dataset. For the 2-class analysis, we considered only positive and negative classes. The comparative analysis presented in Table 5 pertains to the 3-class problem of the Bangla-English-Hindi dataset. For Kannada-English classification, we initially performed binary classification using positive and negative classes, achieving the highest accuracy of 80%. We used two previous studies [5,20] with five different classes for comparative analysis. The highest performance attained was 66% for an ensemble learning model [20] and compared it with our binary classification results. While the accuracy is expected to be lower for the five-class classification, the results remain comparable. For the Sinhala-English dataset, we performed binary classification, achieving an accuracy of 88%. Although accuracy might be expected to decrease with three-class classification, the results remain comparable. The analysis of the Malayalam-English dataset is conducted for a three-class problem, resulting in slightly lower accuracy compared to binary classification. The results of the proposed model outperform existing methods for all datasets except the Malayalam-English dataset, demonstrating its efficacy and superiority for sentiment analysis tasks.

Table 5. Comparative Analysis of Accuracy, Precision, Recall and F1_score of our proposed approach with LSTM model

Model Used	Dataset	Accuracy	Precision	Recall	F1_Score
GPT 3.5 Turbo [18]	Bangla-English-Hindi	-	-	-	0.62
XLM-R [18]	Bangla-English-Hindi	-	-	-	0.59
Proposed Method(LSTM)	Bangla-English-Hindi	**0.89**	**0.91**	**0.89**	**0.89**
Ensemble Learning [20]	Kannada-English	0.66	0.66	0.66	0.66
SSNCSE_NLP [5]	Kannada-English	-	0.64	0.66	0.63
Proposed Method(Bi-LSTM)	Kannada-English	**0.80**	**0.80**	**0.80**	**0.80**
SVM [25]	Sinhala-English	0.71	0.7	0.71	0.69
Linear Regression [25]	Sinhala-English	0.72	0.71	0.72	0.71
Proposed Method(LSTM)	Sinhala-English	**0.88**	**0.87**	**0.88**	**0.87**
SBERT with CBL [3]	Malayalam-English	-	0.71	0.71	0.71
TF-IDF [5]	Malayalam-English	-	0.691	0.69	0.69
Ensemble Learning - CNN [20]	Malayalam-English	-	0.73	0.72	0.72
XLM-RoBERTa [4]	Malayalam-English	-	0.80	0.81	0.80
Proposed Method(Bi-LSTM)	Malayalam-English	**0.71**	**0.71**	**0.71**	**0.71**

6 Conclusion

The cellular automata-based deep learning classifier has demonstrated promising results in sentiment analysis tasks when applied to a code-mixed dataset. With an average accuracy rate of 86%, the proposed approach showcases its effectiveness in accurately classifying sentiment across diverse language contexts. The success of the cellular automata-based classifier can be attributed to its unique architecture, which leverages the dynamic and parallel processing capabilities of cellular automata, coupled with the expressive power of deep learning techniques. Importantly, it surpasses the performance of existing models in this domain, highlighting its potential as a novel approach for sentiment analysis in multilingual or code-mixed environments. Moving forward, further research and refinement of this model hold the potential to enhance the accuracy of sentiment analysis and its applicability across diverse linguistic landscapes.

References

1. Unicode 15.0 character code charts. https://unicode.org/charts/
2. Auer, P.: Code-Switching in Conversation: Language, Interaction and Identity. Routledge, Milton Park (1998)
3. Babu, Y.P., Eswari, R., Nimmi, K.: Cia_nitt@dravidian-codemix-fire2020: Malayalam-English code mixed sentiment analysis using sentence BERT and sentiment features. In: Working Notes of FIRE 2020 - Forum for Information Retrieval Evaluation, Hyderabad, India, 16–20 December 2020. CEUR Workshop Proceedings, vol. 2826, pp. 566–573. CEUR-WS.org (2020)
4. Bai, Y., Zhang, B., Chen, W., Gu, Y., Guan, T., Shi, Q.: Automatic detecting the sentiment of code-mixed text by pre-training model (2021). http://ceur-ws.org
5. Bharathi, B., Samyuktha, G.: Machine learning based approach for sentiment analysis on multilingual code mixing text. In: FIRE (Working Notes), pp. 1038–1043 (2021)
6. Bohra, A., Vijay, D., Singh, V., Akhtar, S.S., Shrivastava, M.: A dataset of Hindi-English code-mixed social media text for hate speech detection. In: Proceedings of the Second Workshop on Computational Modeling of People's Opinions, Personality, and Emotions in Social Media, New Orleans, Louisiana, USA, pp. 36–41. Association for Computational Linguistics (2018). https://doi.org/10.18653/v1/W18-1105
7. Chakravarthi, B.R., et al.: Dravidiancodemix: sentiment analysis and offensive language identification dataset for dravidian languages in code-mixed text. Lang. Resour. Eval. (2022). https://doi.org/10.1007/s10579-022-09583-7
8. Chittaranjan, G., Vyas, Y., Bali, K., Choudhury, M.: Word-level language identification using CRF: code-switching shared task report of MSR India system. In: Proceedings of the First Workshop on Computational Approaches to Code Switching, Doha, Qatar, pp. 73–79. Association for Computational Linguistics (2014). https://doi.org/10.3115/v1/W14-3908
9. Dey, S., Thakur, S., Kandwal, A., Kumar, R., Dasgupta, S., Roy, P.P.: Bharatbhasanet-a unified framework to identify Indian code mix languages. IEEE Access 1 (2024). https://doi.org/10.1109/ACCESS.2024.3396290
10. Elizabeth, M.J., Panda, A.K., Chaudhuri, P.P., Hazari, R.: Cellular automata-based sentiment analysis. In: Das, S., Martinez, G.J. (eds.) ASCAT 2023. AISC, vol. 1443, pp. 53–64. Springer, Singapore (2023). https://doi.org/10.1007/978-981-99-0688-8_5

11. Elizabeth, M., Parsotambhai, S.M., Hazari, R.: Cellular automata enhanced machine learning model for toxic text classification. In: Chopard, B., Bandini, S., Dennunzio, A., Arabi Haddad, M. (eds.) ACRI 2022. LNCS, vol. 13402, pp. 346–355. Springer, Cham (2022). https://doi.org/10.1007/978-3-031-14926-9_31
12. Hidayatullah, A.F., Qazi, A., Lai, D.T.C., Apong, R.A.: A systematic review on language identification of code-mixed text: techniques, data availability, challenges, and framework development. IEEE Access **10**, 122812–122831 (2022). https://doi.org/10.1109/ACCESS.2022.3223703
13. Hochreiter, S., Schmidhuber, J.: Long short-term memory. Neural Comput. **9**(8), 1735–1780 (1997)
14. Kamble, S., Joshi, A.: Hate speech detection from code-mixed Hindi-English tweets using deep learning models (2018). http://arxiv.org/abs/1811.05145
15. von Neumann, J.: The theory of self-reproducing Automata, A. W. Burks ed. Univ. of Illinois Press, Urbana and London (1966)
16. Perera, A., Caldera, A.: Sentiment analysis of code-mixed text: a comprehensive review. J. Univers. Comput. Sci. (JUCS) **30**(2) (2024)
17. Priyadharshini, R., Chakravarthi, B.R., Thavareesan, S., Chinnappa, D., Durairaj, T., Sherly, E.: Overview of the dravidiancodemix 2021 shared task on sentiment detection in Tamil, Malayalam, and Kannada. In: Forum for Information Retrieval Evaluation, FIRE 2021. Association for Computing Machinery (2021)
18. Raihan, M.N., Goswami, D., Mahmud, A., Anastasopoulos, A., Zampieri, M.: Sentmix-3l: a Bangla-English-Hindi code-mixed dataset for sentiment analysis (2023). http://arxiv.org/abs/2310.18023
19. Raja Chakravarthi, B., Jose, N., Suryawanshi, S., Sherly, E., McCrae, J.P.: A sentiment analysis dataset for code-mixed Malayalam-English. arXiv e-prints p. 2006 (2020)
20. Roy, P.K.: Deep ensemble network for sentiment analysis in bi-lingual low-resource languages. ACM Trans. Asian Low-Resour. Lang. Inf. Process. **23**(1) (2024). https://doi.org/10.1145/3600229
21. Sreelakshmi, K., Premjith, B., Soman, K.P.: Detection of hate speech text in Hindi-English code-mixed data. Procedia Comput. Sci. **171**, 737–744 (2020). https://doi.org/10.1016/J.PROCS.2020.04.080
22. Srinivasan, R., Subalalitha, C.N.: Sentimental analysis from imbalanced code-mixed data using machine learning approaches. Distrib. Parallel Databases **41**, 37–52 (2023). https://doi.org/10.1007/s10619-021-07331-4
23. Thara, S., Poornachandran, P.: Code-mixing: a brief survey. In: 2018 International Conference on Advances in Computing, Communications and Informatics (ICACCI), pp. 2382–2388 (2018). https://doi.org/10.1109/ICACCI.2018.8554413
24. Tho, C., Warnars, H.L.H.S., Soewito, B., Gaol, F.L.: Code-mixed sentiment analysis using machine learning approach - a systematic literature review. In: 2020 4th International Conference on Informatics and Computational Sciences (ICICoS), pp. 1–6 (2020). https://doi.org/10.1109/ICICoS51170.2020.9299004
25. Uthpala, D.K., Thirukumaran, S.: Sinhala-English code-mixed language dataset with sentiment annotation, pp. 184–188. Institute of Electrical and Electronics Engineers Inc. (2024). https://doi.org/10.1109/ICARC61713.2024.10499746
26. Wolfram, S.: Theory and Applications of Cellular Automata. World Scientific, Singapore (1986)

Asynchronous Method of Generating Stream Ciphers in a Group of Robots Based on Cellular Automata with Active Cells

Volodymyr Mokhor[1], Stepan Bilan[2(✉)], and Volodymyr Samburskyi[2]

[1] G.E. Pukhov Institute for Modelling in Energy Engineering of the National Academy of Sciences of Ukraine, Kiev, Ukraine
`ipme@ipme.kiev.ua`
[2] Taras Shevchenko National University of Kyiv, Kyiv, Ukraine
`bstepan@ukr.net`

Abstract. The paper solves the problem of effectively organizing the exchange of information in a group of robots, through the use of stream ciphers formed on the basis of cellular automata with active cells. The cellular automata with active cells allows to build a model of a pseudorandom number generator that can simultaneously generate many pseudorandom bit sequences. This makes it possible to organize the transmission of messages from one robot to all robots in the group and vice versa. Messages are exchanged using different self-synchronized stream ciphers generated by one device. Each cellular automata contains the number of active cells, which at the output generate a key gamma and form a stream cipher for the corresponding robot in the group. The number of active cells can either increase or decrease during the functioning of a group of robots, which corresponds to the addition or removal of robots in the group.

Keywords: cellular automata with active cells · asynchronous stream cipher · group of robots · information exchange

1 Introduction

The future is defined by a large number of robots that will interact with each other and with people.

Currently, distributed intelligent systems are increasingly used in human activities. They are especially effective in modern robotic complexes as groups of robots. Recently, the need to develop such groups of robots in the direction of their behavior, navigation, control, interaction and information exchange has become obvious. A group of robots solves complex intellectual problems. A large group of robots (swarm) solves such problems by analogy with living organisms (ants, bees, etc.). In a group of robots, cooperation occurs, which results in collective behavior. Parallelism and multitasking in such a group require effective management or intelligent interaction, which leads to the formation of a swarm of robots operating according to the principles of biological swarms [1, 2]. The collective behavior of biological organisms has been studied in ants [1, 3], bees [4] and other populations with swarming behavior. In papers [5, 6], a swarm was considered as a colony of cells, the behavior of which was modeled using cellular automata with active cells (CAAC). The formation of colonies, their behavior, and the interactions of different colonies are described. These colonies functioned as autonomous groups. Particular attention was paid to the survivability of the colonies, which was determined by the choice of the leader of the colony and their division.

Swarm robotics has defined a new look at robot designs, navigation, path planning, and their information interaction. It is based on the principles of collective intelligence, where each autonomous robot performs one task, and their common solution determines the solution to one more complex intellectual problem. In a non-deterministic environment, the behavior of the group must be determined by counteracting unwanted external influences to maintain the survival of the robot swarm. In such situations, the exchange of information is very important, which depends on the choice of control strategy and methods of encrypting messages transmitted between robots both within the group and between robots of other groups.

2 Formulation of the Problem

The problem of increasing the survivability of a group of robots is an important task in the conditions of group functioning. One of the factors influencing the stability of the high-quality functioning of a group of robots is the information exchange between robots in the group, as well as the effective encryption of messages that are transmitted from robot to robot. For the exchange of information between robots and people in real time, the most effective is the use of stream ciphers, which are generated using a one generator of pseudo-random bit sequences containing multiple outputs. At each output of such a generator, a pseudo-random bit sequence is generated in the form of a key gamma, which differs from the key gammas generated at other outputs of the generator. Such key gammas are involved in generating a stream cipher for each robot in the group, and such a generator, implemented on a single structure, is more reliable. High security is also ensured by the possibility of self-regulating selection of the outputs of such a generator for the formation of a key gamma, which is achieved through the use of an asynchronous method of generating a stream cipher. In this regard, the work solves the problem of increasing the survivability of groups of robots through an asynchronous method of generating stream ciphers for exchanging messages between robots in a group.

3 Information Exchange Between Robots in a Group Based on Asynchronous Stream Ciphers

High-quality autonomous information exchange between robots is a very important task, the solution of which determines the reliable functioning of a group of robots. Information exchange between robots is necessary during swarm navigation and path planning [7, 8]. Two types of information exchange are used here: global and local. Global information exchange involves the use of a common navigation map, with the help of which the path of the entire swarm of robots is planned. As a rule, global navigation is based on centralized control of information exchange. Local navigation is most often based on sensory navigation and data transmitted from the group's robots.

There is also web-based exchange of information between robots [9, 10], which allows increasing the distance between robots that interact with each other. However, this reduces the reliability of the functioning of a group of robots and reduces the speed of data exchange.

Information exchange is used under centralized control [11], which is characterized by centralized collection of information about the state of all robots and the environment on a central device. The central control device can be located on the main robot of the group and transmits control signals to the slave robots of the same group. In this case, for each slave robot, a separate communication channel is allocated from the central device or communication channels are compressed according to frequency, time and other parameters. However, such an organization of information exchange does not guarantee high reliability of the functioning of the entire group of robots, which is vulnerable in the event of an attack on the central control device.

The situation can be improved by information exchange using centralized hierarchical control. In this case, the swarm's vulnerability to enemy attacks is partially improved. However, the reliability of such a strategy leaves much to be desired.

If a leading robot with central control is not in the group, then such a group uses an asynchronous or self-regulating mode of control and behavior of the swarm of robots. Control and exchange of information is determined by the task facing the swarm and the current state of the swarm. In this case, information exchange between robots can occur at any time. Each robot generates a message to other robots at the moment when it needs to report its state and at the moment when its further execution of the task is impossible without information from other robots. The number of robots in the group and the distribution of tasks between the robots determine the complexity of such information exchange. Also, the quality of information exchange is affected by the relative position of robots in space and the radius of the visibility zone of each robot.

In general, organizing information exchange between robots is a complex technical task. At the same time, for complete information protection of a group of robots, it is necessary to use effective encryption of information exchanged between robots. In real time, it is most effective to use stream ciphers, one of the varieties of which are asynchronous stream ciphers [12–14]. Synchronous stream ciphers are formed by bitwise encryption of the input message. Such ciphers are especially effective for real-time encryption of the transmission of secret messages. Using special generators that generate a pseudo-random bit sequence (key gamma); each bit of the input message is encrypted using each bit of the key gamma. In self-synchronizing stream ciphers, a certain number of initial S bits of the input message are analyzed, which indicate the configuration of the structure and initial settings of the generator that generates the key gamma, after receiving these initial S bits. In such an encryption strategy, each robot must have its own channel for transmitting and receiving a ciphergram, and the robot must also have its own generator for generating a key gamma, which is superimposed on the message to generate a stream cipher.

The key gamma generator is a pseudo-random bit sequence generator (PRBSG). With centralized control, the leading robot contains one PRBSG with multiple outputs, the number of which corresponds to the number of slave robots. At each output, a key gamma (pseudo-random bit sequence) is generated, which differs from the key gammas generated at the other outputs (active cells output) of the PRBSG. For reliable autonomy of robots in a group, it is necessary that each slave robot in the group have a PRBSG, the configuration of which differs from the configuration of the generators of other robots. If the configurations of the generators are different, and their initial states are also different, then different key gammas will be formed, which will lead to a violation of the symmetry of the stream cipher. Therefore, key-gamma generators must have identical configurations and equal initial states. For high reliability of operation, such PRBSGs must have many dynamically changing outputs, each of which corresponds to one dynamically changing output in the corresponding robot of the group.

Such a strategy can be achieved by using CAACs, on which a PRBSG with dynamically changing outputs is implemented [5, 15]. CAACs also make it possible to implement efficient navigation and path planning for robots. In paper [16], the cell in which the robot is located at the current moment in time has an active state. The transfer of an active signal to a neighboring cell is carried out by analyzing neighborhood cells, where neighboring cells with logical "1" states represent obstacles, and neighboring cells with logical "0" states indicate free cells to which an active state can be transferred at the next time. Analysis of the states of cells in the vicinity of the active cell at each current time step allows you to select the active cell at the next time step. This cell can only be one of the cells that is in the logical "0" state.

Generators based on CAAC are present in all robots of the group. However, the key gamma is formed for each robot on the active cell output selected by the robot itself. This active cell is determined by the robot itself, and the structure of the cell's neighborhood is also determined, which determines the transmission of the active signal to the cell at the next time step. There may be more active cells on CAAC than there are robots in the group, which makes it difficult to crack the stream cipher and deceives the enemy.

This paper considers the strategy of information exchange between group robots based on asynchronous ciphers generated using CAAC. The use of key-gamma generators based on CAAC increases the survivability of an organized group of robots, regardless of their number in the group.

4 Cellular Automata with Active Cells

Unlike classical cellular automata, in the CAACs only active cells change their state without changing the states of other cells. However, there may be a regime when, simultaneously with active cells, all other (non-active) cells can change their state according to rules different from the rules by which active cells change their state. For a constant CAAC environment, only active cells perform the local transition function (LTF) in accordance with the established neighborhood. Each active cell generates an active signal that is sent to the active inputs of all cells in the neighborhood. At the same time, one of the cells in the neighborhood goes into an active state. All cells in the neighborhood that received an active signal from a neighboring cell analyze their own neighborhood, resulting in one or more cells becoming active.

The second option for transmitting an active signal to one of the cells in the neighborhood is to analyze the states of the cells in the neighborhood by the active cell at the previous time step. In this case, the structure of the neighborhood that is used to transmit the active signal is often different from the structure of the neighborhood that is used to change the information state of the cell.

The work [5] describes models for CAAC with one active cell, with two active cells and with a varying number of active cells. Each cell in such CAAC is described by a local state function (LSF) and LTF. In this case, LSF for all active cells may be the same or may differ. LTFs are selected in accordance with the properties established for the entire CAAC. For CAAC with a varying number of active cells, the number of active cells may decrease or increase.

If, during the functioning of the CAAC, two active cells with the same LTF at one time step transmit an active signal to one cell, then at the next time step there remains one active cell with the same LTF. If the CAAC contains the property of the "birth" of new active cells, then when two active signals from two active cells coincide, which at a given time step have common neighborhood cells, at the next time step another new active cell is "born" (appears), in which LTF is formed as a function of the LTF of matched active cells.

The work [5] presents pseudorandom number generator (PRNG) based on CAAC with a variable number of active cells. To analyze the quality of such generators, NIST tests [17], graphic tests [18] and ENT tests [19] were used. Tests have shown the high quality of operation of such generators. Examples of the functioning of CAs with active cells in Fig. 1 are presented.

Figure 1 shows an example of the transmission of an active signal from cell to cell with different surroundings for the formation of LTF. Such neighborhoods are highlighted in yellow. Three cells of the Moore neighborhood are used. The top row of CAAC shows three time steps of CAAC operation without changes in the shape of the neighborhood. The middle row represents the change in the shapes of the neighborhood at the first time step and the subsequent two time steps of the functioning of the CAAC. How the movement of active cells changes are shown.

At the bottom of Fig. 1, cells that encode the direction of active signal transmission for each active cell are highlighted in yellow. Active cells are respectively highlighted in green and blue. The new "born" cell is highlighted in orange and at the third time step the neighborhood indicating LTF is highlighted in yellow. In the third step, the cells meet and form a new cell with new properties. These properties are determined by a new form of the neighborhood, which determines the direction of transmission of the active signal to the cells of the neighborhood according to LTF. Active cells can also change their state to take into account other shapes of the neighborhoods. If two active cells with the same properties meet, they "die" (disappear) and do not give birth to new cells.

For the "birth" of a new cell, it is not necessary that two active cells coincide in one. Interaction between the neighborhoods of active cells is also possible. The work [5] describes options for the interaction of cells and their neighborhoods, as well as options for the formation of new properties for a new active cell depending on the states of the two interacting cells.

Changing the shapes of the neighborhood to implement the constant movement of active cells can be carried out through a given number of time steps or randomly. In this case, it is not always necessary that all active cells can simultaneously change the shape of the neighborhood to implement LTF. The direction of movement for each active cell is

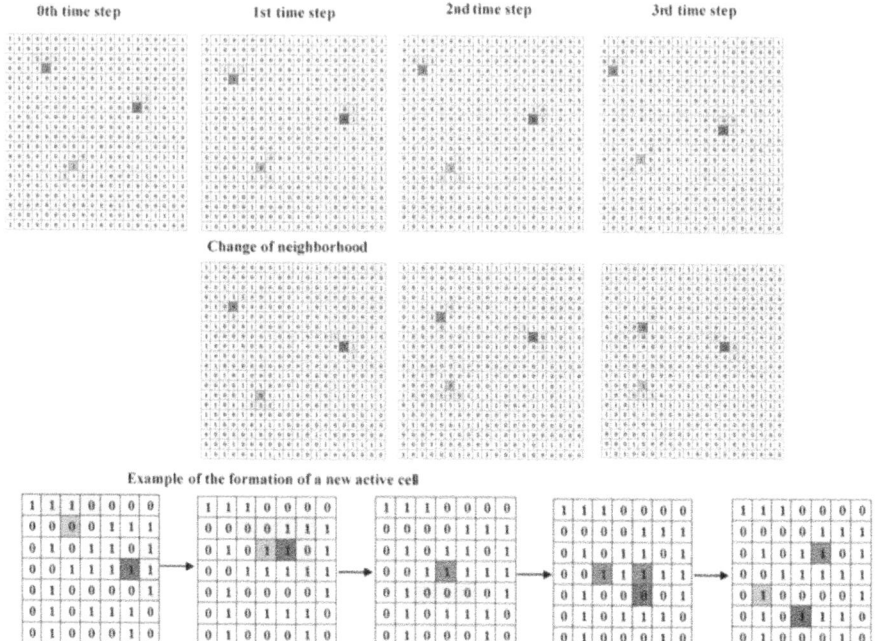

Fig. 1. Examples of the functioning of CAs with active cells (Color figure online)

indicated by the code of the cells in its neighborhood, where the least significant digit of the code is defined on the left, and the most significant is defined as the third, highlighted in a clockwise direction. The direction of movement is encoded by the first character of the code, indicating the upper middle cell of the Moore neighborhood. Then counting continues clockwise through the cells of the neighborhood. An example of such coding of the direction of movement is presented in Fig. 2.

At the top of Fig. 2, there is an active cell (red), the cells in the neighborhood of which indicate the direction of movement (yellow). The bottom part of Fig. 2 shows an example of active signal transmission at the next time step. The direction of transmission of the active signal is indicated by three cells in the neighborhood on the right (green). The code of these three cells indicates the direction of transmission of the active signal according to the coding presented in the upper part of the figure (in this example it is 3). A cell that was active at the previous time step changed its state in accordance with the XOR function for the Moore neighborhood. An additional bit can also be used, which is involved in the implementation of the active signal transmission function or changing the state of the cell [20].

Example of active signal transmission direction encoding

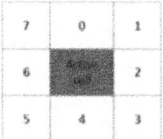

Example of transmission of active signal

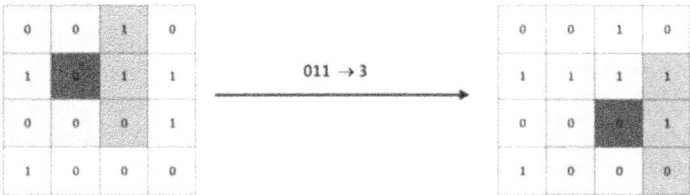

Fig. 2. An example of the functioning of a PRBSG based on CAAC in a group of robots (Color figure online)

5 Formation Method for Encrypting Messages in a Group of Autonomous Robots

A group of self-controlled externally independent driving robots constitutes a distributed intelligent system that is aimed at solving common problems in the process of operation. Each robot in such a group solves a range of problems, the solution of which, in interaction with other robots, allows solving the overall problem of the group.

An important task in the functioning of an autonomous group is the effective organization of information interaction between all autonomous members of this group. To achieve this, effective encryption of all information exchanged between all autonomous members of the group must be ensured. Since a group of autonomous robots consists of leading and slave robots, the leading robot must contain a basic core that generates a key gamma that allows the distribution of individual ciphers for each slave member of the group. Due to this prerequisite and the concentration of all major encryption functions in the lead robot, the entire group can be affected if the lead robot is successfully attacked. In such a situation, the enemy can determine the structure of the key gamma generator and seize control of the entire group. Therefore, it is necessary to distribute encryption processes among all group members.

To increase the reliability of operation and stability of the entire group of robots, it is most effective to use asynchronous stream ciphers generated on the basis of CAAC [5, 17]. Each active cell of the CAAC forms a stable pseudo-random bit sequence, which in a group of autonomous robots is used to form a self-synchronous stream cipher.

Using CAAC technology, a generalized algorithm for exchanging messages between group robots can be represented in the following sequence of steps.

1. The leading robot sets the dimension of the CAAC and the number of active cells, which are determined by the number of slave robots in the group. The number of active cells may be greater than the number of slave robots. A larger number of active

cells is used in case additional robots join the group. In fact, the formation of the swarm is carried out by the leading robot.

2. The initial coordinates of active cells are set that can be set initially in the CAAC of the leading robot, or in asynchronous mode from the slave robots. A more reliable option is to transfer the coordinates of active cells from slave robots. In this case, it becomes necessary to transfer the coordinates of all active cells to all slave robots. This situation arises when a common CAAC is used to encrypt messages for all robots. If you divide into encryption channels for each robot, then there is no need to transfer all the initial coordinates of active cells to all robots.
3. Neighborhood cells are specified for each active cell, which influence the transmission of the active signal to one of the neighborhood cells at the next time step.
4. The key gamma for each slave robot is formed at the output of the corresponding cell to which the slave robot is assigned. The key gamma is superimposed on the message and the generated stream cipher is sent to the input of the slave robot, containing the same CAAC and generating the same key gamma.
5. The transmission of messages from the slave robot is carried out in reverse order according to the rules described in the fourth stage.

The implementation of this algorithm for exchanging messages between robots in a group requires the presence of identical PRBSGs. At the same time, it is possible to introduce additional robots into the group and add new active cells, as well as in real time to delete old active cells and add new ones for each robot in the group.

Setting the initial coordinates of active cells can be done by calculating using a special formula that is used by each robot. All slave robots form some two numbers either randomly or from a predetermined range of numbers. These numbers, generated by each robot, are involved in calculating the coordinates of the active cell for the selected robot. Two numbers are calculated as the remainder of division by numbers that determine the CAAC dimension. For each i-th slave robot, the generated number is different from the other generated numbers for other slave robots.

Each autonomous robot generates two random numbers $M_{i'}$ and $N_{i'}$, which are greater than M and N (the CAAC dimension corresponds to $M \times N$). These numbers are generated to determine the X and Y coordinates of the active cell of the i-th robot. To determine the horizontal coordinate, all numbers generated by other robots are summed up and the remainder of division by M is determined using the formula

$$X_i = \left(\sum_{j=1}^{k} M_{j'} - M_{i'} \right) modM, (j \geq i),$$

where $M_{i'}$ is the first number generated by the i-th robot; $M_{j'}$ is the first number generated by the j-th robot; k is number of robots in the group.

The resulting number indicates the horizontal coordinate value for the active cell of the i-th robot.

The vertical coordinate is determined by the formula

$$Y_i = \left(\sum_{j=1}^{k} N_{j'} - N_{i'} \right) modN, (j \geq i)$$

where $N_i{'}$ is second number generated by the i-th robot; $N_j{'}$ is second number generated by the j-th robot.

This is an example of a possible determination of the coordinates of the active cell for the i-th robot. Other robots may not generate numbers. It is enough that only the i-th robot generates numbers that are much larger than M and N at the corresponding coordinates. Other options for specifying the coordinates of active cells are also possible. For example, the coordinates can be specified by the first bits of the input message from the slave robot, which encode in the binary system the transmitted initial S symbols of the message.

One of the important operations in the asynchronous formation of PRBSG based on CAAC is setting the dimension of CAAC. There can be a large number of options for asynchronous formation of the CAAC dimension.

In asynchronous mode, a selected number of first bits of the message are typically used. With this approach, the first pairs of groups of bits are identified, which encode the locations of active cells. If the number of robots in a team is n, then the first $2bn$ bits of the sequence at the output of the main control robot are used to encode the coordinates of active cells. Here the value of b depends on the dimension of the CAAC in the PRBSG of each robot, i.e. the CAAC dimension is used, corresponding to $2^b \times 2^b$ cells for binary encoding. Also, the dimension of CAAC depends on the message character encoding system. The dimension of the CAAC should determine the full coverage of the coordinates. If we use a binary coding system, then for a 16×16 CAAC, 4 message bits are used to encode one X or Y coordinate. In this case, eight message bits are used to set the coordinates of one active cell.

As recommended by papers [5, 15, 17] the dimension of the CAAC should not be less than 10×10 for one active cell. The dimension can be defined as $Q \times Z$, where Q is the number of logical "1s" in a sequence of n initial bits of the message, and Z is the number of logical "0s".

If during the operation of a group of robots one or more robots are added, then the coordinates of their active cells are received from these robots to the leading robot asynchronously. The leading robot enters the received coordinates into the already functioning PRBSG at the current time step. CAAC states with the new number and location of active cells are transmitted to all robots of the group. If one or more subordinate robots "leave" the group, then their active cells become surrounding cells (inactive) of CAAC. An example of the functioning of a PRBSG based on CAAC in a group of robots is presented in Fig. 3.

Two modes of generator operation is shows in Fig. 3. The first mode implements working without changing the neighborhood of active cells, and the second mode implements a change in the neighborhood every three time steps. The addition (3rd and 5th time steps, respectively, for the first and second modes) and removal (5th and 6th time steps, respectively, for the first and second modes) of slave robots in the group is shown. Each active cell has its own color, which is determined by LTF and corresponds to a slave robot in the group. At the output of this cell, bit k of the key gamma is formed at the corresponding time step. This bit k_j^i encrypts the corresponding bit b_j^i of the message for the j-th robot using the XOR function.

Thus, one CAAC generates many key gammas, the number of which is determined by the number of robots in the group. There may be several leading robots in a group and several PRBSGs.

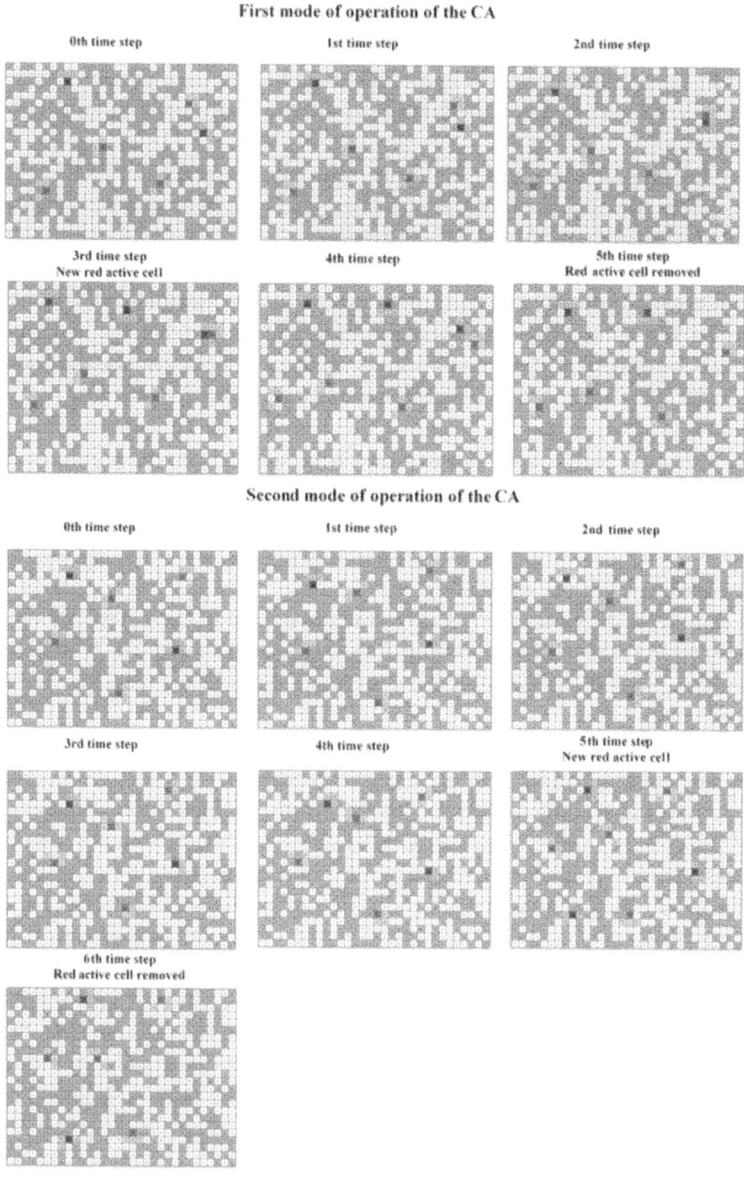

Fig. 3. An example of the functioning of a PRBSG based on CAAC in a group of robots

The described approach is characterized by the fact that all robots in the group know the active cells of the remaining robots, since the same generator is used for all robots.

To increase the reliability of the functioning of a group of robots, it is necessary to divide one generator into many, each of which encrypts only one slave robot.

From the i-th slave robot (R_i) receives a request to exchange messages with the master robot (R_G). The coordinates of the active cell are also transmitted from R_i to R_G. After a specified time, the CAAC$_i$ in the leader and slave robots begin to function. Active cells begin to generate key gamma. Messages are exchanged between the leading R_G and the i-th slave robot. If the initial request for message exchange is made by the leading robot R_G, then before starting the exchange of messages, the leading robot R_G waits for the slave robot to send it the coordinates of the active cell. After receiving the coordinates of the active cell from the slave robot R_i, a key gamma is formed and ciphergrams are exchanged. This organization of interaction between robots in a group makes it possible to separate the slave robots in the group informationally. The vulnerability of the group is determined by attacks on the leading robot.

6 Conclusion

The paper describes and investigates a method for simultaneous generation of multiple asynchronous stream ciphers based on a cellular automata with active cells, that made it possible to organize an effective exchange of information in a group of robots. In this case, the information in this group is transmitted in the form of an asynchronous stream cipher. The reliability of the functioning of a group of robots and resistance to attacks is increased due to the fact that each robot at any time in asynchronous mode itself generates the initial coordinates of its own active cell and transmits them to all robots in the group, and also generates a local state function and a local transition function. The proposed method of information interaction allows a group of robots with a varying number of robots to function reliably. During the operation of a group of robots in real time, robots can be added with the addition of a new active cell, and robots and the corresponding active cells can be removed from the group. The presence of more active cells than robots in a group makes it possible to increase the resistance to hacking of a generator built on the basis of a cellular automaton with active cells. The proposed generator model, unlike generators based on LSFR and other generators, does not allow us to determine the statistical relationships between the ciphers generated by each active cell for each robot.

The authors plan to conduct research in the implementation of multiple asynchronous stream ciphers based on cellular automata with different models of generators on different devices that exchange information.

References

1. Lopes, H.J.M., Lima, D.A.: Surveillance task optimized by evolutionary shared Tabu inverted ant cellular automata model for swarm robotics navigation control. Results Control Optim. **8**, 1–34 (2022)
2. Suárez, P., Iglesias, A., Gálvez, A.: Make robots be bats: specializing robotic swarms to the bat algorithm. Swarm Evol. Comput.Evol. Comput. **44**, 113–129 (2019)
3. Lima, D.A., Oliveira, G.M.: A cellular automata ant memory model of foraging in a swarm of robots. Appl. Math. Model. **47**, 551–572 (2017)

4. Stefanec, M., Szopek, M., Schmickl, T., Mills, R.: Governing the swarm: controlling a bio-hybrid society of bees & robots with computational feedback loops. In: 2017 IEEE Symposium Series on Computational Intelligence, pp. 1–86 . IEEE (2017)
5. Bilan, S.M., Bilan, M.M., Motornyuk, R.L.: New Methods and Paradigms for Modeling Dynamic Processes Based on Cellular Automata, 200 p. IGI-Global, Hershey (2020)
6. Bilan, S., Bilan, M., Motornyuk, R., Yuzhakov, S.: Biometric Data in Smart Cities: Methods and Models of Collective Behavior. CRC Press, Boca Raton, 228 p. (2021)
7. Rubisson, D.L., de Arruda, L.V.R.: Distributed strategy for communication between multiple robots during formation navigation task. Robot. Auton. Syst. **169**, 104509 (2023)
8. Ertug, O., Fabian, S., Boris, L.: Collective navigation of a multi-robot system in an unknown environment. Robot. Auton. Syst.Auton. Syst. **132**, 103604 (2020)
9. Tenorth, M., Beetz, M.: Exchanging action-related information among autonomous robots. In: Lee, S., Yoon, KJ., Lee, J. (eds.) Frontiers of Intelligent Autonomous Systems. SCI, vol. 466, pp. 127–136. Springer, Heidelberg (2013). https://doi.org/10.1007/978-3-642-35485-4_10
10. Waibel, M., et al.: RoboEarth - a World Wide Web for robots. Robot. Autom. Mag. **18**(2), 69–82 (2011)
11. Ivanov, D.: Information exchange in a large group robots. Artif. Intell.. Intell. **4**, 513–521 (2010)
12. Chen, B., et al.: Cryptanalysis of some self-synchronous chaotic stream ciphers and their improved schemes. Int. J. Bifurcat. Chaos 31, 1–26 (2021)
13. Chen, B., Yu, S., Li, D.D.-U., Lü, J.: Cryptanalysis of some self-synchronous chaotic stream ciphers and their improved schemes. Int. J. Bifurcat. Chaos **31**(08), 2150142 (2021)
14. Fan, C., Ding, Q.: A novel image encryption scheme based on self-synchronous chaotic stream cipher and wavelet transform. Entropy **20**(445), 1–13 (2018)
15. Bilan, S., Bilan, M., Bilan, S.: Research of the method of pseudo-random number generation based on asynchronous cellular automata with several active cells. In: MATEC Web of Conferences, vol. 125, pp. 1–6 (2017). 02018
16. Saleem, A., Mahdi, H.M.: Enhancing robot navigation efficiency using cellular automata with active cells. Ann. Emerg. Technol. Comput. **8**(2), 56–70 (2024)
17. NIST SP 800-22: Download Documentation and Software. https://csrc.nist.rip/Projects/Random-Bit-Generation/Documentation-and-Software
18. Chugunkov, I., Muleys, R.: Pseudorandom numbers generators quality assessment using graphic tests. In: Proceedings of the 2014 IEEE NW Russia Young Researchers in Electrical and Electronic Engineering Conference, pp.8–13 (2014)
19. Walker, J.: ENT: a pseudorandom number sequence test program. Fourmilab, Switzerland (2008). https://www.fourmilab.ch/random/
20. Bilan, S.: Formation Methods, Models, and Hardware Implementation of Pseudorandom Number Generators, 301 p. IGI Global, Hershey (2017)

Controlling Desertification Using Cellular Automata and Genetic Algorithms

Alassane Kone[1,2](✉), Samira El Yacoubi[1], and Allyx Fontaine[2]

[1] ESPACE-DEV, University Perpignan Via Domitia, IRD, University Guyane, University Antilles, University Montpellier, University Réunion, Perpignan, France
`alassane.kone@ird.fr, yacoubi@univ-perp.fr`

[2] ESPACE-DEV, University of French Guyana, IRD, University Perpignan Via Domitia, University Antilles, University Montpellier, University Réunion, Guyane, France
`allyx.fontaine@univ-guyane.fr`

Abstract. This work relies on DESERTICAS software, which is designed to model desertification dynamics. Spatiotemporal land changes are represented by cellular automata with continuous states. Their transition function incorporates various factors from the MEDALUS model, fundamental desertification properties, and additional factors such as land use practices, exploitability, and ownership considerations.

The main contribution of this article is the introduction of a control parameter in the DESERTICAS model to study an input-output problem related to the desertification phenomenon and the application of control theory in cellular automata models. The idea is to act directly on a predominant factor by setting its average intensity. This makes it possible to act indirectly on all the other factors in the model and slow down or halt land degradation processes. In this way, we found that the predominant factor in the study of desertification is management, and we determined its intensity using a genetic algorithm (GA) approach. The aim of combining control problem and GAs is to integrate land protection actions into the desertification simulation, i.e. the DESERTICAS software, and turn it into a decision-support tool.

Keywords: Continuous Cellular Automata · Control theory · MEDALUS model · Desertification Assessment · Land Degradation · Genetic algorithms

1 Introduction

Desertification, caused by climatic variability and irrational human activities, is currently one of the most pressing environmental issues [26]. This phenomenon has led to poverty, famine, and displacement, thereby hindering ecological and socio-economic development in developing countries and regions. Consequently, it has captured global attention. Since the establishment of the United Nations Convention to Combat Desertification (UNCCD) in 1994, various efforts have been undertaken worldwide to combat desertification and bring about positive changes [22,26]. In this context, several standardized methodologies have been developed to assess, model, and understand the phenomenon of desertification. Among these, the MEDALUS project (Mediterranean Desertification and Land Use), supported by Europe, stands out as one of the most prominent [6].

In this work, the model used to study desertification dynamics is implemented in the software DESERTICAS [2,3]. The idea behind it is to combine cellular automata and the assessment of MEDALUS project, enhanced by the anthropogenic factors. It has been shown that human actions, represented by the management factor in DESERTICAS, have a huge impact on the desertification process [1]. The management factor is therefore chosen to be a control parameter to integrate land protection actions. Setting an intensity to the management factor index acts implicitly on all desertification factors and thus on the global system to slow down or accelerate the land desertification process. This makes it difficult to isolate this parameter to solve control problems. The desertification simulation provided by DESERTICAS has enabled us to determine the intensity of the predominant factor by extending it with genetic algorithms. Genetics algorithms (GAs) represent a powerful supervised modelling tool that is used in the optimization of complex systems. The peculiarity of GAs results from the fact that they are simple to understand, the required computer code is easy to write and are capable of being applied to an extremely wide range of problems [8]. It has been successfully used to optimize the control parameters in automatic control field [7], to solve Traveling Salesman Problem [5], to design lens [12], to optimize computer network performances [25], in facial recognition [8], etc.

In this article, we present the extension of DESERTICAS by introducing a control parameter, the management, and genetic algorithms in order to turn it into a decision-support tool. It is organized as follows: in the next section, an introduction of the used notions and the theoretical model are given. Section 3 describes the genetic algorithms and the control methods, along with their applications, implementation, and discussions. At the end, a conclusion is given with some perspectives.

2 Desertification Modelling

This section provides a reminder of the desertification modelling used in DESERTICAS, which will serve as a basis for introducing protection measures against desertification. The theoretical model is based on cellular automata with continuous states [3]. Its components are defined according to the MEDALUS assessment extended by anthropogenic factors [1,2] identified during field campaigns and through in-depth knowledge of the functioning of arid zones in the Sahel [15]. Let us start by briefly presenting the basic tools of our study.

2.1 Cellular Automata

A Cellular Automaton (CA) is a discrete dynamical system that is formally defined by a tuple $(\mathcal{L}; \mathcal{S}; \mathcal{N}; f)$ [4,21]. The cellular space $\mathcal{L}$ is a d-dimensional lattice whose elements called cells are arranged depending on their shape and the space dimension d. The state of each cell c belongs to a discrete set $\mathcal{S}$ and updates its value as a function of the current state of the set of its neighbours $\mathcal{N}(c)$ denoted $\mathcal{N}_c$ according to a set of rules, also called the transition function f. If the states of cell c at time t and $t+1$ are respectively denoted by s_c and s_c^+, the transition function is defined by Eq. (1):

$$f: \mathcal{S}^{n+1} \to \mathcal{S} \\ s_{\mathcal{N}_c} \mapsto s_c^+ \quad (1)$$

where n is the number of neighbours except the cell.

2.2 MEDALUS Assessment

The MEDALUS model, developed by the European Commission, assesses land degradation by quantifying factors in four categories: soil, vegetation, climate, and management [6]. Management factor includes agricultural practices, water resource management, soil conservation, forest and pasture management, compliance with land protection policies and community participation [9,10].

Table 1. DSI states

Class of weight	Desertification state
$I_5 = [1.78; 2]$	Very-degraded
$I_4 = [1.53; 1.78[$	Degraded
$I_3 = [1.38; 1.53[$	High
$I_2 = [1.22; 1.38[$	Moderate
$I_1 = [1; 1.22[$	Low

Table 2. Quality indexes of two factors

Class	Quality state	Index of Climate	Index of Management
1	High	$I_1^w = [1; 1.15[$	$I_1^m = [1; 1.25[$
2	Moderate	$I_2^w = [1.15; 1.81[$	$I_2^m = [1.25; 1.50[$
3	Low	$I_3^w = [1.81; 2]$	$I_3^m = [1.50; 2]$

In the MEDALUS model, each factor is assigned a quality index (ranging from 1 to 2) [10]. The degree of land degradation, represented by the Desertification Sensitivity Index (DSI), is calculated as the geometric mean of the quality indexes of soil (l), vegetation (v), climate (w) and management (m) [6] (cf. Eq. (2)).

$$ds = (l \cdot v \cdot w \cdot m)^{\frac{1}{4}} \quad (2)$$

where the DSI is represented by the variable ds. Its values range from 1 to 2, and the corresponding quality states are defined in the MEDALUS model as outlined in Table 1. In the same way, the quality states of climate and management factors of the MEDALUS model, used in our modelling, are given by Table 2.

2.3 Coupling Cellular Automata and MEDALUS Assessment for Desertification Modelling

This section describes the desertification model on which we have based our work characterized by space-time evolution, where desertification spread depends on various degradation factors [6,14]. Presented in previous works and implemented in the software DESERTICAS [1,2], this model combines CA and MEDALUS model with three additional factors: land use type p (in the range $[1,6]$), exploitability e (in range $[\![0,1]\!]$) and ownership o. Land use types include croplands, forests, pasture areas, residential areas, and border areas like roads and water streams. Exploitability factor indicates whether land is used or not, and ownership groups areas by owner.

The study area is represented by a 2-dimensional lattice $\mathcal{L}$ divided into square cells [1,2].

The state of each cell c at time t is given by a tuple $\xi(c)$ defined by $\xi(c) = (l_c, v_c, w_c, m_c, p_c, e_c, o_c)$ [1,2] corresponding, respectively, to the soil, vegetation, climate, management, land use type, exploitability, and ownership indices. We denote by $\xi^+(c) = (l_c^+, v_c^+, w_c^+, m_c^+, p_c, e_c^+, o_c)$ the value of this variable for a cell c at time $t+1$. Therefore, the set of states $\mathcal{S}$ is given by: $\mathcal{S} = [1,2]^4 \times [1,6] \times \{0,1\} \times \{1,2,\ldots,k\}$, where k is the ownership identifier in the study area.

The neighbourhood of a cell c is the Moore neighbourhood of radius r defined as $\mathcal{N}_c = \{c_{i'j'} \in \mathcal{L}, \max(|i-i'|, |j-j'|) \leq r\}$ [1,2] and the state of its neighbourhood is given by $\xi(\mathcal{N}_c) = (\xi(c), \xi(c_1), \xi(c_2), \ldots, \xi(c_n))$.

The most important factors, such as soil (l), vegetation (v), climate (w) and management (m), are characterised for a neighbourhood $\mathcal{N}_c$ by four values calculated as the geometric mean of these factors for all neighbouring cells c_i, $i = 1, \ldots, n$ and give Eq. (3) for each $q \in \{l, v, w, m\}$,

$$q_{\mathcal{N}_c} = (q_{c_1} \cdot q_{c_2} \cdot \ldots \cdot q_{c_n})^{\frac{1}{n}} \qquad (3)$$

The transition function used allows to update the evolution of the state of a cell c according to its current state $\xi(c)$ and the state of its neighbourhood $\xi(\mathcal{N}_c)$. This transition function also integrates the desertification properties such as the combination of factor [6], the irreversibility [6,18], the stress conditions (Properties 1 and 2) [2,19] and the transfer of activities (Property 3).

Property 1. The Desertification Sensitivity Index of a cell c, expressed by the variable ds_c, reaches the upper part of the range of *Degraded* state and its climate and management indexes are *High* i.e. $c \in D_1$:

$$D_1 = \{c \in \mathcal{L} \mid \text{center}(I_4) \leq ds_c < \text{upper}(I_4) \text{ and } w_c \text{ or } m_c \text{ are High }\} \qquad (4)$$

where center(I_i) and upper(I_i) are respectively the center and upper bound of the interval I_i (see Table 1).

Property 2. The Desertification Sensitivity Index of cell c reaches the upper part of the range of *High* state and its climate w_c and management m_c indexes reach the upper part of *High* range values i.e. $c \in D_2$:

$$D_2 = \{c \in \mathcal{L} \mid \text{center}(I_3) \leq ds_c < \text{upper}(I_3) \text{ and } \text{center}(I_3^w) \leq w_c \text{ or } \\ \text{center}(I_3^m) \leq m_c\} \qquad (5)$$

where I_3^w and I_3^m are described in Table 2.

Property 3. [11, 16]
The transfer operation between two lands can only be done by their owners or with their agreements. At the microscopic level, it is possible between two cells c and c', if $(c; c') \in D_3$:

$$D_3 = \{(c, c') \in \mathcal{L} \times \mathcal{L} \mid p_c \in [1; 2; 3], ds_c < \min(I_4), e_c = 0, e_{c'} = 1, ds_{c'} \in I_5,$$
$$p_c = p_{c'} \text{ and } o_c = o_{c'}\} \quad (6)$$

where I_4 and I_5 are described in Table 1 and p_c, e_c, o_c correspond respectively to the indices of land use type, the exploitability, and the ownership factors of the cell c.

The transition function depends on the parameters α and a two generic functions g_α and h [1, 2]. These generic functions allow for the integration of the desertification properties mentioned earlier through the Properties 1, 2 and 3. Let $\alpha = (\alpha_1; \alpha_2; \ldots; \alpha_9)$ be a tuple of factor powers such that $\alpha_i \in [0; 1]$ and $\sum_{i=1}^{9} \alpha_i = 1$. Let g_α be the α-weighted geometric mean defined by Eq. (7).

$$g_\alpha : \mathcal{S}^{n+1} \to [1; 2]$$
$$g_\alpha(\xi(\mathcal{N}_c)) = l_c^{\alpha_1} \cdot l_{\mathcal{N}_c}^{\alpha_2} \cdot v_c^{\alpha_3} \cdot v_{\mathcal{N}_c}^{\alpha_4} \cdot w_c^{\alpha_5} \cdot w_{\mathcal{N}_c}^{\alpha_6} \cdot m_c^{\alpha_7} \cdot m_{\mathcal{N}_c}^{\alpha_8} \cdot ds_c^{\alpha_9} \quad (7)$$

Thus, for $x \in \{l, v, w, m\}$, we define:

$$h_x : [0, 1]^9 \times \mathcal{S}^{n+1} \to [1; 2]$$

$$h_x(\alpha, \xi(\mathcal{N}_c)) = \begin{cases} x_c \text{ if } ds_c \in I_5 \text{ (Irreversibility)} \\ \max(\min(I_5); g_\alpha(\xi(\mathcal{N}_c))) \text{ if } c \in D_1 \text{ or } D_2 \text{ (Property 1 and 2)} \\ (x_c \cdot x_{c'})^{\frac{1}{2}} \text{ if } x = m \text{ and } \exists! \ c' \in \mathcal{L}, (c; c') \in D_3 \text{(Property 3)} \\ g_\alpha(\xi(\mathcal{N}_c)) \text{ otherwise (Normal condition)} \end{cases}$$
$$(8)$$

For a cell c, the transition function f computes in parallel each component of a cell c from its state and the states of its neighbourhood [1, 2]. We have:

$$f : \mathcal{S}^{n+1} \to S$$
$$\xi_c^+ = (f_l(\xi(\mathcal{N}_c)), f_v(\xi(\mathcal{N}_c)), f_w(\xi(\mathcal{N}_c)), f_m(\xi(\mathcal{N}_c)), p_c, f_e(\xi(\mathcal{N}_c)), o_c)$$

where
$$\begin{aligned}
f_l(\xi(\mathcal{N}_c)) &= h_l((\alpha_{l1}; \alpha_{l2}; 0; 0; 0; 0; 0; 0; \alpha_{l9}), \xi(\mathcal{N}_c)), \\
f_v(\xi(\mathcal{N}_c)) &= h_v((\alpha_{v1}; 0; 0; \alpha_{v4}; 0; 0; 0; 0; \alpha_{v9}), \xi(\mathcal{N}_c)), \\
f_w(\xi(\mathcal{N}_c)) &= h_w((\alpha_{w1}; 0; \alpha_{w3}; 0; \alpha_{w5}; \alpha_{w6}; 0; 0; 0), \xi(\mathcal{N}_c)), \\
f_m(\xi(\mathcal{N}_c)) &= h_m((0; 0; 0; 0; 0; 0; \alpha_{m7}; \alpha_{m8}; 0), \xi(\mathcal{N}_c)),
\end{aligned} \quad (9)$$

$$f_e(\xi(\mathcal{N}_c)) = \begin{cases} 1 \text{ if } \exists! \ c' \in \mathcal{L}, (c; c') \in D_3 \\ 0 \text{ if } \exists! \ c' \in \mathcal{L}, (c'; c) \in D_3 \\ e_c \text{ if not} \end{cases}$$

3 Desertification Control

In this section, we present the control theory and the notion of controllability that arises from it. An application of these concepts is made to desertification in order to stop or slow its progress. Thus, through a constructed genetic algorithm, the optimal management value, which is the predominant factor, will be determined, and this value will allow actions throughout the system to protect the land against degradation.

3.1 Control in Cellular Automata

Control theory is a branch of engineering and mathematics that deals with the behavior of dynamical systems with inputs and how to manipulate these inputs to achieve desired outputs. In particular, controllability refers to the property of a system that allows it to be controlled or steered from one state to another within a finite time using inputs. It means the ability to affect the system's behavior through appropriate inputs and to drive it towards targeted objectives. Depending on the objectives, two types of controllability are used in this work and are defined below.

Consider a CA $(\mathcal{L}; \mathcal{S}; \mathcal{N}; f)$ where $\mathcal{L}$ is the cellular space, $\mathcal{S}$ the set of cell states, $\mathcal{N}$ the neighbourhood set and f is the transition function. Let be a given time T, $\mathcal{G}_1$ and $\mathcal{G}_2$ sub-parts of space $\mathcal{L}$ which include respectively $n_{\mathcal{G}_1}$ and $n_{\mathcal{G}_2}$ cells. Let be $s_d \in S^{\mathcal{G}_1}$ a desired configuration on the space $\mathcal{G}_1$ and s_T the configuration of the space $\mathcal{G}_1$ at time T. Let be λ_s a measurement on a configuration s of $S^{\mathcal{L}}$.

We define on the space of configurations $S^{\mathcal{L}}$ a distance Δ by (Fig. 1):

$$\forall (s_1, s_2) \in S^{\mathcal{L}} \times S^{\mathcal{L}}, \Delta(s_1, s_2) = |\lambda_{s_1} - \lambda_{s_2}| \tag{10}$$

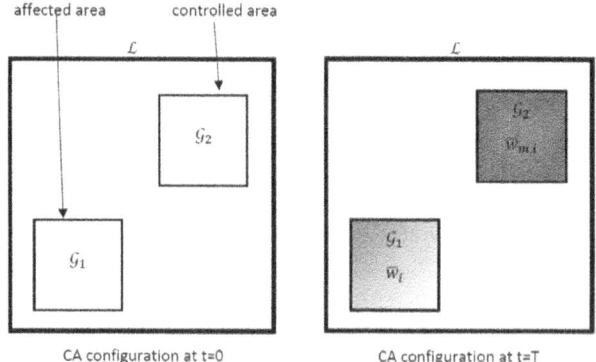

Fig. 1. Control problem

Definition 1. The CA $(\mathcal{L}; \mathcal{S}; \mathcal{N}; f)$ is regionally controllable [20] if, for $s_d \in S^{\mathcal{G}_1}$, there is a control u to apply T times on $\mathcal{G}_2$ to lead $\mathcal{G}_1$ towards a configuration s_T where:

$$s_d = s_T \text{ on } \mathcal{G}_1 \qquad (11)$$

Definition 2. The CA $(\mathcal{L}; \mathcal{S}; \mathcal{N}; f)$ is weakly regionally controllable [20] if, for $s_d \in S^{\mathcal{G}_1}$ and $\epsilon \geq 0$, there is a control u to apply T times on $\mathcal{G}_2$ to lead $\mathcal{G}_1$ towards a configuration s_T where:

$$\Delta(s_d, s_T) \leq \epsilon \text{ on } \mathcal{G}_1 \qquad (12)$$

3.2 Controlling Desertification Using DESERTICAS and Genetic algorithms

The management factor is identified as being the predominant parameter in DESERTICAS [2]. Acting positively on the management factor means acting indirectly on the other desertification factors in order to interrupt the degradation sources and slow down or stop land desertification [1]. The application of the controllability theory described above is achieved from the management parameter. Thus, the management factor is chosen as the control parameter. However, it is difficult to isolate this parameter in the model, simulations are required to determine its intensity. Therefore, we have developed and integrated a genetic algorithm into DESERTICAS whose components are described below.

Genetic Algorithms (GAs). Genetic algorithms (GAs) are numerical optimisation algorithms inspired by both natural selection and natural evolution in genetics [8] with biological terminologies of natural selection, crossover, and mutation [12]. GAs are defined by [17]:

Individual or chromosome is a potential solution to the problem to solve. A chromosome or a individual A of length L is a sequence $A = \{A_1, A_2, ..., A_L\}$, $\forall i \in [\![1, L]\!], A_i \in \{0, 1\}$. A chromosome A is therefore a sequence of bits in binary coding, also called a binary string.

Population. GAs search by simulating evolution, starting from an initial set of solutions or hypotheses called initial population and generating successive generations of solutions [23]. A population is composed of individuals.

Fitness Function. The fitness score of a sequence A is a positive value denoted $u_1(A)$, where u_1 is typically called the fitness function. The fitness function u_1 determines the ability of an individual A to compete with other individuals in the current population [13]. A decoding function $u_2 : \{0,1\}^L \to R$ is used to convert from a binary string to a real-valued digit. The fitness function u_1 is then chosen to transform this value into a positive value, i.e. $u_1 : u_2(\{0,1\}^L) \to R_+^*$. The principle of GAs is to find the sequence that maximizes or minimizes the fitness function u_1 [17].

Selection. The probability that an individual will be selected as the parent for reproduction is based on its fitness score [24]. It creates a new population of chromosomes by using an appropriate selection method (minimum or maximum) [17].

Crossover. This genetic operator takes more than one parent solution and generates children's solutions. Genes from parent chromosomes are taken, and a new offspring is produced. In detail, the operator selects a random crossover point. Everything in the binary string from the first chromosome before this point is copied and everything after this point from the second chromosome is copied to generate a new offspring [8] (Fig. 2).

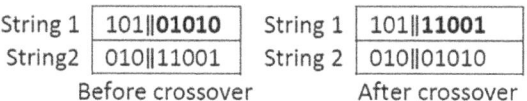

Fig. 2. Crossover operation

Mutation. In certain new offspring formed, some of their genes can be subjected to a mutation with a low random probability. This implies that some of the bits in the bit string can be flipped [24].

Termination. After that selection, crossover, and mutation have been applied to the initial population, a new population is formed and the generational counter is increased by one. These processes of selection, crossover, and mutation continue until a fixed number of generations is reached or convergence criteria are met [8].

Intensity of Management Parameter with a Genetic Algorithm. The index of each desertification factor is in the interval $[1, 2[$. Thus, each management value is a real number whose integer part is 1. The intensity of management depends on the value of its decimal part. The genetic algorithm developed in this present work consists of determining the decimal part of the management which is aggregated to the value 1 to form the management index. Typically, the decimal part of the management ranges from 0 to 99, corresponding to binary values from 0 to 1100011, respectively. So the decimal part has seven digits. For the developed genetic algorithm, each individual (or chromosome) is composed of seven genes and each gene is either 0 or 1.

Initial Population. The initial population $\mathcal{P}$ is composed of 10 individuals uniformly random. For each individual $y_i \in \mathcal{P}$, a management $w_{m,i}$ index is associated by Eq. (13): Let be r_i the decimal corresponding value of the individual y_i and k the binary number of r_i:

$$w_{m,i} = \begin{cases} 1 + \frac{r_i}{10^2} & \text{if } k = 1 \text{ or } k = 2 \\ 1 + \frac{r_i}{10^k} & \text{sinon.} \end{cases} \qquad (13)$$

Fitness Function. Let be $\mathcal{G}_1$ and $\mathcal{G}_2$ two parts of the modelled CA and T a given time. Let be $\bar{w}_d$ the desired mean Desertification Sensitivity Index(DSI) of the part $\mathcal{G}_1$ and $\bar{w}_i$ the mean DSI obtained by apply T times the management $w_{m,i}$ of the ith individual y_i of the genetic algorithm in the part $\mathcal{G}_2$. The fitness function ψ is defined by Eq. (14):

$$\psi : [1,2] \to [0,1] \\ w_{m,i} \mapsto |\bar{w}_i - \bar{w}_d| \tag{14}$$

Selection. At each iteration, three parent individuals are retained and these individuals verify Eq. (15):

$$\min_{w_{m,i}} u(w_{m,i}) = |\bar{w} - \bar{w}_d| \tag{15}$$

Crossover. The crossover is done with a proportion of 0.9, that is to say with a rate of 90%. For each pair of individuals resulting from the selection process, this consists of randomly choosing a real number between 0 and 1. If the chosen number is less than 0.9, the two individuals concerned undergo a crossover from a random crossing point.

Mutation. Some individuals resulting from the crossover carry out a mutation with a rate of $\frac{1}{k_0}$ where k_0 is the bit numbers of each y_i. The bit j of the new individual obtained by mutation from individual y_i is (Fig. 3):

$$y_i[j] = 1 - y_i[j] \tag{16}$$

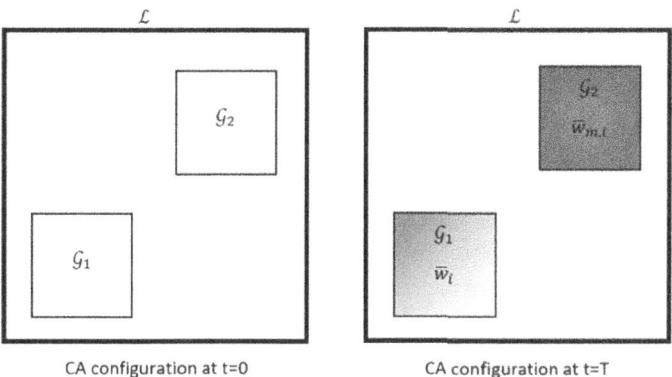

Fig. 3. Developed control problem

New Population Generation. At each iteration, a new population $\mathcal{P}$ is generated with individuals successively resulting from the processes of selection, crossing, and mutation. Indeed, the individual numbers of each population generated vary according to the number of individuals retained after the selection process and according to the individual numbers resulting from the crossover and the mutation.

Termination. The convergence criterion of the GAs developed in this study depends on the type of controllability we aim to implement, with the management factor as the control parameter.

- The developed CA is regionally controllable if, for $\bar{w}_d$ there is an individual $y_j \in \mathcal{P}$ with a management index $w_{m,j}$ applied T times on $\mathcal{G}_2$ given $\bar{w}_j$ a mean DSI on the part $\mathcal{G}_1$, where:

$$\psi(w_{m,j}) = 0 \qquad (17)$$

where ψ is defined by Eq. (14) and y_j is given by GA previously defined.
- The developed CA is regionally controllable if, for $\bar{w}_d$ there is an individual $y_j \in \mathcal{P}$ with a management index $w_{m,j}$ applied T times on $\mathcal{G}_2$ given $\bar{w}_j$ a mean DSI on the part $\mathcal{G}_1$ where:

$$\psi(w_{m,j}) \leq \epsilon \text{ with } \epsilon \geq 0 \qquad (18)$$

Thus, the convergence criterion of the developed GA is defined by Eq. (17) or Eq. (18) according to the type of controllability implemented.

Initial Configuration. The simulations are achieved from the same initial configuration in Fig. 4 representing the studied area. It is divided into four parts where the bottom right part corresponds to a random distribution of desertification factor indexes. The bottom left corresponds to moderate indexes, the top right corresponds to low indexes of the factors and the top left corresponds to high indexes. The cells in red color correspond to the **controlled area**. It corresponds to the part of CA grid on which a control is applied through the management factor. The cells in black color, central to the red cells, correspond to the **affected area**. It corresponds to the part of CA grid where the impact of the control is observed. Thus, in the **controlled area**, different land protection measures will be applied through the management factor index. These measures will have indirect effects on the **affected area**. In the case of our study, the **affected area** will be isolated from other parts of the study area through the **controlled area**. All the factors are assumed to have the same weight $\alpha_i = 1$ in order not to influence the predominance. The considered neighbourhood is Moore's order of 1. The parts of the study area are considered to be exploited: the exploitability factor index is equal to 1 for all CA cells.

3.3 Results and Analysis

In Fig. 5, the initial state of **affected area** is given by the mean DSI index $w = 1.41$. It corresponds to the geometric mean of all cells in the **affected area**. The objective of the study is to search for a management value, using our genetic algorithm, to be applied to the **control area** in order to reduce the

Fig. 4. Initial state of studied area

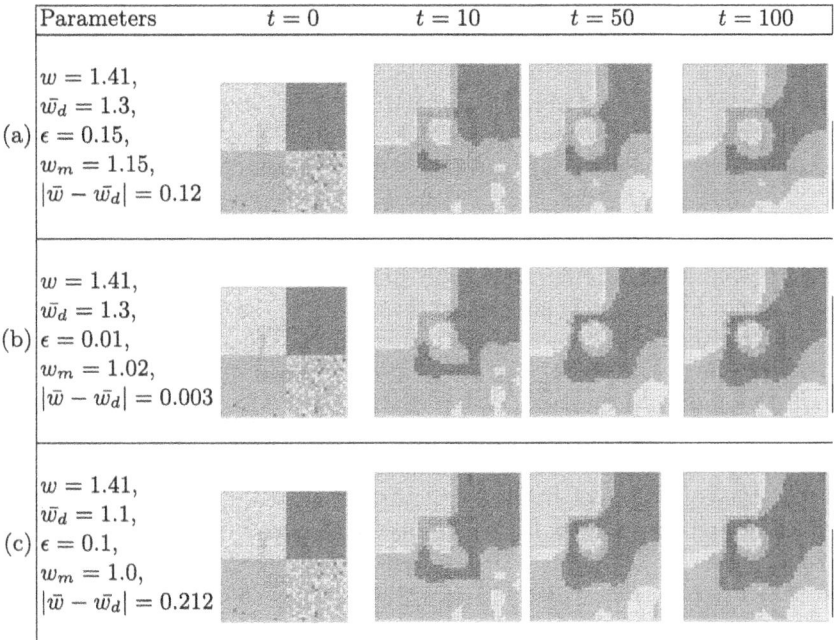

Fig. 5. Evolution of the same area with the management as control parameter

degradation process of the **affected area**. The mean DSI indexes of **affected area** should subsequently be lower by its initial value after this control process and must approach or reach a desired value $\bar{w}_d$.

In Fig. 5(a), the desired mean DSI is $\bar{w}_d = 1.3$ and this value must be approximated to $\epsilon = 0.15$. The implementation of the genetic algorithm allows us to reach this value from $t = 100$ with a management value on **the control area** $w_m = 1.15$. Thus, the fitness functions reaches $\min_{w_{m,i}} \psi(w_{m,i}) = 0.12 < \epsilon$.

This corresponds to a weak regionally controllability case. The genetic algorithm, responsible for determining the value of the control, converges after one iteration. The difference between the desired state and the initial state of the study area is not large enough. Thus, the weak regionally controllability allows for improvement in the state of the area. Indeed, by putting protective measures around an area, the conditions of the central area have been improved and mitigate its degradation process.

In Fig. 5(b), by choosing $\epsilon = 0.01$, the desired state is approached at $t = 100$ and that from 2 iterations. Thus, the value of the fitness function is 0.003 for a management value $w_m = 1.02$. This corresponds practically to a regionally controllability. Indeed, protective actions in the **controlled area** allow to positively improve the condition of the **affected area** and achieve the desired state. This improvement is done without changing the state of the desertified cells which is the degradation threshold. The other cells undergo the positive impacts of the **controlled area** and improve their condition through the neighbouring effects.

In Fig. 5(c), the desired mean DSI is $\bar{w}_d = 1.1$ and this value must be approximated to $\epsilon = 0.1$. From $t = 100$, the genetic algorithm reaches the minimal value of management $w_m = 1.1$ and converges to the corresponding value of the fitness function $\min_{w_{m,i}} \psi(w_{m,i}) = 0.212$. Beyond $t = 100$, the algorithm has almost converged and the values of the fitness function remain identical. This is explained by the fact that the **affected area** contains desertified cells whose conditions cannot be improved by simple control actions. In fact, neighbourhood actions cannot have any real impacts on these cells. In this context, it is necessary to act directly on these cells through restoration or rehabilitation actions. The minimum management value, given by the genetic algorithm, corresponds to the introduction of strong measures in the control area. This has a positive impact on non-desertified cells in the affected area and other neighbouring areas. At the end of this process, the mean index of the cellular automata grid is improved. However, the convergence of the genetic algorithm does not make it possible to reach the desired mean index in the **affected area** but it allows us to reach it at a certain value of ϵ. In this case, it is a weak regionally controllability situation.

In the three cases presented in Fig. 5, the study of controllability is achieved by the action on non-desertified cells. Indeed, the states of the cells in the **affected areas** are improved from the states of neighbouring cells located in the **controlled area**. Thus, the neighbourhood effects improve cell states and subsequently improve the average state of the **affected area**. Although the aim of the study is not the improvement of their states, the areas not located in the **affected area** and in the **controlled areas** also see their states improved when the control value is low.

4 Conclusion

This study highlights the impact of management, that is to say, human actions, on the desertification process. Indeed, acting on management amounts to acting implicitly on all desertification factors and slowing down/accelerating its impacts. This allows to integrate actions into the achieved desertification modelling by acting on the intensity of management.

This study introduces the controllability of the achieved modelling. Indeed, when the difference between the desired state and the initial state of an area is not too great, controllability is achievable. Indeed, the control action allows us to reach the desired state and improve the desertification mean state of an area. It only makes sense if we desire a slight improvement in the state of the study area by putting protective measures in its neighbourhood. Thus, non-desertified cells see their conditions improved and will contribute to the achievement of the improvement objective. However, the control process cannot improve the conditions of desertified cells. On these degraded cells, it is necessary to act directly through actions of rehabilitation, restoration and reassignment which require enormous resources.

In the end, this study shows that genetic algorithms are a good candidate for determining the intensity of the control. We have implemented one into the decision-support tool DESERTICAS. Through different tests, it can determine the intensity of the desired management which represents the control operator.

However, the desertified cells lead to the slow convergence of the algorithm after many iterations because their states cannot be modified and contribute to stabilize the mean state of the study area. Also, the determination of a low state for the study area requires the implementation of the genetic algorithms for a long time. From this perspective, the developed model should be optimized to overcome this insufficiency.

References

1. Koné, A., Fontaine, A., El Yacoubi, S., Loireau, M., Jangorzo, S.N.: Assessing the impact of anthropogenic factors on desertification through deserticas software. In: International Conference Cellular Automata for Research and Industry (ACRI) (2021)
2. Koné, A., Fontaine, A., El Yacoubi, S., Loireau, M., Jangorzo, S.N.: Deserticas, a software to simulate desertification based on medalus and cellular automata. In: International Conference Cellular Automata for Research and Industry (ACRI), pp. 198–208 (2021)
3. Koné, A., Fontaine, A., El Yacoubi, S.: Coupling cellular automata with medalus assessment for the desertification issue. In: The International Conference on Emerging Trends in Engineering & Technology (IConETech) (2020)
4. Chopard, B., Droz, M.: Cellular Automata Modeling of Physical Systems. Collection Ale'a-Saclay: Monographs and Texts in Statistical Physics, Cambridge (1998)
5. Contreras-Bolton, C., Parada, V.: Automatic combination of operators in a genetic algorithm to solve the traveling salesman problem. PLOS ONE **10**(9), 1–25 (2015)
6. Kosmas, C., Tsara, M., Moustakas, N., Karavitis, C.: Identification of indicators for desertification. Ann. Arid Zone **42**, 393–416 (2003)
7. Ying, C., Yong-jie, M., Wen-xia, Y.: Application of improved genetic algorithm in PID controller parameters optimization. TELKOMNIKA Indonesian J. Electr. Eng. **11**, 1524–1530 (2013)

8. Coley, D.A.: An Introduction to Genetic Algorithms for Scientists and Engineers. World Scientific Publishing Co., Inc., USA (1998)
9. Boudjemline, F., Semar, A.: Assessment and mapping of desertification sensitivity with medalus model and GIS - case study: basin of hodna, Algeria. J. Water Land Dev. 17–26 (2018)
10. Lahlaoi, H., Rhinane, H., Hilali, A., Lahssini, S., Moukrim, S.: Desertification assessment using medalus model in watershed oued el maleh, morocco. Geosciences **7**, 2–16 (2017)
11. D'Herbès, J.M., Fezzani, C.: Indicateurs écologiques roselt/oss: Une première approche méthodologique pour la surveillance de la biodiversité et des changements environnementaux. Collection ROSELT/OSS (2004)
12. Höschel, K., Lakshminarayanan, V.: Genetic algorithms for lens design: a review. J. Opt. **48**(1), 134–144 (2018)
13. Namita, K., Anju, R.: Genetic algorithm: a search of complex spaces. Int. J. Comput. Appl. **25**(7), 13–17 (2011)
14. Benslimane, M., Hamimed, A., Wael, E.Z., Abdelkader, K., Mederbal, K.: Analyse et suivi du phénomène de la désertification en algérie du nord. VertigO (2009)
15. Loireau, M., et al.: Guide for the evaluation and monitoring of natural resource exploitation practices roselt/oss programme. ROSELT/OSS Collection (2005)
16. Loireau, M., et al.: Local environmental information system to assess the risk of desertification: circumsaharan compared situations (roselt network). Sci. Planet. Change/Drought **18**, 328–335 (2007)
17. Yildizoglu, M., Vallée, T.: Présentation des algorithmes génétiques et de leurs applications en économie. Revue d'Economie Politique (2004)
18. Dódorico, P., Bhattachan, A., Davis, K.F., Ravi, S., Runyan, C.W.: Global desertification: drivers and feedbacks. Adv. Water Resour. **51**, 326–344 (2013)
19. Shoba, P., Ramakrishnan, S.S.: Modeling the contributing factors of desertification and evaluating their relationships to the soil degradation process through geomatic techniques. Solid Earth **7**, 341–354 (2013)
20. Dridi, S., El Yacoubi, S., Bagnoli, F.: Recent advances in regional controlability of cellular automata. Thesis (2019)
21. El Yacoubi, S., El Jai, A.: Cellular automata modelling and spreadability. Math. Comput. Model. **36**, 1059–1074 (2002)
22. secrétariat de la Convention des Nations unies sur la lutte contre la désertification (CNULD). Désertification: une synthèse visuelle. Centre international UNISFERA (2011)
23. Mathew, T.V.: Genetic algorithm. Indian Institute of Technology Bombay (2009)
24. Mallawaarachchi, V.: Introduction to genetic algorithms. Comput. Genom. (2017)
25. Li, X., Wang, Z., Sun, Y., Zhou, S., Xu, Y., Tan, G.: Genetic algorithm-based content distribution strategy for F-RAN architectures. ETRI J. **41**, 348–357 (2019)
26. Xue, X., Tsunekawa, A., King-Okumu, C.: Editorial: desertification and rehabilitation. Front. Environ. Sci. **10**, 874963 (2022)

Desertification Control Strategies: A Hybrid Approach Using Cellular Automata and Reinforcement Learning

Amira Mouakher[1(✉)], Alassane Kone[1,2], Allyx Fontaine[2], and Samira El Yacoubi[1]

[1] Espace-Dev UMR 228 UPVD, IRD, UM, UA, UG, Perpignan, France
{amira.mouakher,yacoubi}@univ-perp.fr, alassane.kone@ird.fr
[2] Espace-Dev UMR 228 UG, IRD, UM, UA, UPVD, Guyana, France
allyx.fontaine@univ-guyane.fr

Abstract. This paper presents a novel hybrid approach for desertification control that leverages the strengths of cellular automata (CA) modeling and reinforcement learning (RL). We employ the DESERTICAS software, a specifically designed CA model for simulating desertification dynamics. The model incorporates a variety of factors influencing land degradation, including those from the MEDALUS model, fundamental desertification properties, land-use practices, exploitability, and management. Our key contribution is to introduce a control parameter within the DESERTICAS framework. This allows us to formulate desertification control as an input-output problem and apply control theory principles to CA models. By manipulating the average intensity of a dominant factor (identified as management in this study), we can indirectly influence all other factors and potentially decelerate or even halt land degradation processes. Furthermore, we integrate a Reinforcement Learning (RL) agent into the simulation environment. This virtual entity continuously explores different management strategies, dynamically adjusting its actions based on the observed outcomes. This combination of CA modeling and RL constitutes a hybrid approach to desertification control. The experimental results show promising outcomes, with the inclusion of the RL agent leading to a significant reduction in desertified regions. This study paves the way for further exploration of hybrid CA-RL techniques for environmental applications.

Keywords: Cellular automata · Reinforcement learning · Hybrid approach · Desertification control · Management factor

1 Introduction

Desertification, the degradation of land in dry areas (arid, semi-arid, and dry sub-humid environments) [17], is a significant threat arising from the combined effects of human activity and climate fluctuations [17]. Dryland ecosystems, which cover over one-third of the Earth's surface, are particularly vulnerable due to their inherent fragility under excessive use and improper management practices [21]. The environmental, economic, and social consequences of desertification are profound. It compromises regional ecological security and hinders national economic development by reducing soil fertility and degrading vegetation cover [10,26]. The escalating severity of desertification demands immediate attention. Projections estimate that 1.8 billion people will face absolute water scarcity by 2025, and two-thirds of the global population will experience water stress [23]. Additionally, desertification is projected to displace 135 million people by 2045, while over 75% of the Earth's land area is already degraded, with a potential to exceed 90% by 2050 [23]. Therefore, accurate monitoring of desertification dynamics is crucial to understanding its progression and developing frameworks for future prevention.

While previous research has primarily focused on desertification assessment and prediction to capture its driving forces [1,4,8,24,27], this study takes a novel approach by exploring active intervention within the system to mitigate its detrimental effects. This work builds upon the DESERTICAS software [14], which, through a comparative analysis of all factors influencing desertification, highlighted the predominant role of human actions. Our core innovation lies in adapting and applying concepts from systems theory to our cellular automata (CA) model, with the potential to enhance the capabilities of DESERTICAS tool. We focus on two critical aspects directly related to desertification: spreadability analysis, which investigates the expansion patterns of desert zones, and control mechanisms to impede such expansion.

Our approach leverages reinforcement learning to identify the optimal level of intervention for a dominant factor by manipulating its average intensity through rewards and penalties. This will indirectly influence all other factors within the model, potentially slowing down or even halting land degradation processes. Since the Enhanced Model of Desertification identifies management as a key factor, utilizing reinforcement learning to optimize management strategies offers significant promise. Ultimately, we aim to integrate reinforcement learning into the DESERTICAS model to identify and implement effective land protection strategies, transforming it into a valuable decision-support tool.

The paper is structured as follows: Sect. 2 establishes a strong foundation by providing comprehensive definitions of the key concepts used in our approach. Section 3 contextualizes our work by reviewing recent advancements in desertification assessment and monitoring tools. This section then elaborates on the challenges associated with desertification propagation and introduces the methodology employed for mitigation. Section 4 showcases the results achieved through our approach. Finally, Sect. 5 concludes the paper by summarizing the key findings and outlining potential avenues for future research.

2 Preliminaries

This section provides essential background concepts for understanding the proposed approach. We formally define cellular automata (CA), spreadability within the context of dynamical systems and CA, and the concept of reinforcement learning.

Definition 1 (Cellular automata). *A cellular automaton is a mathematical system that models how a grid of cells evolves over time. It is defined as a quadruple $\mathscr{A} = (\mathcal{L}, \mathcal{S}, \mathcal{N}, f)$, where:*

- *$\mathcal{L}$ is the lattice of cells whose dimension is $d \in \mathbb{N}$. Thus, $\mathcal{L} = \mathbb{Z}^d$.*
- *$\mathcal{S}$ is a finite set of possible state values for the cells.*
- *$\mathcal{N}$ is the function that specifies the neighborhood of cell c which locally determines the interaction of cell c with its environment. This neighborhood is usually the same for all cells but it can vary through space and time [6, 20]. The neighborhood function employs the following mathematical notation:*

$$\begin{aligned} \mathcal{N} : \mathcal{L} &\to \mathcal{L}^n \\ c &\mapsto \mathcal{N}(c) = \{c_1, c_2, ..., c_n\} \end{aligned} \quad (1)$$

where the cells c_i, $i = 1, ..., n$ are the neighboring cells and n is the size of $\mathcal{N}(c)$. In most cases, the cell c is in its neighborhood.

- *f is the transition function that calculates the state of a cell at time $t + 1$ using the state of the neighboring cells at time t. The transition function is written as:*

$$\begin{aligned} f : \quad \mathcal{S}^n &\to \quad \mathcal{S} \\ s_t(\mathcal{N}(c)) &\mapsto f(s_t(\mathcal{N}(c))) = s_{t+1}(c) \end{aligned} \quad (2)$$

Where $s_t(c)$ is the state of cell c at time t and $s_t(\mathcal{N}(c)) = \{s_t(c'), c' \in \mathcal{N}(c)\}$ is the state of the neighborhood of c.

This paper delves into the concept of regional control within our proposed approach. The central challenge lies in identifying the optimal control strategy to be applied within a designated area, denoted by ω_{ctrl}. This control aims to achieve a desired configuration within a controlled region, denoted by $\omega_d = \{c_1, \ldots, c_n\}$, such that the state of each cell c_i in ω_d at time T, denoted by $s_d(c_i)$, aligns with a predetermined desired state, $s_T(c_i)$. This objective is visualized in Fig. 1.

Definition 2 (Controllability in cellular automata). *A cellular automaton is considered regionally controllable for a desired configuration $s_d \in S^{\omega_d}$ if there exists a control strategy $u = (u_0, \ldots, u_{T-1}) \in \mathcal{U}$ applicable to the designated area ω_{ctrl} over T time steps, such that the following condition holds:*

$$s_T(c_i) = s_d(c_i) \quad \forall c_i \in \omega_d, \quad (3)$$

where s_T denotes the final configuration of the CA at time T and S^{ω_d} represents the set of all possible configurations for the cells within the controlled region ω_d. The control strategy u belongs to a set of admissible control inputs denoted by $\mathcal{U}$.

In simpler terms, regional controllability implies the existence of a control strategy that can be applied to a specific area of the CA (ω_{ctrl}). This strategy,

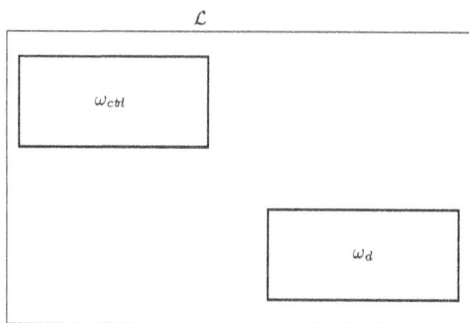

Fig. 1. Regional controllability in a CA.

implemented over a finite number of time steps (T), effectively steers the states of cells within the controlled region (ω_d) to achieve a predetermined desired configuration (s_d).

Definition 3 (Reinforcement learning). *A task is defined [11] by a set of states, $s \in \mathcal{S}$, a set of actions, $a \in \mathcal{A}$, a state-action transition function, $T : \mathcal{S} \times \mathcal{A} \rightarrow \mathcal{S}$, and a reward function, $R : \mathcal{S} \times \mathcal{A} \rightarrow \mathbb{R}$. At each time step, the learner (also called the agent) selects an action, and as a result, is given a reward and its new state. The goal of reinforcement learning is to learn a policy, a mapping from states to actions; the policy is defined as $\pi : \mathcal{S} \rightarrow \mathcal{A}$, which maximizes the sum of its reward over time. We use the infinite-horizon discounted model, where reward over time is a geometrically discounted sum in which the discount, $0 \leq \gamma < 1$ devalues rewards received in the future. Accordingly, when following policy π, we can define the value of each state to be:*

$$V^\pi(s) = \sum_{t=0}^{\infty} \gamma^t r_t \qquad (4)$$

where r_t is the reward received t time steps after starting in state s and following policy π. The optimal policy, denoted as π^*, is the one that maximizes the value, $V^\pi(s)$, for all states s. To learn the optimal policy, we learn its value function, V^*, and its more specific correlate called Q. Let $Q^*(s,a)$ be the value of selecting action a from state s, and thereafter following the optimal policy. This is expressed as:

$$Q^*(s,a) = R(s,a) + \gamma V^*(T(s,a)). \qquad (5)$$

We can now define the optimal policy in terms of Q by selecting from each state the action with the highest expected future reward:

$$\pi^*(s) = \arg\max_a Q^*(s,a). \qquad (6)$$

3 Controlling the Desertification Spread

This section establishes the context by reviewing recent developments in desertification assessment and monitoring tools. It then explores the challenges of desertification spread and introduces the hybrid methodology we propose based on cellular automata and reinforcement learning.

3.1 Related Work

Desertification presents a significant challenge due to its spatio-temporal complexity [25]. The multifaceted nature of this process makes quantification difficult. The attribution of desertification to climate variability, climate change, and human activities varies considerably across space and time. Furthermore, the extent of drylands is not static but can fluctuate based on environmental conditions and, importantly, land use practices. Effective desertification modeling necessitates not only the incorporation of these diverse factors, but also the potential to implement actions aimed at mitigating their impact on global well-being.

The existing literature primarily focuses on land rehabilitation and restoration strategies to combat desertification [3,18]. Several studies address ecological rehabilitation through engineered projects aimed at minimizing environmental impacts on various land types [3,9]. These projects target farmlands, deforested areas, transportation infrastructure, and water conservation efforts threatened by sand and dust storms, as well as other desertification processes. Examples include the rehabilitation of mobile sand dunes and the implementation of water conservation techniques, which have demonstrably improved land conditions in specific regions [16]. Large-scale restoration efforts often combine enrichment planting of native woody and fodder grass species with extensive land preparation techniques to enhance rainwater harvesting and soil permeability [22]. This combined approach promotes long-term land restoration and mitigates the effects of desertification.

Several innovative approaches have emerged in recent years. Expert systems, like the Desertification Risk Assessment Support Tool (DRAST) [12], incorporate comprehensive sets of physical, social, and economic indicators. DRAST guides users in selecting the most pertinent factors for a specific region, facilitating focused and efficient risk assessment. Unlike cumbersome, all-encompassing data approaches, DRAST prioritizes relevant indicators, enabling a holistic understanding of desertification drivers and informed decision-making about land management practices. Additionally, DRAST allows exploration of various strategies and their impact on desertification risk, aiding in the selection of the most suitable mitigation approach for a particular region. In addition to expert systems, other tools leverage complex models. For instance, Assenato et al. presented a scientific approach to combating land degradation, drawing on experiences from six pilot sites within the NL4DL LIFE project [2]. By analyzing both field and satellite data from these southern European locations, the project aimed to develop a transferable method applicable to similar degraded areas. The project's key contribution lies in a decision-making tool that provides a reference procedure for monitoring restoration activities utilizing nature-based solutions. This tool guides users through identifying the specific land degradation processes and selecting the most suitable monitoring indicators and nature-based solutions for the particular degraded land in question. Recently, another study used the MEDALUS model to assess desertification sensitivity within a Moroccan watershed [19]. The authors' objective is to evaluate the susceptibility of different areas to desertification. Their findings reveal that critical and highly sensitive zones encompass 44% of the watershed, primarily concentrated in the northeast and western extremities. Leveraging land use data alongside these results, the authors propose potentially impactful development actions to mitigate the effects of desertification in the identified vulnerable regions.

3.2 DESERTICAS and Spreadability Concept

This paper investigates the desertification modeling framework DESERTICAS [14], which integrates cellular automata with the MEDALUS assessment developed by the European Union. The MEDALUS model quantifies land degradation based on four key factors: soil, vegetation, climate, and management [15]. In this study, the authors have incorporated the MEDALUS model into a continuous cellular automaton framework to capture the spatio-temporal dynamics of desertification. Additionally, anthropogenic factors have been included to enhance the realism of the model.

Table 1. The desertification sensitivity index states.

Class of weight	Desertification state
$I_5 = [1.78; 2]$	Very-degraded
$I_4 = [1.53; 1.78[$	Degraded
$I_3 = [1.38; 1.53[$	High
$I_2 = [1.22; 1.38[$	Moderate
$I_1 = [1; 1.22[$	Low

The Desertification Sensitivity Index (DSI) quantifies the degree of land degradation. It is defined as the geometric mean of four quality indexes: soil quality (l), vegetation cover (v), climatic conditions (w), and land management practices (m). This comprehensive approach allows the DSI to capture the combined influence of these factors on desertification risk. DSI values range from 1 (least sensitive) to 2 (most sensitive). The corresponding quality states associated with these values are detailed in Table 1. The mathematical expression for DSI is given by Eq. 7:

$$DSI = (s_l \times s_v \times s_w \times s_m)^{\frac{1}{4}} \quad (7)$$

Research using the DESERTICAS model has revealed a concerning trend: desertification tends to propagate over time in areas already experiencing it.

We have introduced the concept of "spreadability" to understand this spatial expansion. Spreadability is focused on identifying the dynamics that govern the spatial and temporal increase of a specific property within a distributed parameter system. This concept aligns with the observation that subdomains where a particular spatial property is enforced (e.g., desertification) tend to continuously expand over time [7].

Definition 4 (Spreadability of cellular automata). *Let $\mathscr{A}$ be CA defined by $\mathscr{A} = (\mathcal{L}, \mathcal{S}, \mathcal{N}, f)$ and $\mathcal{P}$ a given property defined by the relationship Eq. (8).*

$$\mathcal{P}s_t(c) \Leftrightarrow s_t(c) = s \tag{8}$$

with $s \in \mathcal{S}$ corresponding to the desired profile and the state s is assumed to be attainable from time t_0. Consider Eq. (9):

$$\theta_t = \{c \in \mathcal{L} \mid \mathcal{P}s_t(c)\} \tag{9}$$

The CA $\mathscr{A}$ is said to be $\mathcal{P}$-spreadable from $\theta_{t_0} = \{c \in \mathcal{L} \mid \mathcal{P}s_{t_0}(c)\}$ if $\theta_t \subset \theta_{t+1}, \forall t \geq t_0$.

Definition 5 (Spreadability in measure). *The CA $\mathscr{A}$ is said to be spreadable in measure or $\mathcal{A}$-spreadability during the time interval $I = [t_0, T]$ if Eq. (10) is verified.*

$$mes(\theta_{t'}/\theta_t) \leq mes(\theta_t/\theta_{t'}) \ \forall t, t', \ t_0 \leq t' \leq t \leq T \tag{10}$$

where $mes(x)$ is the Lebesgue measure of x. When it comes to using the number of elements of the set x, $|x|$ will be used.

The above definition simply means that the surface gained, i.e. the number of cells gained, during spreading is greater than that lost during the time interval $[t', t]$, where t is greater than or equal to t'.

Definition 6 (Spreadability of desertification). *The areas subject to the desertification process in DESERTICAS have states belonging to the interval I_5 corresponding to desertified states (cf. Table 1). Let $\mathscr{A}$ be CA modelling the area subject to the degradation process and $\mathcal{P}$ a property of land desertification defined by the relationship Eq. (11).*

$$\mathcal{P}s_t(c) \Leftrightarrow s_t(c) \in I_5 \tag{11}$$

The CA $\mathscr{A}$ is said to be spreadable in desertification measure if $|\theta_t| < |\theta_{t+1}|, \forall t \geq t_0$. $|\theta_t|$ is the number of cells that satisfy the property $\mathcal{P}$ at time t (desertified areas). So, the desertified areas grow considerably.

Figure 2 depicts a concerning trend: the number of *desertified* cells exhibits a non-decreasing behavior throughout the time interval $[0, 120]$. This observation aligns with the concept of desertification spreadability described in Eq. (11). Essentially, the figure reveals that as the monitored degradation property expands, cells in good states steadily decline in number, while *desertified* cells experience a significant increase. This pattern confirms the presence

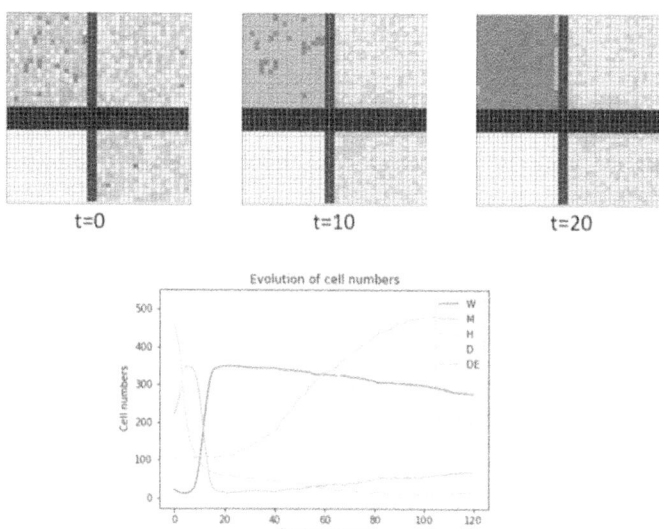

Fig. 2. Non-decreasing of the number of *desertified* cells in the considered study area (W = *Weak*, M = *Moderate*, H = *High*, D = *Degraded* and DE = *Desertified*)

of desertification spreadability within our model, evidenced by the continuous expansion of *desertified* areas.

Our work will focus on deploying appropriate control strategies within the system to either halt or significantly slow down this expansion. We aim to prevent the complete degradation of all cells in the study area and avoid its transformation into a fully desertified landscape. Research by [13] has highlighted the critical role of land management practices. Interventions in this domain can influence not only individual factors but also the system as a whole, potentially leading to desertification mitigation. To curb the observed expansion, we will leverage the theory of controllability and investigate the impact of management strategies using reinforcement learning.

3.3 The Proposed Approach

Desertification arises from a complex interplay of factors that vary both temporally and spatially. These factors include indirect influences such as population growth, socio-economic conditions, policy frameworks, and international trade dynamics, as well as direct drivers like land-use practices and climate-related processes [15].

Our model leverages spatial and temporal data to drive a dynamic simulation of desertification. We categorize this data into five key groups-soil properties, vegetation cover, climatic conditions, exploitability, and current management practices-to initialize and update the states of a cellular automaton across a simulated landscape. At each time step, the CA reevaluates the state of each cell based on its current environmental conditions. Here's where a crucial innovation comes into play: a reinforcement learning agent actively participates in the simulation. This virtual agent continuously explores different management strategies, dynamically adjusting practices based on the immediate outcomes of its actions. We measure the effectiveness of these strategies based on their impact on desertification. A Q-table stores a reward system that enables the agent to learn. The agent rewards actions that slow down desertification, such as improved grazing practices or targeted soil conservation efforts. Conversely, actions that accelerate desertification incur penalties. This feedback loop allows the agent to prioritize optimal management strategies over time, leading to more efficient resource allocation.

The inclusion of reinforcement learning in our desertification model provides significant benefits. Firstly, it enables the agent to dynamically adapt its strategies based on the specific context of each region within the simulation. This dynamic approach targets management efforts at areas most susceptible to desertification, thereby maximizing their effectiveness. Second, the model promotes resource optimization by focusing on optimal strategies. We can direct scarce resources towards interventions that have the highest potential for success, potentially reducing overall costs while achieving greater environmental impact. Finally, the simulation serves as a valuable decision-support tool for policymakers and land managers. By providing insights into how strategic modifications to land management practices can positively influence the environment, the model empowers stakeholders to make informed decisions about combating desertification.

4 Experimental Results

This section details the experimental setup, including the characteristics of the used dataset. It then presents the results obtained by applying our proposed approach, followed by a discussion.

4.1 Data

The DESERTICAS software generates the data used in this study. The process begins with initializing the land state, which forms the foundation of the simulation. This involves setting the initial state of each cell within the study area. Utilizing the graphical interface depicted in Fig. 3, we can generate simulated data for the quality indices of the four key desertification factors (soil, vegetation, climate, and management). At the end of this step, we either generate or import land type, exploitability, and ownership factors. Let's denote the initial quality indices of soil, vegetation, climate, and management for cell c at time $t = 0$ as $l_0(c)$, $v_0(c)$, $w_0(c)$ and $m_0(c)$ respectively. Additionally, let $p_0(c)$, $e_0(c)$ and $o_0(c)$ represent the initial land type, exploitability and ownership factors for the same cell. The combined set of these initial values in DESERTICAS implicitly initializes the state, denoted by $\xi(c)$ of cell c at $t = 0$ as:

$$\xi_0(c) = (l_0(c), v_0(c), w_0(c), m_0(c), p_0(c), e_0(c), o_0(c)) \tag{12}$$

This equation represents a comprehensive state vector encompassing all relevant factors influencing the initial condition of each cell within the simulation.

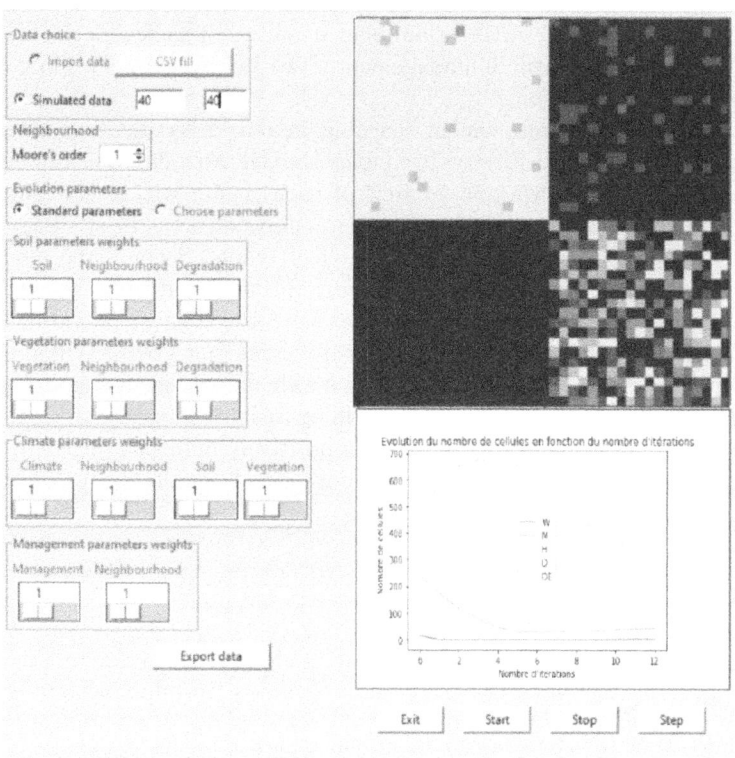

Fig. 3. The graphical interface of DESERTICAS.

4.2 Results and Discussion

This study employs Q-learning [5], a reinforcement learning (RL) technique, to train a model for minimizing desertification risk. The model evaluates desertification zones at three distinct time steps: $t = 10, t = 50, and t = 100$. Desertification risk is calculated using Eq. 7.

The learning process starts with the agent's random selection of an initial action-either increasing or decreasing the management factor. After that, the agent looks at the current state of the environment, which probably includes seeing how the desertification risk values are spread out across the designated area, as shown in Fig. 4. The agent then calculates a reward based on the difference between the previous and current desertification risks. A smaller difference, signifying improvement, translates to a higher reward for the agent. Utilizing this reward and its learned Q-values (expected future rewards for specific actions in specific states), the agent selects the next action (increase or decrease management) that it predicts will lead to the most significant reduction in desertification risk over time. Through this iterative process of action selection, reward evaluation, and state observation, the agent learns from its experiences (state, action, reward, and new state). This learning process allows the agent to update its Q-values, continuously refining its decision-making strategy.

As illustrated in Fig. 4, the desertification risk exhibits a decreasing trend across the three-time steps ($t = 10, t = 50, t = 100$). This trend demonstrates the Q-learning agent's efficacy in learning a policy-a mapping from states to actions-that effectively mitigates desertification risk over time.

To assess the efficacy of reinforcement learning integration, we conducted experiments with a random strategy. Here, the agent disregards the reward and acts randomly. This essentially represents a model without any control strategy. Figure 5 reveals that the desertification risk exhibits a less favorable trajectory over time compared to the Q-learning approach. The contrasting behavior observed in Figs. 4 and 5 underscores the importance of incorporating reinforcement learning for desertification mitigation.

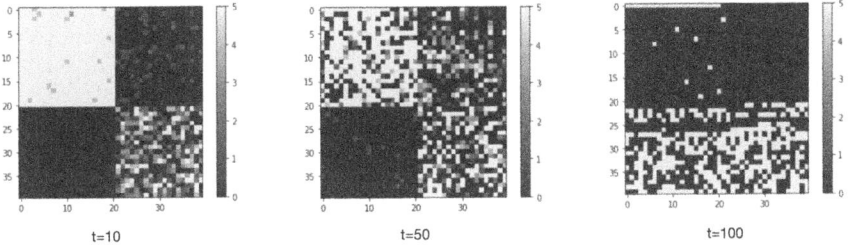

Fig. 4. Reinforcement Learning with a Q-Learning strategy.

The Q-learning curve in Fig. 6 depicts the agent's performance using the Q-learning reinforcement learning technique. As previously described, Q-learning allows the agent to learn a policy-a mapping from states to actions-that minimizes desertification risk. The decreasing trend in Fig. 6 confirms the effectiveness of the Q-learning agent in learning to reduce desertification risk over time. In contrast, the random strategy curve in the same figure shows a slight decrease over time, but it clearly performs below the level of the Q-learning approach. This highlights the limitations of random actions in effectively mitigating desertification. For instance, as shown by Fig. 6, the initial number of desertified cells

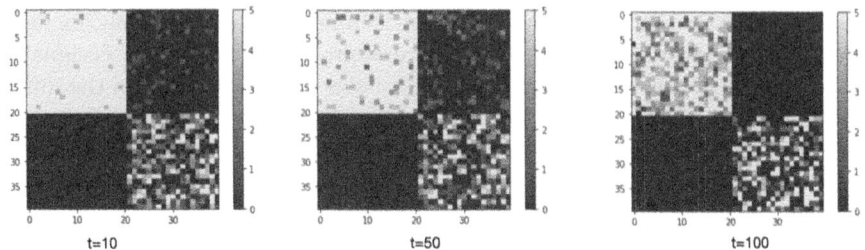

Fig. 5. Reinforcement Learning with a Random strategy.

is 600. Under the Q-learning strategy, this number can be reduced to 200 after 100 iterations. Conversely, the random strategy might only achieve a reduction of around 530 for the same time frame. This highlights the Q-learning strategy's superiority in demonstrably outperforming the random strategy for desertification risk reduction over time.

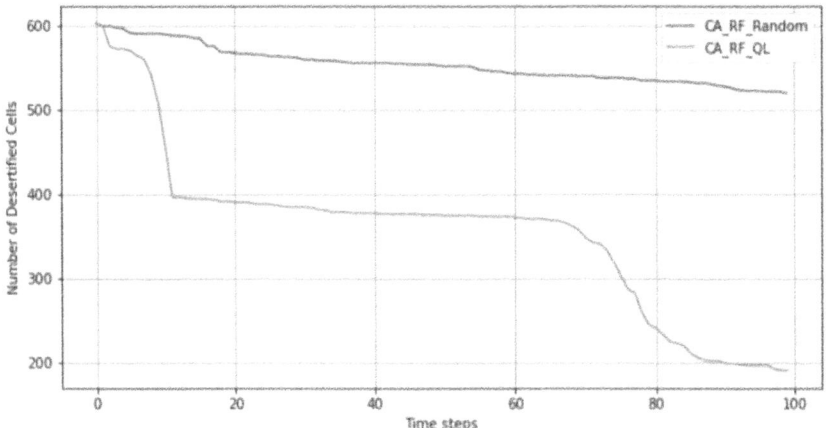

Fig. 6. The evolution of risk desertification over time.

5 Conclusion

This study introduced a novel hybrid approach for desertification control that leverages the strengths of both cellular automata modeling and reinforcement learning. The DESERTICAS CA model, specifically designed for desertification simulations, served as the foundation for incorporating control strategies. We introduced a control parameter within the DESERTICAS framework, allowing us to formulate desertification control as an input-output problem and apply control theory principles. By manipulating the management factor, we demonstrated the potential to influence other desertification factors and potentially decelerate or halt land degradation. In the near future, we plan to pay heed to the development of a Hybrid Digital Twins (HDT) framework for better monitoring of the desertification of the target areas. We model the behavior of target entities using a hybrid model that combines domain-specific physics models and a machine learning approach. We base the decision-making component of the HDT on a first-order ordinary differential equation solver and a transformer-based model for desertification prediction.

References

1. Aldabbagh, Y.A.N., Shafri, H.Z.M., Mansor, S., Ismail, M.H.: Desertification prediction with an integrated 3D convolutional neural network and cellular automata in Al-Muthanna, Iraq. Environ. Monit. Assess. **194**(10), 715 (2022)
2. Assennato, F., Alessi, N., Smiraglia, D., Riitano, N., Labadessa, R., Tarantino, C.: A decision support tool based on field and earth observation to support restoration activities in degraded land based on newlife4drylands pilot sites. In: EGU General Assembly Conference Abstracts, pp. EGU–12747 (2023)
3. Board, M.E.A.: Ecosystems and human well-being: desertification synthesis (2005)
4. Cao, J., Wen, X., Zhang, M., Luo, D., Tan, Y.: Information extraction and prediction of rocky desertification based on remote sensing data. Sustainability **14**(20), 13385 (2022)
5. Dayan, P., Watkins, C.: Q-learning. Mach. Learn. **8**(3), 279–292 (1992)
6. Dridi, S., El Yacoubi, S., Bagnoli, F.: Recent advances in regional controlability of cellular automata. Ph.D. thesis, University of Perpignan Via Domitia (2019)
7. El Yacoubi, S., El Jai, A.: Cellular automata modelling and spreadability. Math. Comput. Model. **36**(9–10), 1059–1074 (2002)
8. Fuxin, Z., Guodong, L., Wenxia, X.: Retracted: Xinjiang desertification disaster prediction research based on cellular neural networks. In: Proceedings of the International Conference on Smart City and Systems Engineering (ICSCSE), pp. 545–548 (2016). https://doi.org/10.1109/ICSCSE.2016.0148
9. Gao, J., Nishio, T., Ichizen, N.: Desertification and rehabilitation in china: an overview. J. Arid Land Stud. **17**(3), 101–112 (2007)
10. Helldén, U., Tottrup, C.: Regional desertification: a global synthesis. Global Planet. Change **64**(3–4), 169–176 (2008)
11. Kaelbling, L.P., Littman, M.L., Moore, A.W.: Reinforcement learning: a survey. J. Artif. Intell. Res. **4**, 237–285 (1996)
12. Karavitis, C.A., et al.: A desertification risk assessment decision support tool (DRAST). CATENA **187**, 104413 (2020)
13. Kone, A., Fontaine, A., El Yacoubi, S., Loireau, M., Jangorzo, S.N.: Assessing the impact of anthropogenic factors on desertification through the DESERTICAS software. J. Cell. Autom. **16** (2022)
14. Koné, A., Fontaine, A., Loireau, M., Jangorzo, S.N., El Yacoubi, S.: DESERTICAS, a software to simulate desertification based on MEDALUS and cellular automata. In: Gwizdałła, T.M., Manzoni, L., Sirakoulis, G.C., Bandini, S., Podlaski, K. (eds.) ACRI 2020. LNCS, vol. 12599, pp. 198–208. Springer, Cham (2021). https://doi.org/10.1007/978-3-030-69480-7_20

15. Kosmas, C., Tsara, M., Moustakas, N., Karavitis, C.: Identification of indicators for desertification. Ann. Arid Zone **42**, 393–416 (2003)
16. Lyu, Y., et al.: Desertification control practices in China. Sustainability **12**(8), 3258 (2020)
17. Ma, H., Zhao, H.: United nations: convention to combat desertification in those countries experiencing serious drought and/or desertification, particularly in Africa. Int. Legal Mater **33**(5), 1328–1382 (1994)
18. Mentis, M.: Environmental rehabilitation of damaged land. For. Ecosyst. **7**(1), 19 (2020)
19. Ourabit, S., Ettaqy, A., Ghachi, M.E.: Assessment of sensitivity to desertification in the Oum Er-Rbia watershed (upstream of Ouled Sidi Driss) using the MEDALUS approach. J. Appl. Life Sci. Environ. **57**(1) (2024)
20. Plénet, T., El Yacoubi, S., Lefèvre, L.: Cellular automata for the observation of complex systems. Ph.D. thesis, University of Perpignan Via Domitia (2022)
21. Programme, U.N.E.: Status of desertification and implementation of the united nations plan of action to combat desertification - report of the executive director (1991). https://wedocs.unep.org/20.500.11822/31094
22. Sacande, M., Parfondry, M., Cicatiello, C.: Restoration in action against desertification: a manual for large-scale restoration to support rural communities' resilience in Africa's great green wall. Africa's Great Green Wall (2020)
23. The World Counts: Global land degradation (2023). https://www.theworldcounts.com/challenges/planet-earth/forests-and-deserts/global-land-degradation
24. Türkeş, M., et al.: Desertification vulnerability and risk assessment for turkey via an analytical hierarchy process model. Land Degradation Dev. **31**(2), 205–214 (2020)
25. Wei, W., et al.: Spatiotemporal changes of land desertification sensitivity in northwest china from 2000 to 2017. J. Geog. Sci. **31**, 46–68 (2021)
26. Xu, D., Zhang, X.: Multi-scenario simulation of desertification in North China for 2030. Land Degradation Dev. **32**(2), 1060–1074 (2021)
27. Zerrouki, N., Dairi, A., Harrou, F., Zerrouki, Y., Sun, Y.: Efficient land desertification detection using a deep learning-driven generative adversarial network approach: a case study. Concurr. Comput.: Pract. Exp. **34**(4), e6604 (2022)

> # Social and Biological Models

Global Analysis of a Lane Merging Strategy for Collaborative Autonomous and Connected Vehicles

Bastien Chopard, Pierre Leone[✉], and Luka Lukic

Computer Science Department, University of Geneva, Geneva, Switzerland
{Bastien.Chopard,Pierre.Leone}@unige.ch, Luka.Lukic@etu.unige.ch

Abstract. In this paper we suggest a simple algorithm for merging the traffic of a main road with on-ramp traffic. We consider collaborative Connected and Autonomous Vehicles (CAVs). We are interested in the global impact of the merging strategy on the input and output flows of the merging area. We refer to the literature for all local aspects of synchronisation of cars. Our analysis focuses on saturated flows and the control of collective behaviour.

1 Introduction

A classical problem in traffic control is the lane merging problem where there is a infinite traffic pattern on a mainstream lane that has to be merged with some exogenous traffic coming from an on-ramp, see Fig. 1. All of us know this configuration as a potential bottleneck leading mainstream cars to stop.

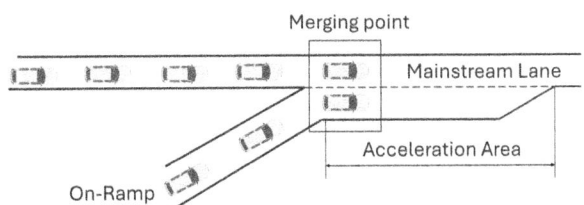

Fig. 1. Illustration of the lane merging problem at on-ramp.

This paper deals with this problem in the context of Autonomous and Connected Vehicles (CAVs). The merging strategy can then make use of some knowledge about the traffic patterns and coordinate the vehicles by exchanging messages. We do not discuss the communication protocols here, we refer to [13] and references therein as well as some other works cited in Sect. 2.

A (one-lane) road section consists in cells that can be occupied by only one car at a time. The distance between 2 cars is the number of empty cells between.

Time is discrete $t \in \mathbb{N}$. If x_a is the cell index of car a and the speed of a, is v_a at time t the position update of the car a is $x_a + v_a$ at time $t+1$. The speed is bounded by v_{max}. Distances, positions and speeds are all natural numbers.

Before the cars apply the merging algorithm, we assume that the traffic is *saturated*. This means that cars on the same lane are moving at the same speed v with head-tail distance $v = d$. This traffic pattern is maximal [2,3]. Intuitively, the condition $v = d$ ensures that the trailing car has the time to brake when the leading car do. Actually, in this condition a trailing car reproduces exactly what did the leading car at the previous step[1], enforcing the condition $v = d$. During the merging algorithm we will assume that cars communicate about their speed. This makes possible that the cars move with head-tail distance shorter that the security distance $d = v$, for a limited time.

2 Related Works

In this section we refer only to merging strategies with CAVs. In particular we do not consider works that try to understand the real traffic patterns due to the human drivers, or make statistical analysis due to the lack of control on the traffic patterns, see for instance [4,11].

The recent review [13] presents a relevant panel of strategies for the coordination of CAVs. We discuss some strategies here with the aim of making clear how our approach is different.

A clever strategy is proposed in [5,7] to turn a car-following regulator to a solution of the lane merging problem, *the virtual car following*. The idea is to map the cars on the other lane to the current lane and let the car-following regulator regulate the distances. Once the distances are large enough merging occurs safely. In [8] the authors refer to the "zipper effect" and show how interaction between cars can be controlled by cellular automata techniques. In our approach we leverage communications between cars to shorten the safety distances. Indeed, we assume that a trailing car is aware of the evolution of the leading car speed.

Another important concept in the literature is the *merging sequence*. Given the traffic on the two lanes it amounts to choosing the order of the cars on the final single lane. In [12] the authors suggest to use a genetic algorithm to compute the optimal merging sequence given a traffic configuration with a complex fitness function. In [9] the authors assume given a merging sequence and coordinate the vehicles accordingly.

In general, the optimization parameters are varied: fuel consumption, average speed in the merging area, total speed of all vehicles, passenger comfort, safety requirements [13].

Model predictive scheme is proposed in [1] for the computation of the optimal merging sequence. Here, the traffic is constraint by communication to a known pattern.

---

[1] The speed of the leading car at time t is the speed of the following car at time $t+1$.

In this work we consider a single merging sequence, see (3), and a single traffic configuration (with parameters). This makes possible to capture the dynamic with closed form formula.

Notice to finish that for many works the problem is solved locally. This means that the focus is on the coordination of vehicles to let one pass before or behind another. The problem amounts to avoiding collision. We refer to these works for this local synchronization and consider the problem globally. Indeed, our strong assumptions on the traffic patterns leads to global understanding of the merging strategy on the traffic.

3 Presentation of the Problem

In this article we consider the problem of On-Ramp lane merging. We assume that we have one target lane on which traffic flows and whose traffic has to be merged with incoming traffic coming from an On-Ramp. As illustrated in Fig. 2 we assume that the cars are located at discrete positions of the roads and move at discrete times $t = 1, 2, \ldots$ This is a similar cellular automata model as in [2,3].

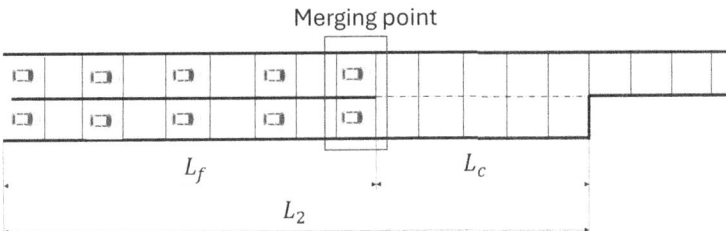

Fig. 2. The lane merging problem. Two lanes have to merge. Up is the target lane. In the initial conditions, cars are assumed to move at the same speed with head-tail distance equals the speed. Such a preconditioning phase is achievable with CAVs.

The originality of our point of view is twofold. First, the lane merging problem can be seen as *making room* in one lane to insert new cars. The usual approach is to make room by slowing down the cars arriving to the road junction. In real scenario this is known to lead to traffic jam. Interestingly this is shown to be optimal in a simplified model in [10]. Our approach consists in *making room by letting the cars accelerate* after reaching the merging point up to a maximal speed v_{max}. Actually, the speed v_{max} is the cruising speed that cars adopt before and after merging. The deceleration of the cars before the junction is controlled to ensure that no car stop moving (speed 0 can be excluded). In summary, we assume that cars arrive at the merging point with speed $0 < v \leq v_{max}$ and head-tail distance $d = v$.

Second, we restrict our analysis to special traffic patterns. On the main road cars moves at the same speed equal to the head-tail distance (this corresponds to a maximal flow traffic pattern, or what we call saturated, see [2,3]). On-ramp are the cars to be merged. For the merging algorithm we first consider a total of N cars see. This is illustrated in Fig. 2. We analyse the merging of the N cars ($N/2$ on the main road and $N/2$ on-ramp) then we analyse the impact of the merging process on the traffic pattern on the main road. Usually, local strategies for lane merging are investigated and it is hard to have the impact of this local perturbation to the global traffic pattern.

Our approach is relevant because the restriction to particular traffic patterns combined with deterministic merging strategy makes possible the computation of precise quantities. For instance The length L_f which represents the length of the ramp is easily seen to be $L_f = \frac{N}{2}(v+1) - v$ where v is the speed of the car (using that for the head-tail distance d it holds that $d = v$). More complex measures are

- The time N_t needed before vehicles reach a maximum vehicle flow[2].
- The length L_c necessary to the application of the merging lane algorithm.
- What is the achievable flow after merging? We start by merging $N/2$ cars with the mainstream and then compute how fast we ca repeat the algorithm.

Before presenting our algorithm and delving to the analysis we propose a rough analysis of some aspects of the problem. We consider two flows entering the merging area:

- The flow on the main road which is $\frac{v}{(v+1)}$, cars move at speed v with head-tail distance $d = v$.
- The flow on-ramp that we denote $\alpha \frac{v}{(v+1)}$, α is then the fraction of the main road flow that can be accepted on the ramp for merging.

The flow out of the merging area is the sum of the two flows and is bounded by the flow of cars at maximum speed, i.e. $v_{max}/(v_{max}+1)$. Hence,

$$(1+\alpha)\frac{v}{v+1} \le \frac{v_{max}}{v_{max}+1}, \text{ i.e. } \alpha \le \frac{v_{max}-v}{v(v_{max}+1)}$$

So we have that *alpha* decreases with v indicating that a small entry speed v allows a larger on-ramp flow. This is confirmed by our analysis, see for instance the results in Fig. 8.

4 Algorithm

We consider a section of the road composed of the target lane n_1 and the second lane n_2 of size $L_2 \in \mathbb{N}$, see Fig. 2 and (1). Vehicles on lane n_2 must move onto lane n_1. In this section we do not consider upstream cars on n_1 upstream, we restrict our attention to the N cars involved in the merging algorithm. We represent a configuration at a time t on the road in the following way:

$$\begin{matrix} n_1 \{\ldots 3 \ldots \ldots \ldots 5 \ldots \ldots \ldots 5 \ldots \ldots \\ n_2 \{\ldots 2 \ldots \ldots 4 \ldots \ldots \ldots \ldots \end{matrix} \quad (1)$$

---

[2] In the sense that the traffic pattern follows the long-term dynamic $v = d = v_{\text{max}}$ as defined and shown to be a recurrent pattern in [2,3].

where a point . denotes an empty cell and a number denotes the speed of a vehicle present in that cell.

4.1 Initial Conditions

The description of the algorithm is limited to the cars involves in the merging algorithm. The cars upstream are assumed to be driven by the car-following speed regulator, i.e. it maintains the head-tail distance $d = v$ with the leading car, and are discussed later in Sect. 5. The speed v is imposed by the algorithm but its value is not yet specified. It is when a car enters one of the section n_1 or n_2 of the road that the regulator switches to the merging mode and applies the algorithm we describe in this section.

We assume that cars are coordinated to an initial configuration[3] in which the vehicles are at equal speed and distance. For example, with $N = 8$ then we would have the following configurations:

$$A \begin{cases} 1.1.1.1............... \\ 1.1.1.1..... \end{cases} \qquad B \begin{cases} 2..2..2..2............ \\ 2..2..2..2.. \end{cases} \qquad (2)$$

Given the initial condition a merging sequence is defined. For example, for configuration **A** the exit order is:

$$A \begin{cases} 7.5.3.1................. \\ 8.6.4.2............ \end{cases} \qquad (3)$$

leading to a final configuration

$$..8...7...6...5...4...3...2...1......$$

where all cars move at speed v_{max} with head-tail distance $d = v_{max}$. We denote by q_a the sequence number of car a.

4.2 Description of the Algorithm

Algorithm 1 works as follows. First, time t is set to 0 at the initial configuration (a traffic pattern like (2)). Let a be a vehicle and q_a its index in the merging sequence. If t is less than q_a, then a maintains its current speed v_a. If t is equal to q_a, then v_a increases by one unit. If, in addition, this position is even, then a changes lanes. Note that the vehicles on lane n_2 always have an even position in the merging sequence. Finally, if t is greater than q_a, then there are three different cases.

- If v_a is less than $v_{max} - 1$ then v_a increases by one unit.
- If v_a equals $v_{max} - 1$ and the head-tail distance with the leading car is smaller ($\leq$) than v_a, then v_a stays constant.

---
[3] Using a strategy similar to one described in Sect. 2.

- If v_a equals $v_{\max} - 1$ and the head-tail distance with the leading car equals $v_a + 1$, then v_a increases by one unit.

It is useful to note that the number of steps a stays at speed $v_{max} - 1$ is

$$\lfloor \frac{q_a}{2} \rfloor (v+1) + 1 \tag{4}$$

Algorithm 1. *Algorithm for vehicles at equal speed and distance with $v \in \llbracket 1, v_{\max} - 2 \rrbracket$*

1: **if** $t < q_a$ **then**
2:     $\tilde{v}_a = v_a$
3: **else if** $t == q_a$ **then**
4:     **if** $q_a \ \% \ 2 == 0$ **then**
5:         $a.\text{changeLanes}()$
6:     **end if**
7:     $\tilde{v}_a = v_a + 1$
8: **else**
9:     **if** $v_a < v_{\max} - 1$ **then**
10:         $\tilde{v}_a = v_a + 1$
11:     **else if** $v_a == v_{\max} - 1$ **then**
12:         **if** $d(a,b) < v_{max}$ **then**   ▷ b is the leading car of a
13:             $\tilde{v}_a = v_a$
14:         **else**
15:             $\tilde{v}_a = v_a + 1$
16:         **end if**
17:     **else**
18:         $\tilde{v}_a = v_a$
19:     **end if**
20: **end if**

4.3 Time Needed to Reach a Maximum Flow

It is relevant to ask how the number of vehicles N influences the time N_t necessary to execute Algorithm 1. The algorithm stops when the last car car involved in the merging process gets to the maximal speed v_{max}. For further reference we call this car *last*. In Fig. 3 at time 9 *last* is boxed. We determine the following formula

$$\boxed{N_t(N, v, v_{\max}) = N\left(1 + \frac{v+1}{2}\right) + v_{max} - v} \tag{5}$$

4.4 Required Lane's Length

We denote L_c the length on lane n_2 necessary to allow all vehicles on this lane, among N vehicles in total, to change lanes. This result is important since related to sizing the road.

$$
\begin{aligned}
&t=0 \begin{cases} 1.1\ldots\ldots\ldots\ldots\ldots\ldots\ldots \\ 1.1\ldots\ldots\ldots\ldots\ldots \end{cases} \quad
t=1 \begin{cases} .1.2\ldots\ldots\ldots\ldots\ldots\ldots \\ .1.1\ldots\ldots\ldots\ldots\ldots \end{cases} \\
&t=2 \begin{cases} ..1.23\ldots\ldots\ldots\ldots\ldots \\ ..1\ldots\ldots\ldots\ldots\ldots \end{cases} \quad
t=3 \begin{cases} \ldots 2..2.3\ldots\ldots\ldots\ldots \\ \ldots 1\ldots\ldots\ldots\ldots \end{cases} \\
&t=4 \begin{cases} \ldots.22..2..3\ldots\ldots\ldots\ldots \\ \ldots\ldots\ldots\ldots\ldots\ldots \end{cases} \quad
t=5 \begin{cases} \ldots\ldots 22..3\ldots 3\ldots\ldots\ldots\ldots \\ \ldots\ldots\ldots\ldots\ldots\ldots \end{cases} \\
&t=6 \begin{cases} \ldots\ldots\ldots 23\ldots 3\ldots 3\ldots\ldots\ldots \\ \ldots\ldots\ldots\ldots\ldots\ldots \end{cases} \quad
t=7 \begin{cases} \ldots\ldots\ldots\ldots 2.3\ldots 3\ldots 3\ldots\ldots \\ \ldots\ldots\ldots\ldots\ldots\ldots \end{cases} \\
&t=8 \begin{cases} \ldots\ldots\ldots\ldots 2.3\ldots 3\ldots 3\ldots 3\ldots \\ \ldots\ldots\ldots\ldots\ldots\ldots \end{cases} \quad
t=9 \begin{cases} \ldots\ldots\ldots\ldots\ldots \boxed{3}\ldots 3\ldots 3\ldots 3 \\ \ldots\ldots\ldots\ldots\ldots\ldots \end{cases}
\end{aligned}
$$

Fig. 3. Example of execution of the merging algorithm with $v_{max} = 3$, $v = 1$ and $N = 4$. At time 9 the algorithm stops when all cars involved in the merging algorithm gets to maximal speed. The last one is boxed at time 9. For further reference we call it *last*. Notice that *last* stays at speed 2 ($= v_{max} - 1$) 5 time units as predicted by (4), i.e. $q_4 = 5$.

In general, for $N > 1$, we have that:

$$\boxed{L_c(N, v) = \frac{N}{2}(v-1) + v}. \tag{6}$$

Indeed, it is the last car that requires the longest road length before changing lane. It changes lane after spanning a distance Nv. The result follows because the last car starts a distance $\frac{N}{2}(v+1) - v$ before the merging point.

5 Repeating the Merging Algorithm

In this Section we address the computation of how many cars car be merged with a one lane saturated traffic pattern as the time goes. Algorithm 1 shows how merging $N/2$ cars. The question now is how frequently bunches of $N/2$ cars can be merged with the incoming flow.

Example in Fig. 3 has to be completed to take into account the upstream flow of cars entering continuously the merging area, see Fig. 4 where we observe at time 2 and 5 the appearance of upstream cars in the mainstream lane. This stream is composed of cars at speed 1 (in general v) and head-tail distance 1 ($d = v$). What is important to notice at time 5 is that the car on the merging point cannot accelerate because of the cars downstream that apply the merging algorithm and are stuck at speed 2 (in general $v_{max} - 1$). Similarly, in Fig. 5 we see when the merging algorithm finishes at time 9, the last car *last* (boxed) is followed by a line of cars stuck at speed 2. To restart the merging algorithm requires to wait for all these cars get to speed v_{max}.

$$t = 2 \begin{cases} 1.1.\underline{2}3............... \\ ..1................ \end{cases} \quad t = 3 \begin{cases} .1.\underline{2}..2.3............. \\ ...1............. \end{cases}$$

$$t = 4 \begin{cases} 1.1.\underline{2}2..2..3............. \\ .... \end{cases} \quad t = 5 \begin{cases} .1.\underline{2}..22..3...3........ \\ .... \end{cases}$$

Fig. 4. Example of Fig. 3 completed to include the flow incoming the merging area. The merging point is underlined. Algorithm 1 can restart at the time when cars arriving at this point can accelerate to v_{max} without restricted by cars down stream at lower speed.

$$t = 8 \begin{cases} .1.\underline{2}..2..2..2..3...3...3...... \\ .... \end{cases}$$

$$t = 9 \begin{cases} ..1.\underline{\,}.2..2..2..\boxed{3}...3...3...3... \\ .... \end{cases}$$

Fig. 5. The cars upstream to *last* (boxed in the figure) cannot accelerate to v_{max} as they enter the merging area. Hence, Algorithm 1 cannot be restarted.

Before turning to the analysis of the performance of Algorithm 1 we describe the traffic dynamics as plotted in Fig. 6. In this figure we see the state of the roads as the merging algorithm is run until the time it can be run again. There $N = 6$ cars participate to the merging process. We plot in magenta the last such car and for further reference we denote it by *last*. The cars plotted on red are the ones not yet in the merging area. Such cars move at speed v and there is an infinite number coming upstream.

On the figure there are 3 phases distinguished. Phase I correspond to the execution of Algorithm 1 as described in Sect. 4.2. We see that at the end of this phase the 6 cars involved in the merging process are at maximal speed, the cars are represented by squares instead of dots when their speed is maximal (v_{max}).

We observe that at the end of phase I there are cars behind *last* that are in the merging area but cannot accelerate to v_{max}. Repeating Algorithm 1 requires to wait that there is no constraint on the acceleration for these cars. These are Phase II and III. Actually, after *last* accelerates the cars behind start to accelerate one after the other and Algorithm 1 can be repeated when the car entering the merging area can accelerate straightaway.

Analysing Phases II and III requires to understand the speed at which the wave of accelerating cars move. It happens that in Phase II the speed is 1 (cell per unit time) - this is plotted as $slope_1$ - and in Phase II it is $(v_{max} - 1)/v$ (cells per unit time) - this is plotted as $slope_2$, see Eq. 8.

The time to complete phases I, II and II are respectively given by Eqs. (5), (7), (8). Due to the space restriction the proofs will be published somewhere else.

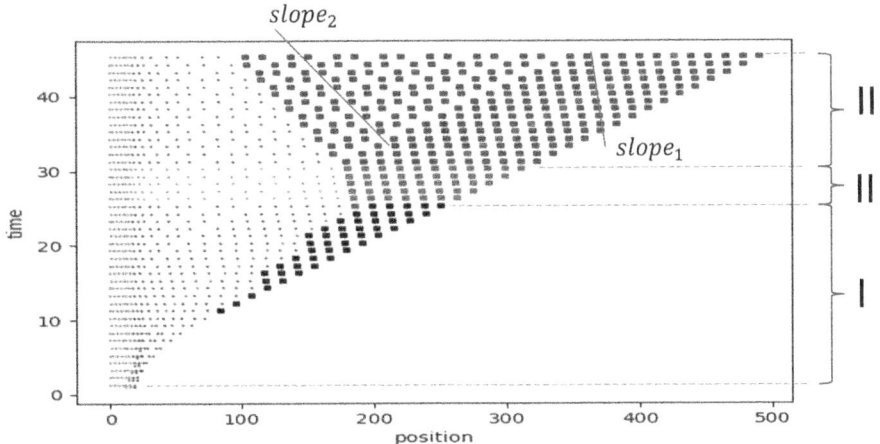

Fig. 6. Plot of the execution of Algorithm 1 until a new batch of cars can be merged with $N = 6$, $v = 2$ and $v_{max} = 12$. In red are the cars before entering the merging area. Cars plotted with a square are the ones at speed v_{max}. Phase I corresponds to the execution of Algorithm 1. The last car involved in the merging algorithm (referred as *last* in the text) is plotted in magenta. Phase II and III correspond to the acceleration phase of the cars behind *last*. At the end of phase III the cars entering the merging area accelerate continuously to v_{max} and a new batch can be merged.

$$k' = \frac{v-1}{v+1}\left(\frac{N}{2}+1\right) + (N\frac{v-1}{2}+v)\frac{v}{v+1} \tag{7}$$

$$\frac{v}{v_{max}-1}(d'_{end} - k'). \tag{8}$$

with

$$d'_{end} = \frac{N}{2}(v-1+(v+1)(v_{max}-1)) + v.$$

Hence, the algorithm can restart after

$$P = N_t + k' + \frac{v}{v_{max}-1}(d'_{end}-k'). \tag{9}$$

This estimation is numerically validated and results are reported in Fig. 7.

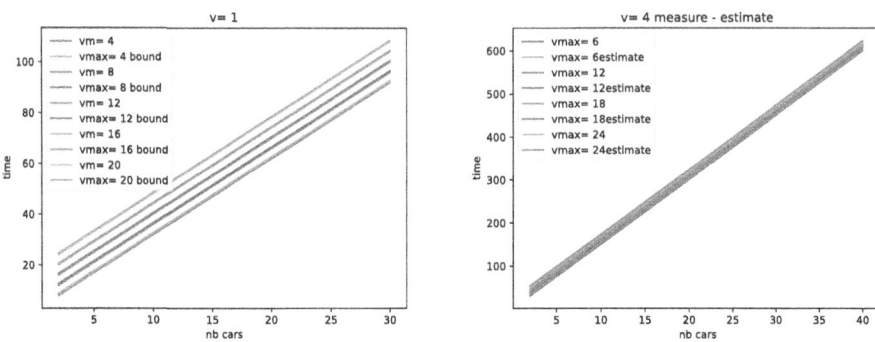

Fig. 7. Numerical validation of (9), vm is v_{max} results of simulations are compared to the bound (9).

Interestingly the three term entering (9) are linear function in N with the slopes increasing with v_{max}.

Using the estimate (9) we can compute the flow after the merging point. Indeed, after a period P, $N/2$ cars are added to the flow. In total there are $\frac{v}{v+1}P + \frac{N}{2}$ cars in the main road for a period P. This leads to a flow of $\frac{v}{v+1} + \frac{N}{2P}$. This total flow can be compared to the flow on-ramp. This is shown in Fig. 8. We observe that increasing the speed of the cars entering the merging area increases the flow but reduces the ability to merge on-ramp traffic.

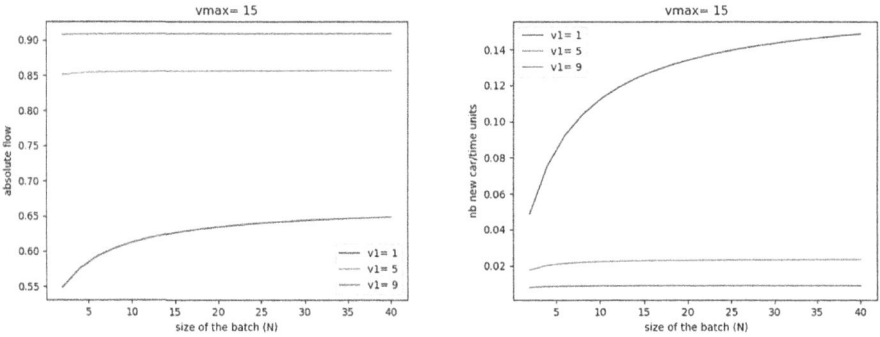

Fig. 8. Left: The total flow after the merging area. Right: The flow on-ramp.

6 Conclusion

The analysis conducted in this paper shows that the performance of merging algorithms in a global setting involve many aspects of the traffic pattern dynamics that are not easy to grasp all together. But, from the point of view of optimizing the process it is mandatory to understand the global process. If our analysis does not lead to a comprehensive rule of thumb we believe it is relevant in showing a way to analyse the problem. Moreover, we think our approach to consider the merging of bunches of cars and repeat the process relevant.

Our analysis leads to relevant conclusions. In particular, it shows that it is important to control the speed of the cars arriving to the merging area. This speed v should be as low as possible. Notice, that Eq. (6) leads to the same conclusion since the length of the double road infrastructure increases with v.

From a technical point of view, the derivation of the estimate (9) shows that it is possible to express the speed of the front wave of acceleration/deceleration of traffic patterns. These speeds are functions of the head-tail distances between the cars. This is an improvement with respect to [2,3].

As for criticism, our model should consider the more realistic situations where cars occupy more than one cell. This is paramount since it makes possible to consider realistic acceleration. Moreover, it is important to add this new parameter and see whether is reinforce or not our conclusions.

We mention that we do not assert that Algorithm 1 is optimal. Actually, before doing its complete analysis it was not clear in what respect this algorithm should be optimised. This paper suggests that what should be optimized is the period P given by (9).

From an heuristic point of view, to optimize Algorithm 1 we may exploit the fact that more than one car at a time can increase its distance. More generally, formalising the problem as an optimisation problem for solution would be as well relevant. We will consider such directions for further works.

Finally, we qualified this work as global to stress that we quantify the impact of the lane merging algorithm on entering and output flows. Our analysis shows that merging cars is only part of the challenge. On a long road the interaction between the cars may have long range effect that need to be understood. This was already pointed out in the pioneering work [6].

References

1. Cao, W., Mukai, M., Kawabe, T., Nishira, H., Fujiki, N.: Cooperative vehicle path generation during merging using model predictive control with real-time optimization. Control. Eng. Pract. **34**, 98–105 (2015)
2. Cohen, N., Chopard, B., Leone, P.: Maximum traffic flow patterns in interacting autonomous vehicles. In: Chopard, B., Bandini, S., Dennunzio, A., Arabi Haddad, M. (eds.) ACRI 2022. LNCS, vol. 13402, pp. 281–291. Springer, Cham (2022). https://doi.org/10.1007/978-3-031-14926-9_25
3. Cohen, N., Chopard, B., Leone, P.: Optimal safe driving dynamics for autonomous interacting vehicles. Nat. Comput. 1–13 (2023)
4. Guzmán, H., Lárraga, M., Alvarez-Icaza, L., Huerta, F.: On-ramp traffic merging modeling based on cellular automata. In: 2015 IEEE European Modelling Symposium (EMS), pp. 103–109. IEEE (2015)

5. Huang, Z., Zhuang, W., Yin, G., Xu, L., Luo, K.: Cooperative merging for multiple connected and automated vehicles at highway on-ramps via virtual platoon formation. In: 2019 Chinese Control Conference (CCC), pp. 6709–6714 (2019)
6. Lighthill, M.J., Be Whitham, G.: On kinematic waves i. flood movement in long rivers. Proc. Roy. Soc. London. Ser. A Math. Phys. Sci. **229**(1178), 281–316 (1955)
7. Milanes, V., Godoy, J., Villagra, J., Perez, J.: Automated on-ramp merging system for congested traffic situations. IEEE Trans. Intell. Transp. Syst. **12**(2), 500–508 (2011)
8. Nishi, R., Miki, H., Tomoeda, A., Nishinari, K.: Achievement of alternative configurations of vehicles on multiple lanes. Phys. Rev. E **79**(6), 066119 (2009)
9. Ntousakis, I.A., Nikolos, I.K., Papageorgiou, M.: Optimal vehicle trajectory planning in the context of cooperative merging on highways. Transp. Res. Part C: Emerg. Technol. **71**, 464–488 (2016)
10. Petig, T., Schiller, E.M., Suomela, J.: Changing lanes on a highway. In: 18th Workshop on Algorithmic Approaches for Transportation Modelling, Optimization, and Systems (ATMOS 2018). Schloss Dagstuhl-Leibniz-Zentrum fuer Informatik (2018)
11. Sun, S., An, X., Zhao, J., Li, P., Shao, H.: Modeling and simulation of lane-changing management strategies at on-ramp and off-ramp pair areas based on cellular automaton. IEEE Access **9**, 35034–35044 (2021)
12. Xu, L., Lu, J., Ran, B., Yang, F., Zhang, J.: Cooperative merging strategy for connected vehicles at highway on-ramps. J. Transp. Eng. Part A: Syst. **145**(6), 04019022 (2019)
13. Zhu, J., Easa, S., Gao, K.: Merging control strategies of connected and autonomous vehicles at freeway on-ramps: a comprehensive review. J. Intell. Connect. Veh. **5**(2), 99–111 (2022)

Binary Hiking Optimization Algorithm

Tahir Sağ

Selçuk University, 42075 Konya, Turkey
tahirsag@selcuk.edu.tr

Abstract. The Hiking Optimization Algorithm (HOA) is a newly designed metaheuristic optimization algorithm that stands out for its simplicity and problem-solving capability. Inspired by hiking, HOA draws parallels between the search landscapes of optimization problems and the rugged terrains encountered in nature. In this study, HOA is adapted for binary optimization problems, resulting in the proposal of the Binary Hiking Optimization Algorithm. The proposed algorithm, utilizing S-shaped and V-shaped transfer functions, is applied to Uncapaciated Facility Location Problems, and its performance is assessed. Key metrics such as mean, standard deviation, and GAP values are computed to evaluate the algorithm's effectiveness. Furthermore, an improvement method for the location update strategy is proposed for HOA, inspired by the Moore neighborhood used in cellular automata. The experimental results demonstrate that the binary HOA, strengthened by appropriate transfer functions, serves as a robust approach to tackle complex binary optimization challenges.

Keywords: HOA · BinHAO · UFLP

1 Introduction

The pursuit of efficient optimization algorithms has long been a focal point in the realm of computational intelligence, driven by the ever-growing complexity of real-world problems across various domains. Among the myriad of approaches, metaheuristic algorithms stand out for their ability to navigate complex search spaces, offering promising solutions to optimization challenges [1].

In this context, the Hiking Optimization Algorithm (HOA) [2] emerges as a novel metaheuristic approach, drawing inspiration from the adaptive behavior of hikers traversing rugged terrains. Mimicking the strategic decisions made by hikers to navigate challenging landscapes, HOA embodies a balance between exploration and exploitation, making it particularly adept at solving optimization problems.

While HOA targeted continuous optimization tasks, this study extends its utility by proposing the Binary Hiking Optimization Algorithm (BinHOA). BinHOA caters specifically to binary optimization problems, where decision variables are constrained to binary values, offering a tailored solution to a wide array of practical challenges.

The focus of this work lies in adapting and evaluating BinHOA for solving Uncapaciated Facility Location Problems (UFLPs) [3], a class of binary optimization challenges with significant theoretical and practical implications. UFLPs find application across diverse domains such as logistics, transportation, and facility management, underscoring the importance of developing effective optimization techniques to address them.

Many optimization algorithms have been proposed to solve continuous problems; however, many problems are binary in nature. Continuous optimization algorithms cannot directly solve binary optimization problems, necessitating modifications to their operational structures. Nevertheless, various methods have been proposed in the literature to solve binary optimization problems like the UFL problem. Some of the recently suggested algorithms include the Binary Battle Royale Optimizer (BinBRO) [4], the Binary Tree Seed Algorithm (SimLogicTSA-ELSM) [5], and the Binary Archimedes Optimization Algorithm (BAOA) [6], An enhanced group-theory optimization algorithm (EGTOA) [7], binary differential evolution algorithm (T-NBDE) [8], and etc.

By leveraging S-shaped and V-shaped transfer functions, BinHOA transforms continuous solutions into binary representations, facilitating its application to UFLPs. Through an experimentation, this study investigates the performance of BinHOA across a range of UFLP instances, assessing its efficacy in terms of mean fitness, standard deviation, and GAP values. This study presents BinHOA as a promising approach for tackling complex binary optimization challenges, demonstrating its adaptability and effectiveness in solving UFLPs. The findings underscore the significance of transfer function selection in optimizing algorithm performance, paving the way for future research to explore BinHOA's scalability, robustness, and applicability across diverse optimization domains. Moreover, a novel improvement method for the location update strategy in HOA is proposed, inspired by the Moore neighborhood concept utilized in cellular automata. This version of binary HOA algorithm is named as BinHOA_CA. Likewise, BinHOA_CA was applied to UFL problems under the same conditions and the effect of the proposed neighborhood method on performance was evaluated.

The structure of paper is organized as follows: Sect. 2 provides an explanation of the standard Hiking Optimization Algorithm. Section 3 gives an overview of the UFL problems. Section 4 elaborates on the proposed binary algorithms. Following that, Sect. 5 presents the experimental results and discussions. Finally, Sect. 6 offers concluding remarks and suggestions for future research directions.

2 The Hiking Optimization Algorithm

The Hiking Optimization Algorithm (HOA) [2] is inspired by the behavior of hikers navigating mountainous terrains. Hikers adjust their paths based on the steepness of the terrain to maintain speed, avoiding overly steep areas that slow them down. In other words, steep terrains slow hikers down and extend the duration of the hike. Hikers aware of the terrain can estimate the time to reach the peak. This is similar to agents finding local or global optima in optimization problems, where the search space resembles the mountainous terrains. Agents may get bogged down due to the problem's complexity, prolonging the search for the global optima, akin to hikers' experiences. This behavior is analogous to agents navigating a search space to find local or global optima in optimization problems. Figure 1 shows the pseudocode of the algorithm.

```
01  INPUT: problem parameters( ObjFun, lb, ub, nVar, nPop, MaxIter )
02  PRE-ALLOCATE storage for fitness values and best iteration fitness
03  INITIALIZE population of hikers
04      Pop ← random binary matrix of size (nPop, nVar)
05  EVALUATE initial fitness of each hiker
06  Best.iteration(1) ← minimum value in fit
07  MAIN LOOP for iterations
08  FOR each iteration i in 1 to MaxIter
09      FIND the hiker with the best fitness
10          ind ← index of minimum value in fit
11          Xbest ← Pop(ind, :)
12      FOR each hiker j in 1 to nPop
13          Xini ← Pop(j, :)
14          theta = random integer between 0 and 50
15          s ← tan(theta)
16          SF ← random integer between 1 and 2
17          velocity ← 6 * exp(-3.5 * abs(s + 0.05))
18          new_velocity ← velocity + random array of size (1, nVar) * (Xbest - SF * Xini)
19          newPop ← Pop(j, :) + new_velocity
20          fnew ← ObjFun(newPop)
21          IF fnew < fit(j) THEN
22              Pop(j, :) ← newPop
23              fit(j) ← fnew
24          END IF
25      END FOR
26      UPDATE the best fitness value for the current iteration
27      Best.iteration(i+1) ← minimum value in fit
28  END FOR
29  STORE global best fitness and position
30  OUTPUT: bestObj, bestParams, iters
```

Fig. 1. Pseudocode of HOA.

The mathematical foundation of HOA is based on Tobler's Hiking Function (THF) [9], which models a hiker's speed as an exponential function of terrain steepness. The THF is given by Eq. (1).

$$velocity_{i,t} = 6e^{-3.5|S_{i,t}+0.05|} \qquad (1)$$

where $velocity_{i,t}$ is the velocity of hiker i at iteration t, and $S_{i,t}$ is the slope of the terrain defined in Eq. (2).

$$S_{i,t} = \frac{dh}{dx} = \tan\theta_{i,t} \qquad (2)$$

where dh and dx indicate the difference in elevation and the distance covered by the hiker, respectively. $\theta_{i,t}$ is the angle of inclination of the terrain, lies within the range of [0, 50°]. The hiker's velocity is updated considering their initial velocity, the lead hiker's position, and a sweep factor that ensures they stay relatively close to the lead hiker. This is described by Eq. (3).

$$new\_velocity_{i,t} = velocity_{i,t-1} + \gamma_{i,t}(\beta_{best} - \alpha_{i,t}\beta_{i,t}) \qquad (3)$$

where $\gamma_{i,t}$ is a uniformly distributed random variable within [0, 1], β_{best} is the position of the lead hiker, and $\alpha_{i,t}$ is the sweep factor ranging from 1 to 3. The updated position $\beta_{i,t+1}$ of hiker i is given by Eq. (4).

$$\beta_{i,t+1} = \beta_{i,t} + new\_velocity_{i,t} \qquad (4)$$

At the start of the algorithm, the positions of the hikers are initialized randomly within the bounds of the problem. The initialization is given by Eq. (5).

$$\beta_{i,t} = \phi_1 + \delta_j(\phi_2 - \phi_1) \tag{5}$$

where ϕ_1 and ϕ_2 represent the lower and upper bounds of the decision variables, respectively, and δ_j is a uniformly distributed random variable within [0, 1].

In each iteration, the fitness values of all hikers are calculated by using the objective function of the optimization problem considered, and the hiker with the best fitness is designated as the lead hiker. The process is repeated for the number of iterations (*MaxIter*). In summary, HOA leverages the adaptive behavior of hikers to solve optimization problems by balancing exploration and exploitation. Adjusting parameters such as the sweep factor and terrain slope guides the search process effectively, mirroring the decision-making of hikers on a trail.

3 Uncapacitated Facility Location Problems

The Uncapacitated Facility Location Problem (UFLP) stands as a significant binary optimization challenge and holds a prominent place among NP-hard problems in location theory [3]. With its broad theoretical implications and practical relevance, the UFLP finds application across diverse domains such as warehouse management, network design, logistics, transportation, and public facility location selection.

In essence, the UFL problem entails the allocation of a given set of potential facilities to a corresponding set of customers. It assumes that the facilities have no capacity constraints, and each customer is served by a single facility. The decision regarding the operation of each facility is binary in nature, signifying its status as either open or closed. The objective in solving UFLP is to minimize the total cost associated with both facility operation and customer service. This cost is composed of two key components: the service cost incurred in serving customers and the operational cost of opening facilities. Mathematically, UFLP is formally expressed through Eq. (6).

$$\text{minimize } f(W, X) = \sum_{i=1}^{m}\sum_{j=1}^{n} d_{ij}.w_{ij} + \sum_{j=1}^{n} g_j.x_j \tag{6}$$

$$s.t. \sum_{j=1}^{n} w_{ij} = 1 \tag{7}$$

$$w_{ij}, x_j \in \{0,1\}, \ w_{ij} \leq x_j, \ i = 1,\ldots,m \text{ and } j = 1,2,\ldots,n \tag{8}$$

where d_{ij} is the service cost that *ith* customer receives from the *jth* facility, and g_j is the opening cost required by the opening of the *jth* facility. The variable w_{ij} takes on a value of 1 if customer i is served by facility j, otherwise, it remains 0. The variable x_j takes a value of 1 if facility j is open, otherwise it takes a value of 0.

The OR-Library [10] provides a comprehensive dataset for UFLPs. This study utilized a selection of 12 problems from the small-sized and medium-sized categories, which are detailed in Table 1. The small-sized Cap71–74 problems consist of 16 facilities and 50 customers, offering simplicity ideal for initial algorithm testing. Moving to the medium-sized Cap101–104 problems, with 25 facilities and 50 customers, presents a higher complexity level, demanding algorithms to manage more facilities while retaining the same number of customers. Finally, the large-sized Cap131–134 problems, with 50 facilities and 50 customers each, escalate in scale and complexity, posing challenges for efficient resource allocation and solution optimization.

Table 1. Properties of UFLPs

Instance	$m \times n$	Optimum Value
Cap71	16×50	9.3261575E+05
Cap72	16×50	9.7779940E+05
Cap73	16×50	1.01064145E+06
Cap74	16×50	1.03497698E+06
Cap101	25×50	7.9664844E+05
Cap102	25×50	8.547042E+05
Cap103	25×50	8.9378211E+05
Cap104	25×50	9.2894175E+05
Cap131	50×50	7.9343956E+05
Cap132	50×50	8.5149533E+05
Cap133	50×50	8.9307671E+05
Cap134	50×50	9.2894175E+05

4 Binary Hiking Optimization Algorithm (BinHOA)

The Binary Hiking Optimization Algorithm (BinHOA) is an adaptation of the continuous HOA for solving binary optimization problems. BinHOA is designed to address problems in binary search spaces, where variables take on values of either 0 or 1. This involves transforming the continuous solutions generated by HOA into binary solutions using transfer functions. Two types of transfer functions are used to convert the updated position of hiker i $(\beta_{i,t+1})$, from continuous values to binary values: S-shaped and V-shaped transfer functions [11]. The shapes and mathematical definitions of the functions are given in Fig. 2.

The eight different versions of BinHOA derived from transfer functions are proposed in this paper. In this approach, new generated position is transformed into {0, 1} values by using Eq. (9).

$$\beta_{i,t+1} = \begin{cases} 0 & \text{if rand} < tf(\beta_{i,t+1}) \\ 1 & \text{if rand} \geq tf(\beta_{i,t+1}) \end{cases} \quad (9)$$

where $tf(\beta_{i,t+1})$ is a transfer function that is used to calculate the probability values from the value of new generated position $\beta_{i,t+1}$ of hiker i. rand is a rando number in [0, 1].

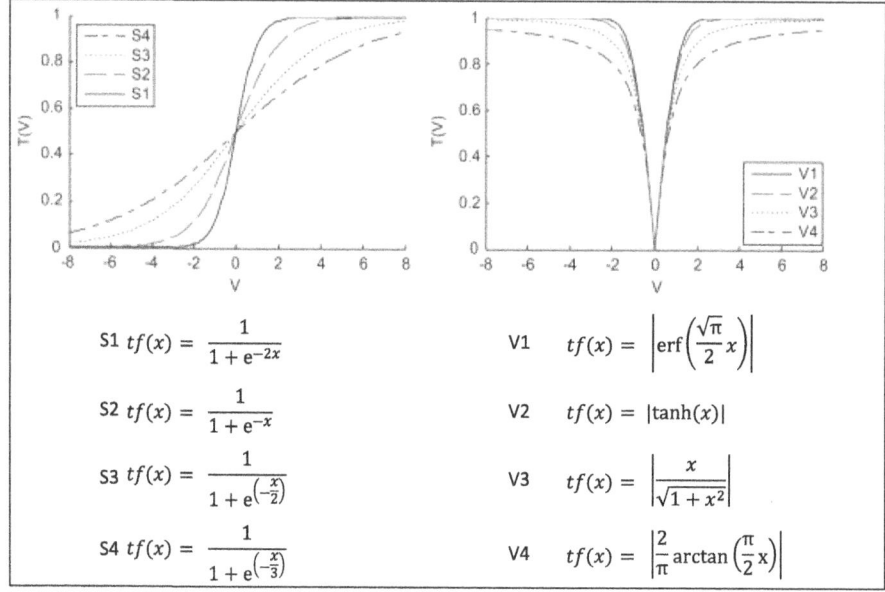

Fig. 2. Transfer functions.

Cellular automata (CA) can be utilized in various ways within metaheuristic algorithms: Exploration and Diversity Maintenance of the Solution Space, Local Search and Update Mechanisms, and Parallel Computing. In this study, inspired by the Moore neighborhood used in cellular automata [12], an improvement method for position update strategy has been proposed. The Moore neighborhood comprises 8 neighboring cells surrounding a cell in a two-dimensional grid. This neighborhood is used to determine interactions or state changes by considering a specific cell and its immediate surrounding cells. In the proposed approach, a cellular BinHOA algorithm was modeled and named BinHOA_CA. Unlike the standard algorithm, the cellular HOA algorithm arranges individuals on a grid and allows them to interact only with neighboring individuals. This helps maintain diversity and avoid local minima. The selected neighboring solution is used with equal probability in place of the lead hiker β_{best} in Eq. (3), and the velocity value is calculated as in Eqs. (10) and (11).

$$\beta_{neighbour} = \begin{cases} \beta_{best}, & rand < 0.5 \\ \beta_{Moore}, & otherwise \end{cases} \quad (10)$$

$$new\_velocity_{i,t} = velocity_{i,t-1} + \gamma_{i,t}(\beta_{neighbour} - \alpha_{i,t}\beta_{i,t}) \quad (11)$$

where the position β_{Moore} is a candidate solution randomly selected from among the 8 solutions in the Moore neighborhood of the candidate solution to be updated. Assuming that the candidate solutions forming the population are arranged in a grid layout, the neighbors of any solution can be determined. For a run with $nPop = 100$, the arrangement of candidate solutions is shown on a 10x10 grid as depicted in Fig. 3. For instance, the

Moore Neighborhood of the 25th candidate solution consists of the set of solutions *neighbors* = {14, 15, 16, 24, 26, 34, 35, 36}, highlighted in yellow in the Fig. 3.

1	11	21	31	41	51	61	71	81	91
2	12	22	32	42	52	62	72	82	92
3	13	23	33	43	53	63	73	83	93
4	14	24	34	44	54	64	74	84	94
5	15	25	35	45	55	65	75	85	95
6	16	26	36	46	56	66	76	86	96
7	17	27	37	47	57	67	77	87	97
8	18	28	38	48	58	68	78	88	98
9	19	29	39	49	59	69	79	89	99
10	20	30	40	50	60	70	80	90	100

Fig. 3. 10 × 10 Grid layout representation of the population.

5 Experimental Results

Eight different binary versions with S-shaped and V-shaped transfer functions were developed for both the standard HOA algorithm and the BinHOA_CA algorithms, whose position update strategy was modified using Moore neighborhood. To evaluate the algorithms, each of the 8 binary versions of the BinHOA algorithm and the BinHOA_CA algorithm was tested on 12 UFL problems. This evaluation involved 10 independent experiments, each running for a maximum of 500 iterations with a population size of 40. Ten runs were conducted for comparison. The results, including the mean, standard deviation, and GAP metrics, are shown in Table 2 for BinHOA and in Table 3 for BinHOA_CA. The GAP values, as defined in Eq. (12), represent the difference between average fitness values and optimum values.

$$\text{GAP}(\%) = \frac{\textit{Mean Fitness} - \textit{Optimum Value}}{\textit{Optimum Value}} \times 100 \qquad (12)$$

where Mean Fitness represents the average of results from 10 independent runs. Optimum Value means the solution of the problem as global optimum value given in Table 1 for each UFLP.

Table 2 presents the performance of various versions (V1 to V8) of BinHOA on UFLPs. Across multiple instances (Cap71 to Cap134), V5 and V6 consistently demonstrate superior performance, evidenced by mean values closely aligning with optimal values and minimal standard deviations, indicating consistent results. These versions also show the lowest GAP values, underscoring their effectiveness in reaching optimal solutions.

In contrast, V2, V3, and V4 frequently exhibit higher mean values and greater variability, reflecting less reliable optimization. Specifically, V3 often shows the highest mean values and GAP, indicating significant deviations from the optimal solutions.

For instance, in Cap71 and Cap72, V1, V5, V6, and V7 achieve mean values at or near the optimum with low variability, whereas V2, V3, and V4 show higher means and standard deviations. This pattern persists in subsequent instances such as Cap73, Cap74, and Cap101, where V5 and V6 maintain superior performance with lower GAP values compared to other versions.

Overall, V5 and V6 are identified as the most effective versions of BinHOA, demonstrating robust optimization capabilities across different UFLP instances. Their consistent performance highlights the importance of selecting appropriate transfer functions to enhance the algorithm's efficiency in solving binary optimization problems. Conversely, the higher variability and mean values in V2, V3, and V4 suggest that these versions are less reliable, emphasizing the need for careful choice of transfer functions to achieve optimal results.

Table 2. Statistical results of all versions of BinHOA on UFLPs

		BinHOA-V1	BinHOA-V2	BinHOA-V3	BinHOA-V4	BinHOA-V5	BinHOA-V6	BinHOA-V7	BinHOA-V8
Cap71	Mean	9,32616E+14	9,32997E+14	9,38837E+14	9,33798E+14	9,32616E+14	9,32616E+14	9,32616E+14	9,33092E+14
	Std	0,00000E+00	4,92205E+14	2,48350E+14	1,45997E+14	0,00000E+00	0,00000E+00	0,00000E+00	5,02354E+14
	GAP	0,00000E+00	4,08807E−02	6,67094E−01	1,26732E−01	0,00000E+00	0,00000E+00	0,00000E+00	5,11009E−02
Cap72	Mean	9,78230E+14	9,79838E+14	9,81587E+14	9,80690E+14	9,77799E+14	9,77799E+14	9,77799E+14	9,77799E+14
	Std	5,56109E+14	1,57780E+14	1,44130E+14	1,65390E+14	1,22713E+04	1,22713E+04	1,22713E+04	1,22713E+04
	GAP	4,40540E−02	2,08536E−01	3,87368E−01	2,95648E−01	1,19058E+00	1,19058E+00	1,19058E+00	1,19058E+00
Cap73	Mean	1,01357E+14	1,01375E+14	1,01215E+14	1,01331E+14	1,01119E+14	1,01138E+14	1,01119E+14	1,01119E+14
	Std	1,83414E+14	2,22202E+14	1,33421E+14	1,52935E+14	8,86643E+14	9,47861E+14	8,86643E+14	8,86643E+14
	GAP	2,89899E−01	3,07294E−01	1,49365E−01	2,63632E−01	5,44859E−02	7,26479E−02	5,44859E−02	5,44859E−02
Cap74	Mean	1,05004E+14	1,04663E+14	1,03859E+14	1,04375E+14	1,03498E+14	1,03498E+14	1,03498E+14	1,03498E+14
	Std	4,90481E+14	6,00594E+14	2,48685E+14	4,79705E+14	1,22713E+04	1,22713E+04	1,22713E+04	1,22713E+04
	GAP	1,45578E+14	1,12586E+14	3,48791E−01	8,48040E−01	−1,12481E+00	−1,12481E+00	−1,12481E+00	−1,12481E+00
Cap101	Mean	8,00355E+14	8,02372E+14	8,07515E+14	8,02125E+14	7,97323E+14	7,97500E+14	7,98692E+14	8,00285E+14
	Std	1,88873E+14	1,39999E+14	3,43073E+14	2,06288E+14	7,54421E+14	9,69010E+14	1,74499E+14	1,81707E+14
	GAP	4,65272E−01	7,18408E−01	1,36403E+14	6,87506E−01	8,46686E−02	1,06926E−01	2,56560E−01	4,56434E−01
Cap102	Mean	8,68038E+14	8,63039E+14	8,65274E+14	8,66109E+14	8,55408E+14	8,55544E+14	8,55132E+14	8,57073E+14
	Std	2,23034E+14	2,44871E+14	3,43814E+14	2,73019E+14	7,49759E+14	9,70575E+14	7,65253E+14	1,64068E+14
	GAP	1,55999E+14	9,75224E−01	1,23664E+14	1,33438E+14	8,23421E−02	9,82868E−02	5,00003E−02	2,77166E−01
Cap103	Mean	9,15343E+14	9,12716E+14	9,05490E+14	9,10568E+14	8,94472E+14	8,94258E+14	8,94576E+14	8,94676E+14
	Std	8,59591E+14	4,44845E+14	3,73052E+14	3,24926E+14	5,90533E+14	5,27832E+14	1,17934E+14	5,21406E+14
	GAP	2,41237E+14	2,11839E+14	1,30997E+14	1,87810E+14	7,72386E−02	5,31928E−02	8,88097E−02	1,00042E−01
Cap104	Mean	9,81851E+14	9,73030E+14	9,47278E+14	9,62813E+14	9,28942E+14	9,28942E+14	9,28942E+14	9,28942E+14
	Std	1,46350E+14	8,09427E+14	6,87003E+14	8,07244E+14	1,22713E+04	1,22713E+04	1,22713E+04	1,22713E+04
	GAP	5,69562E+14	4,74609E+14	1,97394E+14	3,64623E+14	2,50641E+00	2,50641E+00	2,50641E+00	2,50641E+00
Cap131	Mean	8,49556E+14	8,41142E+14	8,30174E+14	8,37498E+14	7,97373E+14	7,97592E+14	7,98657E+14	8,02399E+14
	Std	4,71193E+14	4,92587E+14	6,09428E+14	3,88652E+14	2,90399E+14	1,14364E+14	2,52155E+14	3,14113E+14
	GAP	7,07251E+14	6,01209E+14	4,62981E+14	5,55286E+14	4,95779E−01	5,23356E−01	6,57573E−01	1,12915E+14
Cap132	Mean	9,52096E+14	9,36804E+14	8,97535E+14	9,27601E+14	8,52066E+14	8,53494E+14	8,54632E+14	8,54145E+14
	Std	1,07716E+14	1,42182E+14	9,32979E+14	7,24744E+14	1,03016E+14	1,96905E+14	2,57628E+14	2,11214E+14
	GAP	1,18146E+14	1,00187E+14	5,40693E+14	8,93784E+14	6,69653E−02	2,34754E−01	3,68396E−01	3,11235E−01
Cap133	Mean	1,05201E+14	1,02830E+14	9,55882E+14	1,00520E+14	8,94080E+14	8,94136E+14	8,94529E+14	8,94761E+14
	Std	1,97870E+14	1,42320E+14	1,46728E+14	1,04773E+14	4,55642E+14	6,95473E+14	6,15573E+14	1,32839E+14
	GAP	1,77960E+14	1,51409E+14	7,03245E+14	1,25546E+14	1,12325E−01	1,18660E−01	1,62569E−01	1,88538E−01
Cap134	Mean	1,18725E+14	1,14072E+14	1,03389E+14	1,11962E+14	9,28942E+14	9,29102E+14	9,28942E+14	9,28942E+14
	Std	2,15215E+14	2,51172E+14	1,06148E+14	1,42214E+14	1,22713E+04	2,58822E+14	1,22713E+04	1,22713E+04
	GAP	2,78071E+14	2,27976E+14	1,12976E+14	2,05266E+14	2,50641E+00	1,73040E−02	2,50641E+00	2,50641E+00

Table 3 presents the performance of eight versions (V1 to V8) of BinHOA_CA on UFLPs. BinHOACA-V1 to BinHOACA-V8 show consistent mean performance in most cases, but versions with adaptive components (BinHOACA-V5 to BinHOACA-V7) often display reduced standard deviation, indicating more stable performance. For instance, Cap71 shows zero standard deviation for BinHOACA-V5 to V7, contrasting with higher variability in earlier versions.

Table 3. Statistical results of all versions of BinHOA_CA on UFLPs

		BinHOACA-V1	BinHOACA-V2	BinHOACA-V3	BinHOACA-V4	BinHOACA-V5	BinHOACA-V6	BinHOACA-V7	BinHOACA-V8
Cap71	Mean	9,32711E+14	9,32932E+14	9,38785E+14	9,34073E+14	9,32616E+14	9,32616E+14	9,32616E+14	9,32963E+14
	Std	3,01412E+14	5,16729E+14	1,04772E+14	1,05750E+14	0,00000E+00	0,00000E+00	0,00000E+00	5,65629E+14
	GAP	1,02202E−02	3,39566E−02	6,61489E−01	1,56303E−01	0,00000E+00	0,00000E+00	0,00000E+00	3,72527E−02
Cap72	Mean	9,78122E+14	9,79659E+14	9,82438E+14	9,79585E+14	9,77799E+14	9,77799E+14	9,77799E+14	9,77799E+14
	Std	5,20192E+14	1,83905E+14	2,11745E+14	1,41046E+14	1,22713E+04	1,22713E+04	1,22713E+04	1,22713E+04
	GAP	3,30405E−02	1,90165E−01	4,74423E−01	1,82565E−01	1,19058E+00	1,19058E+00	1,19058E+00	1,19058E+00
Cap73	Mean	1,01415E+14	1,01312E+14	1,01283E+14	1,01252E+14	1,01121E+14	1,01083E+14	1,01064E+14	1,01119E+14
	Std	1,99888E+14	1,57847E+14	1,35495E+14	1,65185E+14	8,76650E+14	5,80444E+14	0,00000E+00	8,86643E+14
	GAP	3,47373E−01	2,45374E−01	2,16982E−01	1,85970E−01	5,61355E−02	1,81620E−02	0,00000E+00	5,44859E−02
Cap74	Mean	1,05119E+14	1,04540E+14	1,03822E+14	1,04182E+14	1,03498E+14	1,03498E+14	1,03498E+14	1,03498E+14
	Std	3,93353E+14	2,81235E+14	2,24677E+14	3,18250E+14	1,22713E+04	1,22713E+04	1,22713E+04	1,22713E+04
	GAP	1,56621E+14	1,00752E+14	3,13335E−01	6,60898E−01	−1,1248E+00	−1,12481E+00	−1,12481E+00	−1,12481E+00
Cap101	Mean	8,00336E+14	8,02591E+14	8,07763E+14	8,03398E+14	7,97251E+14	7,97352E+14	7,98159E+14	8,00213E+14
	Std	1,66392E+14	1,02826E+14	2,46395E+14	2,27958E+14	4,15558E+14	3,73442E+14	1,29811E+14	1,87097E+14
	GAP	4,62827E−01	7,45925E−01	1,39512E+14	8,47300E−01	7,55918E−02	8,82554E−02	1,89586E−01	4,47465E−01
Cap102	Mean	8,66378E+14	8,63524E+14	8,64606E+14	8,63970E+14	8,55037E+14	8,54907E+14	8,55632E+14	8,56462E+14
	Std	3,69309E+14	3,61721E+14	3,98178E+14	3,74188E+14	5,54748E+14	4,44241E+14	1,24163E+14	8,60700E+14
	GAP	1,36584E+14	1,03190E+14	1,15854E+14	1,08412E+14	3,89657E−02	2,37532E−02	1,08506E−01	2,05610E−01
Cap103	Mean	9,18360E+14	9,10928E+14	9,04894E+14	9,10662E+14	8,94710E+14	8,94269E+14	8,94348E+14	8,94586E+14
	Std	6,18679E+14	6,14571E+14	2,75128E+14	4,63610E+14	4,99991E+14	4,80538E+14	4,43522E+14	5,15915E+14
	GAP	2,74989E+14	1,91838E+14	1,24325E+14	1,88855E+14	1,03857E−01	5,44355E−02	6,33082E−02	8,99263E−02
Cap104	Mean	9,82001E+14	9,71568E+14	9,47522E+14	9,58121E+14	9,28942E+14	9,28942E+14	9,29506E+14	9,28942E+14
	Std	8,28441E+14	8,31174E+14	8,56440E+14	7,69317E+14	1,22713E+04	1,22713E+04	1,78518E+14	1,22713E+04
	GAP	5,71176E+14	4,58872E+14	2,00013E+14	3,14109E+14	2,50641E+00	2,50641E+00	6,07705E−02	2,50641E+00
Cap131	Mean	8,54288E+14	8,42474E+14	8,29276E+14	8,38313E+14	7,95157E+14	7,97395E+14	7,99518E+14	8,02071E+14
	Std	4,20772E+14	5,54007E+14	4,74011E+14	3,61803E+14	1,74100E+14	2,44133E+14	2,79262E+14	3,44095E+14
	GAP	7,66890E+14	6,18003E+14	4,51659E+14	5,65553E+14	2,16406E−01	4,98541E−01	7,66025E−01	1,08780E+14
Cap132	Mean	9,60194E+14	9,37814E+14	8,95973E+14	9,28838E+14	8,53141E+14	8,52181E+14	8,54174E+14	8,56579E+14
	Std	1,31368E+14	7,07685E+14	9,92825E+14	3,03094E+14	1,98781E+14	6,95520E+14	2,51089E+14	1,98183E+14
	GAP	1,27656E+14	1,01373E+14	5,22353E+14	9,08317E+14	1,93317E−01	8,05360E−02	3,14610E−01	5,97047E−01
Cap133	Mean	1,04910E+14	1,03133E+14	9,52663E+14	1,00325E+14	8,93880E+14	8,94279E+14	8,94195E+14	8,94730E+14
	Std	2,26270E+14	1,15492E+14	1,69320E+14	1,07704E+14	4,27549E+14	1,22697E+14	5,29881E+14	1,51067E+14
	GAP	1,74703E+14	1,54811E+14	6,67198E+14	1,23363E+14	8,99707E−02	1,34659E−01	1,25262E−01	1,85142E−01
Cap134	Mean	1,19233E+14	1,15264E+14	1,03400E+14	1,11608E+14	9,28995E+14	9,28942E+14	9,28995E+14	9,28942E+14
	Std	2,70915E+14	1,49389E+14	1,53797E+14	1,52928E+14	1,69439E+14	1,22713E+04	1,69439E+14	1,22713E+04
	GAP	2,83532E+14	2,40813E+14	1,13093E+14	2,01448E+14	5,76799E−03	2,50641E+00	5,76799E−03	2,50641E+00

The GAP values, representing the deviation from optimal solutions, further highlight the efficacy of adaptive versions. For example, the GAP for Cap71 is zero in BinHOACA-V5 to V7, suggesting optimal or near-optimal solutions. Conversely, earlier versions exhibit larger gaps, especially BinHOACA-V3, indicating less efficient solutions.

Standard deviation patterns also suggest that newer versions (BinHOACA-V5 to V8) provide more consistent performance. In Cap73, zero standard deviation is noted in BinHOACA-V7, contrasting sharply with the high variability in BinHOACA-V1 to V3. This pattern repeats across other capacities, such as Cap72 and Cap104, demonstrating the enhanced reliability of later versions. In conclusion, the adaptive versions of BinHOA_CA (V5 to V7) consistently outperform earlier versions in terms of solution quality and stability.

Table 4 provides a comparison of the mean values for BinHOA and BinHOA_CA across various versions and capacities, where a value of 1 indicates that BinHOA outperforms BinHOA_CA, 2 indicates that BinHOA_CA outperforms BinHOA, and 0 indicates that both algorithms perform equally well. The results demonstrate the impact of modifying the position update strategy in BinHOA, inspired by Moore Neighborhood, on the performance of BinHOA_CA.

Across all instances and versions, BinHOA_CA shows a clear advantage. Specifically, BinHOA_CA outperforms BinHOA in more instances, with the following counts of better performances (indicated by 2s): V1 (4), V2 (6), V3 (7), V4 (7), V5 (4), V6 (6), V7 (5), and V8 (6). Conversely, BinHOA only outperforms BinHOA_CA in the following counts: V1 (8), V2 (6), V3 (5), V4 (5), V5 (4), V6 (2), V7 (4), and V8 (1). The number of instances where both algorithms perform equally (indicated by 0s) varies, with V1 (0), V2 (0), V3 (0), V4 (0), V5 (4), V6 (4), V7 (3), and V8 (5). The analysis reveals that BinHOA_CA generally offers better performance, particularly in versions V6 and V8, where it clearly outperforms BinHOA in more instances.

In summary, Table 4 underscores the success of the BinHOA_CA algorithm by demonstrating its superiority in achieving better mean values compared to the original BinHOA algorithm. This improvement can be attributed to the modified position update strategy inspired by the Moore Neighborhood, which enhances the algorithm's capability to explore and exploit the solution space more effectively.

Table 4. Comparison of mean values of BinHOA and BinHOA_CA for all versions.

	V1	V2	V3	V4	V5	V6	V7	V8
Cap71	1	2	2	1	0	0	0	2
Cap72	2	2	1	2	0	0	0	0
Cap73	1	2	1	2	1	2	2	0
Cap74	1	2	2	2	0	0	0	0
Cap101	2	1	1	1	2	2	2	2
Cap102	2	1	2	2	2	2	1	2
Cap103	1	2	2	1	1	1	2	2
Cap104	1	2	1	2	0	0	1	0
Cap131	1	1	2	1	2	2	1	2
Cap132	1	1	2	1	1	2	2	1
Cap133	2	1	2	2	2	1	2	2
Cap134	1	1	1	2	1	2	1	0
	8/0/4	6/0/6	5/0/7	5/0/7	4/4/4	2/4/6	4/3/5	1/5/6

6 Conclusions and Future Works

The Binary Hiking Optimization Algorithm (BinHOA), proposed and evaluated in this study, has demonstrated significant potential in addressing binary optimization problems, particularly the Uncapacitated Facility Location Problems (UFLPs). By incorporating various S-shaped and V-shaped transfer functions, BinHOA has effectively adapted the continuous Hiking Optimization Algorithm (HOA) to the binary domain, facilitating its application to a broader range of optimization challenges.

The experimental results underscore the critical role of transfer functions in determining the algorithm's performance. Among the tested versions, V5 and V6 consistently outperformed others, showing lower mean values, standard deviations, and GAP values, thus indicating their superior ability to find optimal or near-optimal solutions. These findings highlight the necessity of selecting appropriate transfer functions to enhance the efficiency of BinHOA in binary optimization tasks.

A novel enhancement method for the location update strategy was proposed, drawing inspiration from the Moore neighborhood concept used in cellular automata. This led to the development of BinHOA_CA, an advanced version of BinHOA. The performance comparison between BinHOA and BinHOA_CA across various UFLP instances revealed that BinHOA_CA generally offers better performance, particularly in versions V6 and V8, where it outperformed BinHOA in more instances. The modified position update strategy enabled BinHOA_CA to explore and exploit the solution space more effectively, thereby achieving better mean values and more consistent results.

The detailed analysis in this study provides a comprehensive understanding of the strengths and limitations of BinHOA and its enhanced version, BinHOA_CA. The superiority of BinHOA_CA in most instances highlights the effectiveness of the proposed improvement method. Moreover, the comparative analysis revealed that while some versions of BinHOA performed adequately, the adaptive versions (V5 and V6) consistently delivered better results across different UFLP instances, emphasizing the importance of algorithmic adaptability and strategic enhancements.

In conclusion, this study contributes significantly to the field of binary optimization by introducing and validating BinHOA and BinHOA_CA. The success of these algorithms in solving UFLPs suggests their potential applicability to other binary optimization problems. Future work aims to integrate different aspects of cellular automata techniques, similar to the Moore neighborhood, into the optimization algorithm to enhance performance further.

Acknowledgments. The author would like to thank Selçuk University Scientific Research Projects Coordination Office for their institutional support and the use of the laboratory established within the scope of the Data Intensive and Computerized Vision Research Laboratory Infrastructure Project.

References

1. Martí, R., Sevaux, M., Sörensen, K.: 50 years of metaheuristics. Eur. J. Oper. Res. (2024)
2. Oladejo, S.O., Ekwe, S.O., Mirjalili, S.: The hiking optimization algorithm: a novel human-based metaheuristic approach. Knowl.-Based Syst. **296**, 111880 (2024)
3. Kuehn, A.A., Hamburger, M.J.: A heuristic program for locating warehouses. Manag. Sci. **9**, 643–666 (1963)
4. Akan, T., Agahian, S., Dehkharghani, R.: Battle royale optimizer for solving binary optimization problems. Softw. Impacts **12**, 100274 (2022)
5. Karakoyun, M., Ozkis, A.: A binary tree seed algorithm with selection-based local search mechanism for huge-sized optimization problems. Appl. Soft Comput. **129**, 109590 (2022)
6. Çınar, A.C.: A comprehensive comparison of binary archimedes optimization algorithms on uncapacitated facility location problems. Düzce Üniversitesi Bilim ve Teknoloji Dergisi **10**, 27–38 (2022)
7. Zhang, F., He, Y., Ouyang, H., Li, W.: A fast and efficient discrete evolutionary algorithm for the uncapacitated facility location problem. Expert Syst. Appl. **213**, 118978 (2023)
8. He, Y., Zhang, F., Mirjalili, S., Zhang, T.: Novel binary differential evolution algorithm based on Taper-shaped transfer functions for binary optimization problems. Swarm Evol. Comput. **69**, 101022 (2022)
9. Goodchild, M.F.: Beyond Tobler's hiking function. Geogr. Anal. **52**, 558–569 (2020)
10. Beasley, J.E.: OR-library: distributing test problems by electronic mail. J. Oper. Res. Soc. **41**, 1069–1072 (1990)
11. Mirjalili, S., Lewis, A.: S-shaped versus V-shaped transfer functions for binary particle swarm optimization. Swarm Evol. Comput. **9**, 1–14 (2013)
12. Zaitsev, D.A.: A generalized neighborhood for cellular automata. Theoret. Comput. Sci. **666**, 21–35 (2017)

A Spatial Daisyworld Model

Franco Bagnoli[1,2,3], Marco Bosi[1], and Tommaso Matteuzzi[1]

[1] Department of Physics and Astronomy and CSDC, University of Florence,
via G. Sansone 1, 50019 Sesto Fiorentino, Italy
{franco.bagnoli,tommaso.matteuzzi}@unifi.it, marco.bosi@edu.unifi.it
[2] INFN, Florence sect., Florence, Italy
[3] Associate Member of the UMR Espace-Dev, University of Perpignan via Domitia, Perpignan, France

Abstract. We present a simple spatial implementation of the classic Daisyworld model, similar to a cellular automaton. The spatial version reproduces well the standard scenario with hysteresis cycles, and allows for a more physical implementation of the diffusion of temperature. It allows also to examine spatial behaviors, in particular the difficulty of colonization of territory by a more adapted species. This element, coupled to a finite velocity of the variation of luminosity of the sun, induces the appearance of new hysteresis cycles that reduce the adaptation capacity of the system. Mutations do not increase stability and may enhance the spurious hysteresis cycles.

Keywords: Daisyworld · phase transitions · hysteresis cycles

1 Introduction

The Daisyworld model was introduced to support Gaia hypothesis [1], i.e., that life forms interact with each other and with inorganic matter, forming a single system that evolves and self-regulates to maintain the conditions of the Earth's environment such that life can thrive.

The Gaia hypothesis was strongly criticized, but it sparked many debates and encouraged several scientists to study models, such as the Daisyworld and its increasingly complex derivatives, which have proven useful for the study of ecosystem dynamics. Some examples are covered in Ref. [2].

In the original model there are two species of daisies, "white" and "black" ones, characterized by different albedo (the fraction of sunlight reflected by a body). Both daises survive only within a certain temperature range, and the local temperature is given by a balance between the incoming radiation (modulated by the albedo), dissipation to space, and heat diffusion.

The variable proportion of white and black daises is able to modulate the total albedo of the planet and therefore to regulate its temperature beyond the survival limits of an empty planet, when the sun intensity changes.

The original model is formulated in mean-field (differential or difference equations) terms. A certain number of spatial versions have been developed [3,4]. In this paper we want to show that a cellular automaton version of the model reproduces qualitatively the mean-field results and shows additional features (hysteresis cycles), that reduce the adaptability of the system, not present in the mean-field models and not reported in the spatial versions.

The outline of the paper is the following. The mean-field model is presented in Sect. 2. We then introduce the spatial cellular automaton version in Sect. 3. The numerical results are shown in Sect. 4, and an extension of the spatial model including mutations is presented in Sect. 5. Conclusions are drawn in the last section.

2 Mean-Field Model

Let us recall the main features of Lovelock and Watson's original Daisyworld model [5], using dimensionless variables.

The original mean-field model is based on a set of consistency equations plus a set of differential equations which we reformulate using unit time steps, so that differential equations become discrete maps. This is essentially also what was done by Watson and Lovelock, who integrated the equations using the Euler procedure. The prime denotes quantities at the following time step.

In the original model there are two species of daisies, "white" ($k = 1$) and "black" ($k = -1$) ones, characterized by different albedo (A_k), which is the fraction of sunlight reflected by a body. It is measured on a scale that goes from 0 (no reflection), corresponding to a black body, to 1 (reflection of all incident radiation) corresponding to a perfect white body (or a mirror). We assume that $A_1 = 0.75$ and $A_{-1} = 0.25$, while the albedo of the empty ground is $A_0 = 0.5$.

The fraction of the planet surface covered by species k is indicated by x_k. Clearly $\sum_k x_k = 1$. The average albedo $\overline{A}$ of the planet is

$$\overline{A} = \sum_k x_k A_k$$

The temperature T of the planet is given by the balance between the energy received by the sun, which irradiates a power S converted into heat according with the albedo $\overline{A}$, and the energy radiated by the planet, according with the Stefan-Boltzmann law. Assuming immediate equilibrium,

$$RT^4 = S(1 - \overline{A}), \qquad (1)$$

where R is the radiance of the planet. The model does not include the effects of an atmosphere, but the radiance R can model some effects due to cloud shading or greenhouse ones. The temperature T is given by the ratio S/R and in the following we shall put $R = 1$. The temperature T_e of the bare planet is similarly given by

$$RT_e^4 = S(1 - A_0).$$

The evolution of the populations of daisies is modeled by two coupled logistic equations (reduced to maps using the Euler approximation)

$$x'_k = x_k + [x_k(x_0 \beta_k - \gamma) + b]\Delta t, \qquad (2)$$

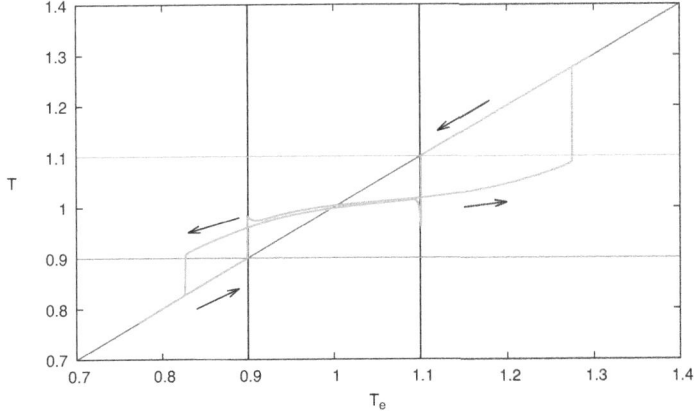

Fig. 1. A simulation of the mean-field version of the Daisyworld, showing the temperature of the planet with daisies, T, versus the temperature of the "empty planet" T_e when the sun intensity, S, is raised up to extinction of the daisies and then lowered. Arrows shows the cycle over S. The straight purple line marks the temperature of the empty planet. The yellow line marks the temperature of the planet with daises (repopulated after extinction). The vertical and horizontal lines denoted the comfort zones (for the empty $T_e = T^* \pm K$ and populated planet $T = T^* \pm K$, respectively). The presence of daisies stabilizes the temperature until it reaches the limit of the comfort zone, after which daisies disappear. By reverting the direction of the sun intensity, daises reappear when the temperature of the empty planet reaches the comfort zone. At this transition, there is a sudden increase or drop of the temperature, due to a quick increase of a population of single-color daises, followed by the restoration of a mixed population (see also Fig. 2). Simulation parameters are: $T^* = 1$, $K = 0.1$, $q = 0.01$, (almost equilibrium) $b = 4 \cdot 10^{-5}$, $\gamma = 0.01$ $\delta t = 0.01$ and $\eta = 10^{-5}$. The comfort zone $T_e = T^* \pm K$ and $T = T^* \pm K$ are marked.

for $k \neq 0$, where γ is the death rate, β_k is the survival rate (depending on temperature), x_0 is the empty fraction of the planet, and b is a small constant needed to remove the presence of absorbing states, i.e., to repopulate the planet in case of extinction (it may represent the presence of dormant seeds). The logic of these equations is that the variation of the population is given by the deaths (proportional to γ) and the reproduction after survival (β). Reproduction occurs if empty space (x_0) is available. Notice that the two equations are coupled by $x_0 = 1 - x_1 - x_{-1}$.

The survival function β_k establishes the temperature range compatible with life

$$\beta_k = \beta(T_k) = \begin{cases} 1 - \left(\frac{T_k - T^*}{K}\right)^2 & \text{for } |T_k - T^*| \leq K, \\ 0 & \text{otherwise,} \end{cases} \tag{3}$$

where T^* is the optimal temperature, and K is the survival temperature range. The comfort zone is therefore $T = T^* \pm K$.

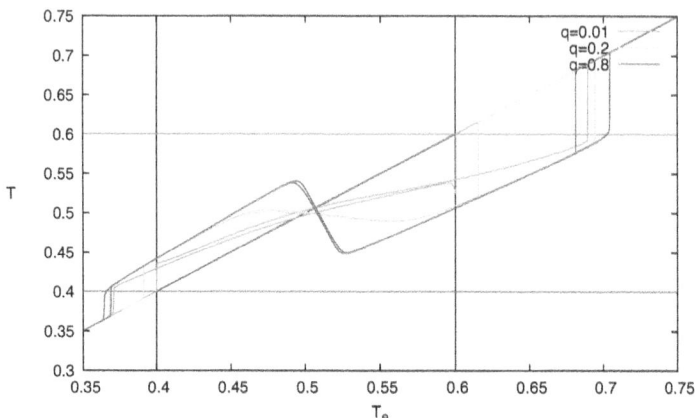

Fig. 2. Three simulations of the mean-field version of the Daisyworld model for different values of the diffusion parameter q, $q = 0.01$, (almost equilibrium), $q = 0.2$ and $q = 0.8$ (almost isolation), showing the average temperature T (thick line) vs the "empty world" temperature T_e. Notice that increasing the parameter q (i.e., reducing the diffusion of heat among different daisies species), more extended single-species intervals are present, marked by the slanted straight lines corresponding to a planet with a total albedo equal to that of black or white daisies. For large diffusion ($q = 0.01$) these intervals reduce to the small peaks at the transition. Other parameters are: $T^* = 0.5$, $K = 0.1$, $b = 4 \cdot 10^{-5}$, $\gamma = 0.01$ $\delta t = 0.01$ and $\eta = 10^{-5}$, The comfort zones $T^* \pm K$ and $T = T^* \pm K$ are marked by vertical and horizontal lines, resp.

Finally, there is an equation ruling the diffusion of heat among the zones occupied by the different species. In the original paper [5] the temperature of daisies was heuristically set to

$$T_k^4 = T^4 + q(\overline{A} - A_k)^{1/4},$$

where q is a parameter that controls diffusion. If $q = 0$, we have immediate relaxation $T_k = T_e$, i.e., infinite diffusion. On the other hand, if $q = S$, taking into account Eq. (1) we have

$$T_k = S(1 - A_k),$$

i.e., no diffusion.

This approach is questionable, since in the simulations the sun intensity $S = S(t)$ is being changed, so we prefer to define the temperature of daisies as

$$T_k^4 = T^4 + qS(\overline{A} - A_k)^{1/4},$$

where now q goes from 0 (immediate equilibrium) to 1 (no diffusion).

Finally, the sun intensity is varied according to some scheduling, for instance

$$S(t + \Delta t) = S(t) + \alpha \eta \Delta t,$$

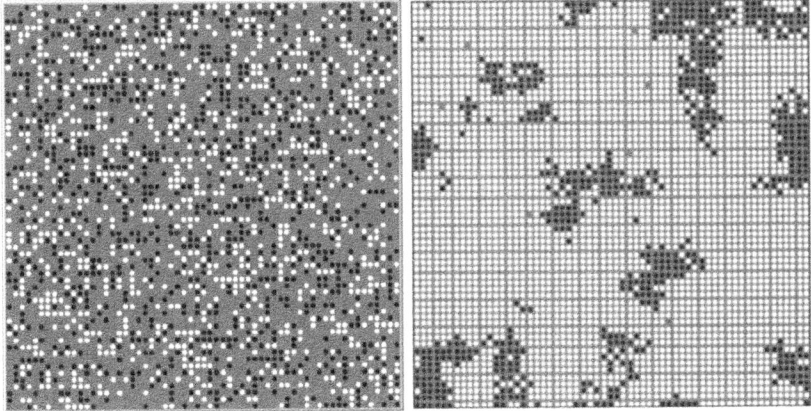

Fig. 3. Spatial Daisyworld model (Sect. 3). (left) A typical setup configuration, black and white dots marks the location of daises, the red color marks the empty cells. (right) A configuration for $D = 1$, $\Delta t = 0.01$. The planet is partitioned into a 60×60 grid (shown by the red lines on the right)

where $\alpha = \pm 1$ gives the increasing or decreasing of the radiation, and η is the variation rate, which in the original model is infinitesimal.

A typical simulation of T versus T_e when the sun intensity is changed is reported in Fig. 1. One can notice that the temperature of the simulated world follows that of the empty one until it arrives within the comfort zone. At this point, daisies colonizes the planet stabilizing the temperature, and this stabilization lasts well outside the comfort zone of the empty planet, up to $T_e \simeq 1.28$, where the effective temperature T is almost equal to $T^* + K$. The difference is due to the mortality rate of daisies.

At this point daisies get extincted and the temperature returns to that of the empty planet. We can then reverse the sign of α, decreasing the sun intensity. Daisies start to appear when the temperature of the empty planet enters again the comfort zone, stabilizing the temperature. An essentially similar scenario appears at the opposite limit of the comfort zone. At transition there is a peak corresponding to a single-species colonization of the planet, followed by a quick re-equilibrium among species due to a large diffusion of heat $q = 0.01$, as we shall see in the following.

The effects of the factor q, related to diffusion, is reported in Fig. 2. For large values of q, i.e., limited ($q = 0.2$ or almost no ($q = 0.8$) diffusion, the appearance of daises occurs for values of temperature of the empty planet T_e which are almost the same for which daisies get extincted (when α is reversed). This is due to the fact that daises of the appropriate color reduce or increase the temperature of their location without being hindered by the temperature or nearby empty locations. Therefore the first colonization of the planet is due to single-color daisies, and this is marked by the straight slanted lines intervals,

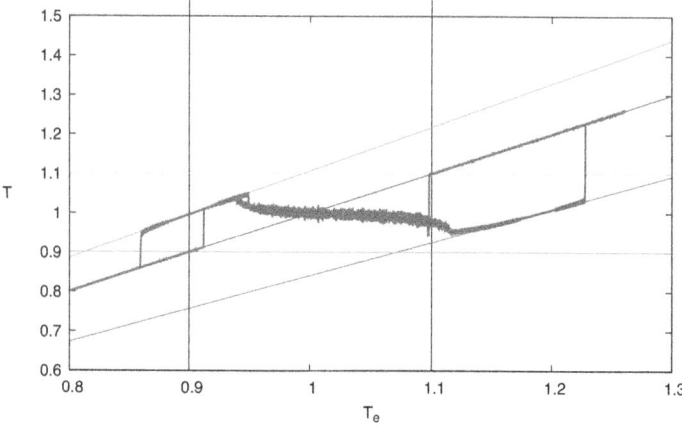

Fig. 4. The temperature diagram of the spatial Daisyworld model. T vs T_e when the sun intensity is changed in a cycle. Horizontal and vertical lines show the comfort zone. The continuous oblique lines, above and below the line $T = T_e$, show the temperature of the planet completely covered by black or white daisies. Other simulation parameters: $N = 60$, $T^* = 1$, $D = 1$, $\Delta t = 0.01$, $K = 0.1$ $\eta = 10^{-6}$, $\gamma = 0.01$, $b = 4 \cdot 10^{-5}$.

particularly evident for $q = 0.8$. At $T_e = 0.5$ there is a sudden switch of the population.

On the contrary, for $q = 0.01$ (almost equilibrium) the temperature is always constant and therefore the population is always mixed (except for a tiny interval at the transition). When the planet is populated, the small value of q is still able to make a difference between the survival probability of the two species and the population can survive, changing its composition, up to essentially the same limits of other values of q, but the appearance of daisies after extinction is severely hindered by the fact that almost immediately also daisies of opposite color may appear. Therefore, for $q = 0.01$ re-population occurs when $T_e = T^* \pm K$ and the single-species interval is limited to the tiny interval corresponding to the peak at the transition.

Two hysteresis cycles are clearly present, with the main result that the presence of daisies stabilizes the temperature of the planet well beyond the comfort zone of the empty planet.

We shall see how this behavior changes in a spatial model.

3 Spatial Model

Our aim was to define a spatial version of the model as a 2D cellular automaton (or non-movable agents, Fig. 3), see also Refs. [3,4].

Each cell s_{ij} of the automation (i and j run from 1 to $N = 60$) can take three values, $s_{ij} \in \{-1, 0, 1\}$, which correspond to the presence of a black (-1) or white (1) daisy, or empty patch (0). We can define the albedo $a_{ij} = A(s_{ij})$ imposing as in the mean-field model that $A(1) = 0.75$, $A(0) = 0.5$ and $A(-1) = 0.25$.

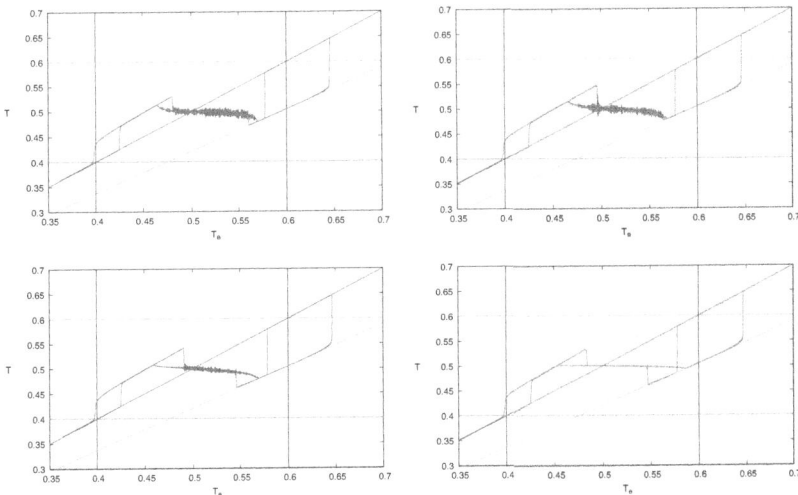

Fig. 5. The temperature diagram of the spatial Daisyworld model for different values of the diffusion coefficient D. (top left) $D = 0$, (top right) $D = 1$, (bottom left) $D = 5$, (bottom right) $D = 20$. Continuous lines as in Fig. 4. Other parameters: $N = 60$, $T^* = 0.5$, $\Delta t = 0.01$, $K = 0.1$, $\eta = 10^{-6}$, $\gamma = 0.01$, $b = 4 \cdot 10^{-5}$.

Each cell has also its own temperature T_{ij}, which evolves according with the energy balance, which we choose to make a bit more realistic, defining

$$T_{ij}\left(t + \frac{1}{2}\Delta t\right) = T_{ij} + \left[(S * (1 - a_{ij}) - (T_{ij})^4\right]\Delta t, \qquad (4)$$

followed by diffusion

$$T_{ij}(t + \Delta t) = (1 - D\Delta t)T_{ij}\left(t + \frac{1}{2}\Delta t\right) + \overline{T}_{ij}\left(t + \frac{1}{2}\Delta t\right)D\Delta t, \qquad (5)$$

where $\overline{T}_{ij}$ is the temperature averaged over the 8 neighbors. The quantity $\beta(T_{ij})$ of Eq. (3) is now interpreted as a survival probability.

The logistic equations Eq. (2) are replaced by a rule for empty cells: an empty cell selects one of its 8 neighbors, and if it is occupied, its state is copied. Notice that temperature does not change, so too hot or too cold patches cannot be occupied by reproduction, until eventually cooled by diffusion or emission.

4 Simulations

Simulations reproduce well the expected behavior (see Fig. 4). By changing the diffusion rate D (the actual diffusion coefficient is $D\Delta t$), we observe however a different scenario with respect to the mean-field approximation presented in Sect. 2.

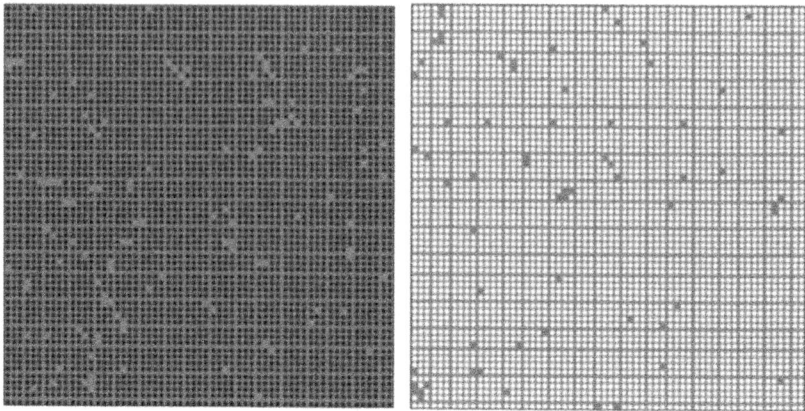

Fig. 6. Typical configurations of the spatial Daisyworld in the additional hysteresis cycles. The appearance of the opposite type of daisies is hindered by the small number of empty cells.

With no diffusion ($D = 0$, Fig. 5 top left), the scenario is similar to the previous one, except that the hysteresis cycles are smaller and shifted inside the comfort zone. This is due to the "inertia" of the surface, couples to a finite (albeit small) variation rate of the sun. Differently from the mean-field case, in the spatial model the temperature follows an evolution equation (Eq. (4)).

By increasing the diffusion rate, there appear two new hysteresis cycles, which are quite evident varying D, see Fig. 5 showing the hysteresis diagram for $D = 0$ (top left), $D = 1$ (bottom-left), $D = 5$ (top-right) and $D = 10$ (bottom-right). This is again due to the temperature inertia, in this case due to diffusion, Eq. (5), so that when the sun intensity decreases or increases the surface remains too hot or too cold (resp.) for a longer time.

In these new hysteresis cycles, the appearance of the opposite type of daisies is not very probable due to the small number of empty cells, see Fig. 6.

5 Mutations

The model with only two species of daisies is not really quite biological, since in general mutations and adaptations occur. It is possible to add mutation to the spatial Daisyworld model during the reproduction phase, see also Ref. [6]. We allow daisies to mutate their albedo with a probability $\mu = 10^{-4}$ (order one mutation per time step). The mutation range was limited to the same interval (0.25–0.75) of the original one, to avoid to introduce spurious control capacities.

Results are reported in Fig. 7 for different values of the diffusion rate D. By comparison with Fig. 5, one sees that mutation does not stabilizes the system more. It has the effect, for moderate values of D, of lowering the control capacity in the extreme cases (only white daisies, $T_e \simeq 0.8$), since mutations only decrease the number of surviving daisies.

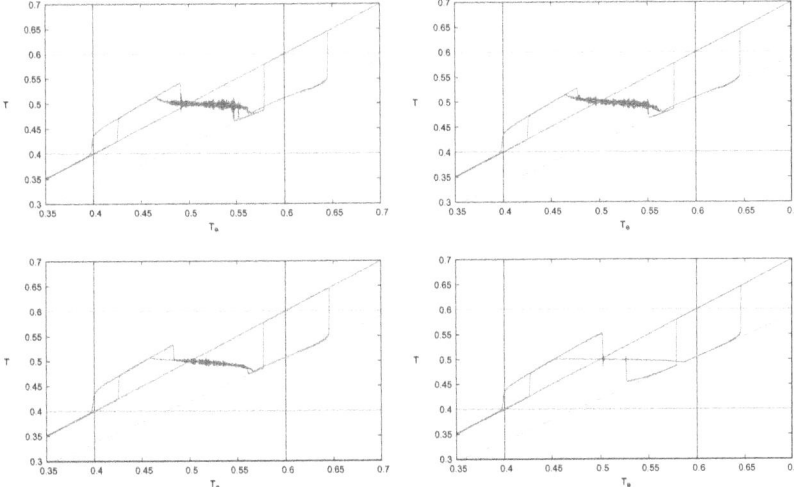

Fig. 7. The temperature diagram of the spatial Daisyworld model in presence of mutations, $\mu = 10^{-4}$. $D = 0$ (top left), $D = 1$ (top right), $D = 5$ (bottom left) and $D = 20$ (bottom right). Other parameters: $N = 60$, $T^* = 0.5$, $\Delta t = 0.01$, $K = 0.1$, $\eta = 10^{-6}$, $\gamma = 0.01$, $b = 4 \cdot 10^{-5}$.

The most conspicuous effect is that for large diffusion rates, for instance $D = 20$ (Fig. 7 bottom right), where the extra hysteresis cycles are enhanced. This is due to the survival of mutants of intermediate albedo, which are not so efficient for temperature control as daisies of opposite albedo replacing the dying ones, but still able to survive with respect to the latter.

6 Conclusions

We presented a simple spatial implementation of the classic Daisyworld model. We revised the standard mean-field model inserting a consistent mechanism for the diffusion of temperature, compatible with the energy balance. We show that this mechanism promotes unexpected effects, in particular an inversion in the temperature curve.

The spatial version reproduces well the mean-field scenario with hysteresis cycles, and allows for a more physical implementation of the diffusion of temperature.

In this model, spatial effects like the difficulty of colonization of territory by a more adapted species are easily reproduced. This difficulty, coupled to a finite velocity of the variation of luminosity of the star, induces the appearance of new hysteresis cycles that reduce the adaptation capacity of the system.

Finally, we observe that mutations do not increase the system stability while they may enhance the amplitude of the spurious hysteresis cycles.

Acknowledgments. This publication was produced with the co-funding of European Union - Next Generation EU, in the context of The National Recovery and Resilience Plan, Investment 1.5 Ecosystems of Innovation, Project Tuscany Health Ecosystem (THE), CUP: B83C22003920001.

Disclosure of Interests. The authors declare no conflict of interest.

References

1. Dutreuil, S.: James Lovelock's Gaia hypothesis: "a new look at life on earth"... for the life and the earth sciences. In: Dreamers, Visionaries, and Revolutionaries in the Life Sciences, pp. 272–287 (2018)
2. Lenton, T.M., Lovelock, J.E.: Daisyworld revisited: quantifying biological effects on planetary self-regulation. Tellus b **53**(3), 288–305 (2001)
3. Punithan, D., McKay, R.I.B.: Evolutionary dynamics and ecosystems feedback in two dimensional daisyworld. Artif. Life **13**, 91–98 (2012)
4. Wood, A.J., et al.: Daisyworld: a review. Rev. Geophys. **46**(1) (2008)
5. Watson, A.J., Lovelock, J.E.: Biological homeostasis of the global environment: the parable of daisyworld. Tellus B: Chem. Phys. Meteorol. **35**(4), 284–289 (1983)
6. von Bloh, W., Block, A., Schellnhuber, H.-J.: Selfstabilization of the biosphere under global change: a tutorial geophysiological approach. Tellus B **49**(3), 249–262 (1997)

A Reaction-Diffusion Cellular Automata Model for Mycelium-Based Engineered Living Materials Evolution

Ioannis Tompris[1]($\boxtimes$), Ioannis K. Chatzipaschalis[1,2],
Theodoros Panagiotis Chatzinikolaou[1], Iosif-Angelos Fyrigos[1],
Michail-Antisthenis Tsompanas[3], Andrew Adamatzky[3], Phil Ayres[4],
and Georgios Ch. Sirakoulis[1]

[1] Democritus University of Thrace, DUTH University Campus, 67100 Xanthi, Greece
itompris@ee.duth.gr
[2] Universitat Politècnica de Catalunya, 08034 Barcelona, Spain
[3] University of the West of England, Bristol, UK
[4] Chair for Biohybrid Architecture, Institute of Architecture and Technology, Royal Danish Academy, 1435 København K, Denmark

Abstract. Engineered living materials (ELMs) and, more specifically, mycelium-based ELMs have been proposed as a solution to address the escalating societal pressures related to human-induced environmental disruption, scarcity of resources, and the anticipated increase in material demand. However, due to the complex biological mechanisms they emulate, their environmental sensitivity, slow supply chain and regulations, these devices present significant challenges for reproduction. Consequently, modeling the phenomena underlying such devices becomes critically important. In this context, we introduce a comprehensive mycelium-based ELM framework that incorporates reaction-diffusion processes and the modeling tool of Cellular Automata (CA). This framework successfully simulates the ELM's unpredictable growth mechanisms and closely resembles the mycelium's biological structure through the exploitation of the reaction-diffusion activator-inhibitor system. Finally, an augmented 3D version is presented that enhances the realism of our findings and strives to provide a deeper understanding of such materials.

Keywords: Cellular Automata · Reaction-Diffusion · Mycelium-based ELMs · 2D and 3D Modeling

1 Introduction

Engineered Living Materials (ELMs) represent an innovative and burgeoning area of research, distinguished by their remarkable capacity to incorporate the self-healing, regenerative, and adaptive properties inherent to biological systems with the robust and versatile structural capabilities, emblematic of materials science [21]. This interdisciplinary approach enables the development of dynamic materials that can adapt, evolve, and respond to environmental stimuli, mirroring the inherent functionalities of living organisms while maintaining the strength and reliability of engineered substances. Additionally, ELMs are utilized in designing biosensors that offer sensitive, selective, and biocompatible solutions for health and environmental monitoring [20,23,26]. This demonstrates ELMs' versatility and their revolutionary potential in various applications.

Out of all the ELMs, mycelium-based ones stand out and are the primary focus of this work [2,10,14]. Mycelium, the vegetative part of a fungus, is renowned for its rapid growth and sustainability, presenting an eco-friendly alternative to traditional materials while also being in abundance in nature [4]. Mycelium-based ELMs are being increasingly applied across a wide range of sectors, showcasing their ability to innovate upon traditional materials science. Their uses include creating self-healing biofilms that improve durability [10], developing microgels for drug delivery and soft robotics [19], enhancing 3D printing for complex bioactive constructs [5], and producing nanofibers through electrospinning for biomedical applications and electronics [13].

However, their practical implementation is proven to be a difficult task, due to their intricate mechanisms at a microscopic level and more specifically the mycelium's tips, known as the hyphae, which exhibit behaviors such as extension, apical and lateral branching, anastomosis, nutrient uptake, and substrate diffusion, among others [11]. These mechanisms are the exact ones that we aim to replicate using reaction-diffusion (RD) systems capable of generating fractal-like patterns akin to natural mycelium networks [9,25]. Unlike diffusion systems, which depict growth without accounting for losses and involve individually incorporating and managing each mechanism through alterations in the corresponding equations, constituting a forced growth process, reaction-diffusion systems compensate for losses via reaction processes. This results in more spontaneous growth, offering a more natural and realistic simulation of such organisms.

Towards this direction, significant research has focused on Cellular Automata (CA) over the past decades [1]. CA are computational models that utilize parallel processing capabilities and can act as distributed computational systems [3]. In CA, complex global phenomena emerge from the local interactions of simple, identical units. CA integrate computational power with mathematical and physical concepts to explore complex systems arranged as evolving cell grids with local interconnections [22]. A defining feature of CA models is their ability to demonstrate emergent behavior; this is where complex patterns and behaviors develop from simple inter-cellular rules. Moreover, the parallel computing structure inherent in CA supports a versatile and robust tool for modeling and simulating large and complex systems [7,8,29], providing insights into these systems' dynamics that might be unattainable through other modeling techniques and methodologies.

This work is focused on modeling the complex dynamics found in Mycelium-based ELMs, especially at the Mycelium's tip, which includes tip extension, branching, anastomosis, and nutrient uptake from the ELM's substrate. The approach involves utilizing an established RD system based on activator-inhibitor pairs [27] and modifying it to generate patterns that exhibit these dynamics. The result is a generic model that accurately represents growth in Mycelium-based

ELMs. In addition, the system is constructed on a CA framework, providing a discrete model that enhances parallelism while also providing control over the ELM's diffusion. Consequently, this results in 2D and 3D general Mycelium-based ELM models that can also be fitted to experimental data to represent the growth of biological organisms by adjusting the corresponding RD and CA parameters.

2 Mycelium-Based ELM Reaction-Diffusion Mechanisms

Reaction-diffusion systems are widely used as mathematical models that describe the spatio-temporal changes and interactions between multiple substances, and as such, they are instrumental in producing complex patterns [18] and play a crucial role in understanding the formation of biological patterns [12,15]. Moreover, these models are scalable to any desired dimension because they are based on partial differential equations, which are inherently multidimensional. This characteristic is notably advantageous, as it enables the design of both two-dimensional and three-dimensional models based on these principles.

Moreover, in order to improve the understanding behind pattern formation, it is crucial to examine the details of the reaction-diffusion system employed [27]. This system outlines the dynamics between two abstract key substances depicted in Fig. 1(a), that affect the model's state: the activator and the inhibitor, and their corresponding influence on Mycelium formation. The activator encourages the Mycelium's growth on the condition that its concentration exceeds a certain threshold. On the other hand, the inhibitor acts to reduce the concentration of the activator, thus having a suppressive effect. Moreover, in Fig. 1(b) a more sophisticated point of view on the inhibitor's effect on the system is depicted, in which the inhibitor shapes the growth of the Mycelium, by managing the uncontrolled growth of the activator.

Fig. 1. (a) Activator-inhibitor dynamics. (b) The inhibitor's effect on the Mycelium's growth.

In reaction-diffusion models, activator and inhibitor pairs make it possible to simulate complex mycelial growth behaviors such as tip expansion, branching, and anastomosis. Within this framework, inhibitors manage and limit growth, whereas activators promote expansion. The outward diffusion of the activator drives the extension of mycelial hyphae by effectively replicating tip extension. Branching emerges naturally, caused by the inhibitor's increased concentration, which gives shape to the model by preventing the contact of the activator fronts. On the other hand, when the activator's concentration is high enough, the activator fronts merge, which is similar to anastomosis, which is when hyphal tips join together.

The mathematical expressions of the RD system are given in Eqs. (1)–(4):

$$\frac{\partial u}{\partial t} = \nabla^2 u + \zeta(\kappa u + u^2 - \lambda uv), \quad where\ u \in \Omega_c \tag{1}$$

$$\frac{\partial v}{\partial t} = d\nabla^2 v + \zeta(\mu u^3 - v), \quad where\ v \in \Omega \tag{2}$$

$$\frac{dc}{dt} = \zeta \nu c(a(u) - c)(c - 1), \quad where\ c \in \Omega \tag{3}$$

$$a(u) = \begin{cases} \xi, & if\ u \geqslant threshold \\ \xi - \theta(u - threshold), & otherwise \end{cases} \tag{4}$$

where u and v represent the concentrations of the activator and the inhibitor, respectively. In Eq. (1) and Eq. (2), the first term on the right includes the Laplacian operator that describes how the solutions propagate through space. This term is called the diffusion term. Subsequently, the second term, which is a polynomial of u and v, elucidates the interaction between the solutions u and v, known as the reaction term. Equation (3) describes the derivative of the mycelium state c over time, with negative values of c to highlight the presence of a mycelial tip, while $|c|$ describes the concentration of the mycelium. Tips are formed due to the extremely high and condensed concentration of the activator in a specific area, which significantly increases c. According to Eq. (3), this will result in the derivative of c to be highly negative, as ξ, which will ultimately be the value of $a(u)$, is a constant selected to be smaller than 1, thus explaining the negative value of the tips. Finally, Eq. (4) provides the function $a(u)$, which describes the dependence of the mycelium on the activator.

It is important to remember that v and c can take on values anywhere in the grid-plane $\Pi_{i,j}$, which is shown as Ω. But u can only take values in the space bounded by Ω_C, which shows a region close to c, as illustrated in Fig. 1(a). Finally, ζ, κ, λ, μ and ν are all reaction parameters that affect the corresponding interactions and ultimately influence the pattern generated. The parameters ξ and θ directly influence $a(u)$ and enable the achievement of both positive and negative gradients in the calculation of c.

3 Reaction-Diffusion CA Towards 2D and 3D Simulation of Mycelium-Based ELM

3.1 CA Grid Configuration

In order to use the reaction-diffusion model mentioned earlier in a discrete time and space framework that allows for a lot of parallelism and a simple, modular simulation approach, a CA framework is employed. The proposed CA-based model uses a discrete square grid where each CA cell contains the values of u, v and c that are continuous. The activator is able to diffuse only within the space denoted by Ω_c. Ideally, u would diffuse radially adjacent to the Mycelium, to support a more natural and realistic diffusion. However, given the computational and geometric limitations of creating a circle within a discrete grid, the activator was conditionally updated, taking into account neighbors in a circular-like fashion, as depicted in Fig. 2, where in dark blue is the center cell and in light blue its corresponding neighbors.

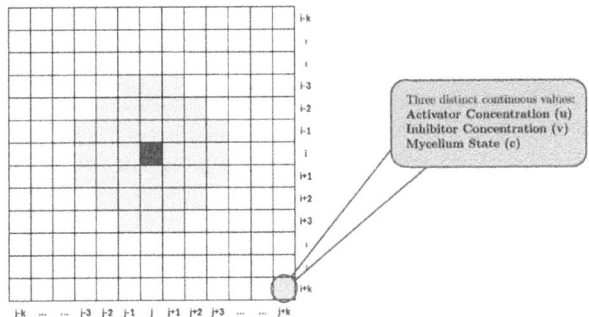

Fig. 2. Variable spaces Ω (whole grid-plane $\Pi_{i,j}$) and Ω_c (dark and light blue colors describing the referenced cell and its neighborhood accordingly) with a depth of 3 cells in the CA grid employed. In gray, a CA cell is shown containing the three continuous values that correspond to the state equations of the activator, inhibitor and mycelium. (Color figure online)

3.2 2D and 3D Representation of the Discrete Reaction-Diffusion Process

Before we can use the diffusion processes (Eqs. 1 and 2), which are usually described by partial differential equations, in a discrete CA framework, we need to use the discrete Laplacian operator. The application of this operator is equivalent to a convolution operation with a 3 × 3 kernel [16].

Furthermore, by modifying the kernel's weights, the spatial growth of the activator and the inhibitor can be controlled. More specifically, it was observed that by applying two different kernels, we can achieve two different diffusion patterns in our model. The first one, which is ∇_A^2, shown in Eq. 5, results in a denser diffusion because of the increased central cell weight and ultimately in a more condensed mycelium. The second one, ∇_B^2, because of the increased neighbor interactions, describes a finer and more precise diffusion, resulting in the mycelium's tips. Alternating at the appropriate moments can create a more realistic growth pattern, depicting a condensed biomass at the center with emerging tips at the edges, similar to the morphology of biological fungi [17].

$$\nabla_A^2 = \frac{1}{20}\begin{pmatrix} 0 & 1 & 0 \\ 1 & -4 & 1 \\ 0 & 1 & 0 \end{pmatrix} \quad \nabla_B^2 = \frac{1}{100}\begin{pmatrix} 10 & 35 & 10 \\ 35 & -11 & 35 \\ 10 & 35 & 10 \end{pmatrix} \quad (5)$$

Furthermore, we can further expand the dimensionality of the model from two dimensions, which is projected on one plane $\Pi_{i,j}$, to three dimensions, projected on three perpendicular planes: $\Pi_{i,k}$, $\Pi_{i,j}$, and $\Pi_{j,k}$. Similarly with the two dimensions, a convolution operation is performed on the planes $\Pi_{i,k}$, $\Pi_{i,j}$, and $\Pi_{j,k}$ with the three kernels $\nabla_{i,k}^2$, $\nabla_{i,j}^2$, and $\nabla_{j,k}^2$, respectively. These kernels, shown in Eq. 6, are introduced in [24] and are the equivalent of the 3D discrete Laplacian operator. It is noteworthy that all other parameters remain unchanged for the 3D model. Finally, it was observed that the kernel alternation didn't affect the model's growth in three dimensions, as the number of neighbors increases exponentially, rendering a small change in some of the neighbors' weights negligible.

$$\nabla_{i,k}^2 = \frac{1}{26}\begin{pmatrix} 2 & 3 & 2 \\ 3 & 6 & 3 \\ 2 & 3 & 2 \end{pmatrix} \quad \nabla_{i,j}^2 = \frac{1}{26}\begin{pmatrix} 3 & 6 & 3 \\ 6 & -88 & 6 \\ 3 & 6 & 3 \end{pmatrix} \quad \nabla_{j,k}^2 = \frac{1}{26}\begin{pmatrix} 2 & 3 & 2 \\ 3 & 6 & 3 \\ 2 & 3 & 2 \end{pmatrix} \quad (6)$$

3.3 Algorithm for Implementing the Mycelium-Based ELM Model

The algorithm detailing the model is depicted as Algorithm 1. Initially, the setting of hyperparameters (including total iterations and grid size) and the reaction-diffusion parameters takes place. Following, the concentrations of the activator, the inhibitor, and the Mycelium are initialized. The initialization of those three components is restricted to a small region in the center of the grid in order to observe and grasp the Mycelium's growth in all possible directions. Furthermore, all three components must have an initial value: the Mycelium requires the activator to grow, the activator can only propagate within the region defined by the Mycelium, and the inhibitor must be introduced into the system for the RD processes to function. It should be noted that the initialization within the central area is conducted randomly, and given the sensitivity of reaction-diffusion systems [28], the initial randomness is propagated throughout the developing pattern with each iteration.

Moreover, the simulation incorporates a substrate that varies in its capacity to support mycelium's growth, being either more or less conducive. This is integrated as a matrix which ranges from 0 to 1, that is multiplied to the reaction part u (line 19) and either promotes or suppresses it, affecting in turn c. As a result, more fertile areas are instantiated with a higher value, while less fertile areas get a rather small non-zero value. The model also randomly introduces obstacles, such as rocks within the substrate, which are clusters of zeros inside the matrix that impede the activator's (u) development and, consequently, the growth of the Mycelium in those specific areas.

The CA model then initiates a loop phase where its state is continuously updated. Initially, Ω_c is established based on the neighborhood defined in Fig. 2. Then, Ω_c is iteratively updated at each simulation step to represent the confined area in which the activator can take values, resulting in the activator gradually expanding in a circular pattern. The "for loops" in lines 9 and 10 do not iterate over the whole grid to prevent overflow, due to the considerable size of Ω_c. Next, in line 18, a kernel between ∇_A^2 and ∇_B^2 is chosen to control the diffusion of u and v. This is equivalent to a change in the CA's rule update. Subsequent processes include the conditional and unconditional update of the activator and the inhibitor, respectively (lines 17–20). Following, function $a(u)$ is computed to determine the concentration of the mycelium c at each time step (lines 21–25). Lastly, concentration limiters are used to make sure that both the activator and the suppressor diffuse properly, as discussed in Sect. 3.1.

Algorithm 1. Mycelium-based ELM Model

```
 1: grid ← 800
 2: iterations ← totalIterations
 3: {θ, ζ, κ, λ, μ, ν, ξ, amax, imax, threshold} ← {2.5, 625, 0.5, 0.8, 2.6, 1, 0.49, 20, 30, 0.5}
 4: u, v, c ← and(size(10))
 5: substrate ← rand(size(grid))
 6: step, i, j (counters) ← 0
 7: Ω_c ← 0
 8: for step < iterations do
 9:     for i < (grid-length(Ω_c)) do
10:         for j < (grid-length(Ω_c)) do
11:             if c(i,j) > 0 then
12:                 Ω_c(i,j) ← 1
13:             end if
14:         end for
15:     end for
16:     for i < grid do
17:         for j < grid do
18:             rule alternate
19:             if Ω_c(i,j) == 1 then
20:                 u_new(i,j) ← u_old(i,j) + ∇²u_old(i,j) + substrate · ζ(κu_old(i,j) + u_old(i,j)² − λu_old(i,j)v_old(i,j))
21:             end if
22:             v_new(i,j) ← v_old(i,j) + +d∇²v_old(i,j) + ζ(μu_old(i,j)³ − v_old(i,j))
23:             if u_new(i,j) ≥ threshold then
24:                 a(u_new(i,j)) ← ξ
25:             else
26:                 a(u_new(i,j)) ← ξ − θ(u_old(i,j) − threshold)
27:             end if
28:             c_new(i,j) ← c_old(i,j) + Ω_c ζ ν c_old(i,j)(a(u_new(i,j)) − c_old(i,j))(c_old(i,j) − 1)
29:         end for
30:     end for
31:     if u_new(i,j) > amax then
32:         amax ← u_new(i,j)
33:     end if
34:     if v_new(i,j) > imax then
35:         imax ← v_new(i,j)
36:     end if
37: end for
```

In conclusion, the model's growth is succinctly described by a three-way interaction, in which the space Ω_c permits values for the activator, which subsequently stimulates the growth of the Mycelium. Following this, Ω_c is updated and expands in regions where c is non-zero. This process is repeated and establishes the use of the variable range Ω_c.

4 Simulation Results

The results of the 2D model's simulations are presented in this Section. It is clear that the 2D model can accurately replicate fundamental1 mycelium behaviors like tip extension, branching, and anastomosis, confirming its ability to mimic fungal-based ELMs. By simulating these processes, the model looks like a fungus, especially by reflecting the patterns found in fungi, especially those in the Rhizoctonia Solani family [6]. This lends credibility to the modeling results. It is noteworthy that the employed model is not used to represent specific chemical species. Instead, it serves as a mathematical framework designed to broadly and effectively capture the dynamics of the system, aiming to accurately simulate the evolution of mycelium. The involved abstract variables are employed to model the emergent patterns and behaviors seen in the ELM bio-inspired system, rather than directly representing particular chemical entities. In addition, simulation results for the 3D model, an extension of the 2D model, are presented.

4.1 2D Model Results

In Fig. 3, the CA-based mycelium's growth is depicted, which is shown at different simulation steps. One can observe the condensed biomass at the center, while the hyphae are visible at the periphery. This is accomplished by alternating between the two kernels, ∇_A^2 and ∇_B^2, in a 3:1 ratio, respectively, every 100 iterations, following in a sense the natural process. This means that, every 100 iterations, diffusion is performed with ∇_A^2 for 75 iterations and with ∇_B^2 for the remaining 25 iterations.

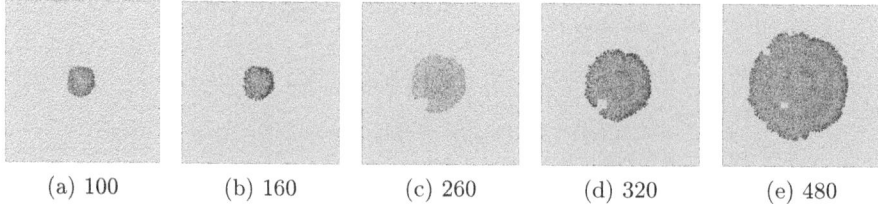

(a) 100   (b) 160   (c) 260   (d) 320   (e) 480

Fig. 3. Mycelium concentration c at different simulation steps.

Focusing on Fig. 4, the patterns of the activator, inhibitor, and mycelium are very similar. This is reasonable given that the activator's growth, which the inhibitor shapes, directly affects the mycelium's state. Furthermore, their concentration ranges are represented on the corresponding color bars. Furthermore, in areas where the substrate is more prosperous (closer to 1), there is a

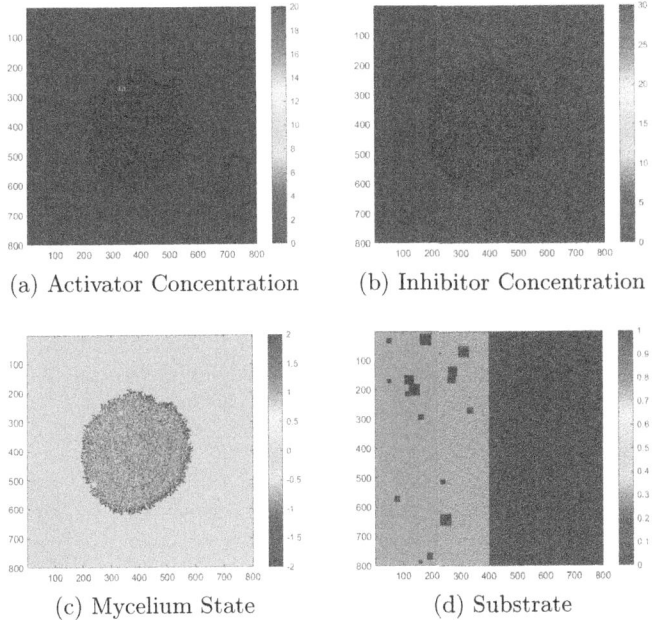

Fig. 4. (a–c) Final patterns of the activator, inhibitor concentrations and the mycelium state respectively after 500 timesteps, and (d) substrate initialization for the 2D model.

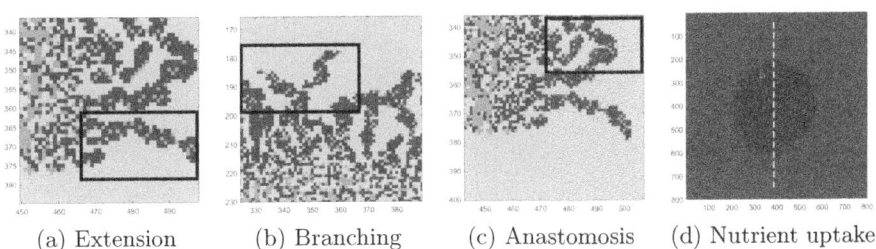

Fig. 5. (a–c) Simulated biological mechanisms of the mycelium's hyphae depicted in the mycelium's concentration c. (d) Nutrient uptake and mycelium's density depicted in the activator's concentration u.

higher concentration of the activator, which results in a thicker cell structure. The model also randomly introduces obstacles, for example rocks, within the substrate, which impede the growth of the ELM in those specific areas.

Moreover, the fundamental mycelium-based ELM mechanisms which include hyphae extension, apical and lateral branching, anastomosis and nutrient uptake are observed and showcased in Fig. 5, which is a magnification of the Mycelium's State in Fig. 4(c). Extension, branching, and anastomosis are readily observable in the mycelium's hyphae, while nutrient uptake is indicated by the density of the activator. The substrate morphology results in a denser right region, as shown in Fig. 4(d).

4.2 3D Model Results

In Fig. 6, the results of the 3D model simulation are presented. The augmentations consistently maintain the integrity of the 2D model's design, effectively showcasing the condensed biomass along with the precise spherical morphology of the fungi. These enhancements support the spatial and structural fidelity of the model and also underline the robustness of the model in replicating complex biological forms. The detailed visualization aids in understanding the dynamic interactions and growth patterns within the modeled environment.

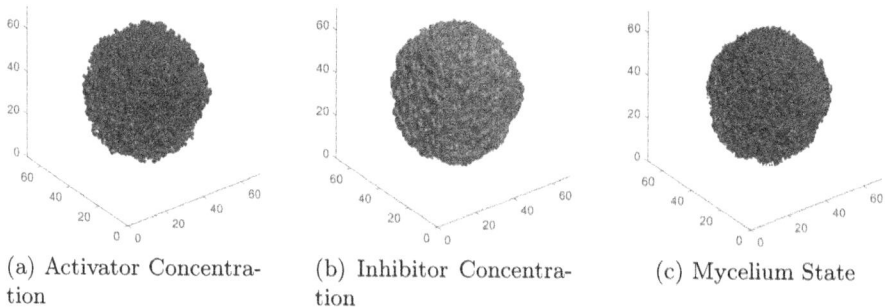

(a) Activator Concentration (b) Inhibitor Concentration (c) Mycelium State

Fig. 6. Final patterns of the activator, inhibitor concentration and the mycelium state, respectively, after 100 time steps for the 3D model.

5 Conclusions and Future Work

This paper introduces a mycelium-based ELM model, utilizing CA within their framework. The model is grounded in RD processes, which are prevalent in natural systems, providing enhanced insights into the dynamics of such organisms. It employs an activator-inhibitor interaction, capable of generating a diverse array of patterns while also extending to three dimensions. Overall, the model effectively replicates the fundamental mechanisms of mycelium, including tip extension, branching, anastomosis, and nutrient uptake, and it also exhibits characteristic bio-inspired patterns. The model alternates between different Laplacian operators in its diffusion processes in order to capture the Mycelium's essence, which includes a condensed central biomass and discernible tips at its edges. Both 2D and 3D simulation results reflect the pattern observed in the biological fungi, validating the results of the proposed reaction-diffusion CA model.

Future work will involve inclusion of external factors such as temperature and electrical stimuli, among others, using the reaction-diffusion's capabilities for dynamic interactions. Such a system could be enhanced by moving it to a Field Gate Programmable Array (FPGA) board, which can provide significant simulation parallelism and a real-time interaction with those external factors by utilizing the board's Analog to Digital Converters (ADCs). Additionally, a mycelium-based ELM with bacterial symbiosis will be developed to improve control and structural properties, advancing the study of biologically inspired systems.

Acknowledgments. This work has been supported by the framework of the FUNGATERIA project, which has received funding from the European Union's HORIZON-EIC-2021-PATHFINDER CHALLENGES program under grant agreement No. 101071145.

References

1. Adamatzky, A.: Cellular Automata: A Volume in the Encyclopedia of Complexity and Systems Science. Springer, Heidelberg (2018). https://doi.org/10.1007/978-1-4939-8700-9
2. Adamatzky, A.: Fungal Machines: Sensing and Computing with Fungi, vol. 47. Springer, Heidelberg (2023). https://doi.org/10.1007/978-3-031-38336-6
3. Ahangaran, M., Taghizadeh, N., Beigy, H.: Associative cellular learning automata and its applications. Appl. Soft Comput. **53**, 1–18 (2017)
4. Angelova, G.V., Brazkova, M.S., Krastanov, A.I.: Renewable mycelium based composite - sustainable approach for lignocellulose waste recovery and alternative to synthetic materials - a review. Zeitschrift für Naturforschung C **76**(11–12), 431–442 (2021)
5. Antinori, M.E., et al.: Advanced mycelium materials as potential self-growing biomedical scaffolds. Sci. Rep. **11**(1), 12630 (2021)
6. Boswell, G.P.: Modelling mycelial networks in structured environments. Mycol. Res. **112**(9), 1015–1025 (2008)
7. Chatzinikolaou, T.P., et al.: Wave cellular automata for computing applications. In: 2022 IEEE International Symposium on Circuits and Systems (ISCAS), pp. 3463–3467. IEEE (2022)
8. Chatzipaschalis, I.K., Chatzinikolaou, T.P., Fyrigos, I.A., Adamatzky, A., Rubio, A., Sirakoulis, G.C.: Memristor-based cellular automata for natural language processing. In: 2023 30th IEEE International Conference on Electronics, Circuits and Systems (ICECS), pp. 1–4. IEEE (2023)
9. Davidson, F., Sleeman, B., Rayner, A., Crawford, J., Ritz, K.: Large-scale behavior of fungal mycelia. Math. Comput. Model. **24**(10), 81–87 (1996)
10. Elsacker, E., Zhang, M., Dade-Robertson, M.: Fungal engineered living materials: the viability of pure mycelium materials with self-healing functionalities. Adv. Func. Mater. **33**(29), 2301875 (2023)
11. Fricker, M.D., Heaton, L.L., Jones, N.S., Boddy, L.: The mycelium as a network. Fungal Kingdom 335–367 (2017)
12. Gierer, A., Meinhardt, H.: A theory of biological pattern formation. Kybernetik **12**, 30–39 (1972)
13. Heide, A., Wiebe, P., Sabantina, L., Ehrmann, A.: Suitability of mycelium-reinforced nanofiber mats for filtration of different dyes. Polymers **15**(19) (2023)

14. Karana, E., Blauwhoff, D., Hultink, E.J., Camere, S.: When the material grows: a case study on designing (with) mycelium-based materials. Int. J. Design **12**(2) (2018)
15. Kondo, S., Miura, T.: Reaction-diffusion model as a framework for understanding biological pattern formation. Science **329**(5999), 1616–1620 (2010)
16. Kong, H., Akakin, H.C., Sarma, S.E.: A generalized Laplacian of Gaussian filter for blob detection and its applications. IEEE Trans. Cybern. **43**(6), 1719–1733 (2013)
17. Krull, R., Cordes, C., Horn, H., Kampen, I., Kwade, A., Neu, T.R., Nörtemann, B.: Morphology of filamentous fungi: linking cellular biology to process engineering using aspergillus niger. Biosyst. Eng. II: Linking Cell. Netw. Bioprocess. 1–21 (2010)
18. Landge, A.N., Jordan, B.M., Diego, X., Müller, P.: Pattern formation mechanisms of self-organizing reaction-diffusion systems. Dev. Biol. **460**(1), 2–11 (2020)
19. Liu, A.P., et al.: The living interface between synthetic biology and biomaterial design. Nat. Mater. **21**(4), 390–397 (2022)
20. Liu, S., Xu, W.: Engineered living materials-based sensing and actuation. Frontiers in Sensors **1** (2020)
21. Mora-Boza, A., Acosta, S., Puertas-Bartolomé, M.: Chapter 9 - biopolymers for the development of living materials for biomedical applications. In: Sessini, V., Ghosh, S., Mosquera, M.E. (eds.) Biopolymers, pp. 263–294. Elsevier (2023)
22. Neumann, J.V.: Theory of Self-reproducing Automata. University of Illinois Press (1966)
23. Nguyen, P.Q., Courchesne, N.M.D., Duraj-Thatte, A., Praveschotinunt, P., Joshi, N.S.: Engineered living materials: prospects and challenges for using biological systems to direct the assembly of smart materials. Adv. Mater. **30**(19), 1704847 (2018)
24. O'Reilly, R.C., Beck, J.M.: A family of large-stencil discrete Laplacian approximations in three-dimensions. Int. J. Numer. Methods Eng. 1–16 (2006)
25. Regalado, C., Crawford, J., Ritz, K., Sleeman, B.: The origins of spatial heterogeneity in vegetative mycelia: a reaction-diffusion model. Mycol. Res. **100**(12), 1473–1480 (1996)
26. Rodrigo-Navarro, A., Sankaran, S., Dalby, M.J., del Campo, A., Salmeron-Sanchez, M.: Engineered living biomaterials. Nat. Rev. Mater. **6**(12), 1175–1190 (2021)
27. Sugimura, K., Shimono, K., Uemura, T., Mochizuki, A.: Self-organizing mechanism for development of space-filling neuronal dendrites. PLOS Comput.l Biol. **3**(11), 1–12 (2007)
28. Van Gorder, R.A.: A theory of pattern formation for reaction-diffusion systems on temporal networks. Proc. Roy. Soc. A: Math. Phys. Eng. Sci. **477**(2247), 20200753 (2021)
29. Wolfram, S.: Cellular Automata and Complexity: Collected Papers. CRC Press (2018)

Mycelium-Based ELM Digital Twin Implemented in FPGA

Ioannis K. Chatzipaschalis[1,2(✉)], Ioannis Tompris[1], Konstantinos Rallis[1], Theodoros Panagiotis Chatzinikolaou[1], Iosif-Angelos Fyrigos[1], Michail-Antisthenis Tsompanas[3], Andrew Adamatzky[3], Phil Ayres[4], Antonio Rubio[2], and Georgios Ch. Sirakoulis[1]

[1] Democritus University of Thrace, DUTH University Campus, 67100 Xanthi, Greece
ichatzip@ee.duth.gr
[2] Universitat Politècnica de Catalunya, 08034 Barcelona, Spain
[3] University of the West of England, Bristol, UK
[4] Chair for Biohybrid Architecture, Institute of Architecture and Technology, Royal Danish Academy, 1435 København K, Denmark

Abstract. Engineered Living Materials (ELMs) based on fungal mycelium offer a promising solution to the challenges posed by environmental disruption, resource scarcity, and rising material demands, while also exhibiting interesting computational capabilities. These materials are cost-effective, widely available, and environmentally beneficial due to their biodegradable nature. Using a Field-Programmable Gate Array (FPGA) to create a digital twin could make it much easier to predict growth patterns, find the best conditions for development, and look into possible uses for mycelial ELMs. Digital twins with advanced hardware integration are particularly effective, providing deep insights into the physical and mathematical aspects of these materials. This paper describes the use of Cellular Automata (CAs) and reaction-diffusion systems to model these processes due to their ability to handle the behavior of complex systems with scalability and parallelism. This approach has enabled the high-fidelity simulation of ELMs behaviors and the successful prototype implementation on a FPGA, making it a significant step towards practical applications of mycelial ELMs.

Keywords: Cellular Automata · Reaction-Diffusion · Mycelium-based ELMs · Reconfigurable Hardware · FPGA

1 Introduction

Engineered Living Materials (ELMs), particularly those based on fungal mycelium, represent a novel class of materials inspired by living organisms. These mycelium-based ELMs offer a sustainable solution to the escalating societal challenges posed by human-induced environmental disruption, resource scarcity, and increasing material demands. They stand out for their affordable manufacturing process, natural ubiquity, and significant environmental benefits like biodegradability [1,11]. However, due to the complex biological mechanisms they encompass, their environmental sensitivity, the slow supply chain, and regulations, these materials present significant challenges for reproduction [15].

A digital twin would help scientists and engineers accurately predict growth patterns, find the best conditions for mycelial ELM development, and look at possible uses [21]. A digital twin [19] is a virtual representation that serves as the nearly identical digital equivalent of a real-world physical object, system, or process. This model is used for practical purposes including simulation, integration, testing, monitoring, and maintenance, effectively mirroring its real-world counterpart.

In this context, to effectively design a digital twin of the ELMs, it is crucial to select a modeling framework that can be easily ported into hardware while also taking full advantage of the available resources, as analyzed in Sect. 2. More specifically, Cellular Automata (CAs) have been strategically selected for this task due to their robust ability to simulate complex systems, as well as being scalable, modular, and showcasing massive parallelism [7,8]. CAs have been extensively utilized in the literature, from unconventional computing [17,20] to intelligent applications [5,14]. In this framework, CAs are specifically utilized to create a detailed grid that models the reaction-diffusion processes, providing a dynamic mapping of the morphological activities of the ELMs [4]. Leveraging this hardware-compatible modeling framework, along with their robust ability to simulate complex systems, has enabled the successful implementation of a mycelium-based ELM prototype on a Field-Programmable Gate Array (FPGA), elaborated in Sect. 3. The simulation results are presented in Sect. 4, while Sect. 5 concludes this work.

2 Hardware-Compatible ELM Model

2.1 Reaction-Diffusion Processes

The Reaction-Diffusion (RD) system employed consists of activator-inhibitor pairs that engage in dynamic interactions throughout the designated space, denoted as u and v, respectively [18]. These interactions significantly influence a third component, c, which characterizes the concentration of ELM in the specified region. In more detail, the activator u facilitates the growth of ELM by promoting its concentration c in regions where the activator is sufficiently present. On the other hand, the inhibitor v impedes the growth of ELM by constraining the concentration of the activator.

It is important to note that the reaction-diffusion model used, which includes the activator and inhibitor mentioned above, is not meant to be a representation of real chemical species. Instead, it is a mathematical framework used to show how the system works in a way that is generic enough and fits the mycelium evolution. Instead of directly modeling specific chemical entities, the involved abstract variables are used to mimic the patterns and behaviors that are seen in the ELM bio-inspired system.

This system has been utilized to model various processes and patterns by leveraging the complex dynamics of the activator-inhibitor pairs. In this study, the variable c denotes the concentration of the ELM. It signifies the degree of presence or absence of the material formed through the interplay between the

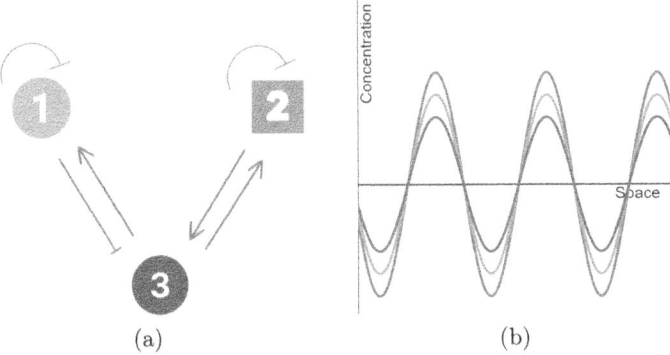

Fig. 1. (a) Activator (3), inhibitor (1) and main body (2) dynamics. Blue arrows signify promotion, while red flat-head links indicate suppression. (b) Concentration-Space graph for the concentrations of the activator (blue), the inhibitor (orange) and the main body (red). Adapted from [12]. (Color figure online)

activator and the inhibitor. In this case, the activator u is strategically limited to a certain radius from the ELM's main body c. This radius is in the space shown as Ω_c, which shows the mycelium's plasma membrane. The inhibitor, on the other hand, is free to spread out in this space. The RD system employed, comprising of three components, describes a three-way interaction between the activator, the inhibitor, and the main body. More specifically, the activator diffuses within the main body and promotes its growth when its concentration reaches a set amount indicated by Eq. 4. Once the main body is generated, the activator's morphology is no longer affected directly by the inhibitor, which diffuses freely and demotes the activator's concentration. This interaction leads to an in-phase evolution of the system's components, as depicted in Fig. 1(b), resulting in a Mycelium-like pattern rather than a traditional Turing pattern generated by Turing instability. The presence of the main body component disturbs the phenomenon of Turing instability [23]. This results to a final pattern that depends on the combined effects of diffusion, growth rates, and the dynamics of the components. The dynamics of the activator-inhibitor system are illustrated in Fig. 1 and are mathematically expressed in Eqs. (1)–(4), which refer to a reaction-diffusion model for bio-inspired evolution of structures:

$$\frac{\partial u}{\partial t} = \nabla^2 u + n\zeta(\kappa u + u^2 - \lambda uv) \; \epsilon \; \Omega_c \tag{1}$$

$$\frac{\partial v}{\partial t} = d\nabla^2 v + \zeta(\mu u^3 - v) \; \epsilon \; \Omega \tag{2}$$

$$\frac{dc}{dt} = \zeta \nu c(a(u) - c)(c - 1) \; \epsilon \; \Omega \tag{3}$$

$$a(u) = \begin{cases} \xi, & \text{if } u \geqslant threshold \\ \xi - \theta(u - threshold), & \text{otherwise} \end{cases} \tag{4}$$

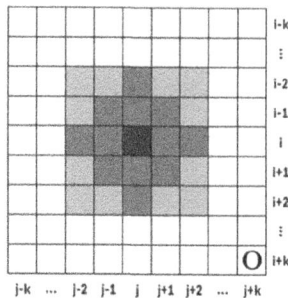

$$\boxed{O} = \begin{cases} u(i+k, j+k), & u \in \Omega_c \\ v(i+k, j+k), & v \in \Omega \\ c(i+k, j+k), & c \in \Omega \end{cases}$$

Fig. 2. Variable spaces Ω (whole grid) and Ω_c (colored cells) in the CA grid employed. The cells of the evolution enable matrix N are categorized in orange cells that have reduced weights (equal to 1/10), in light blue colored cells (equal to 7/20), and in the central dark blue colored cell (with weight equal to 1/9) to support circular growth. (Color figure online)

On the right side of Eq. (1) and Eq. (2), the initial term featuring the Laplacian operator delineates how the solutions u and v diffuse. The subsequent term, a polynomial of u and v, explicates the interactions between these solutions, referred to as the reaction term. Equation (3) describes the time derivative of the ELM's main body c, which correlates with the previous state of c and the function $a(u)$, indicating the dependence of the cell's growth on the activator. This dependency is detailed in Eq. (4). Additionally, the parameters d, ζ, κ, λ, μ, and ν are reaction coefficients that modify the corresponding interactions and significantly impact the resulting pattern, while n represents the nutrient concentration of the substrate. Finally, the parameters ξ and θ exert direct control over $a(u)$, indirectly affecting the evolution of c.

2.2 CA Framework

The foundational principles of CA concerning the cells within the utilized grid are the CA cell's state, its rule, and its neighborhood [2]. Also, the grid employed in this implementation is illustrated in Fig. 2 as a square grid, consisting of i rows and j columns. Each cell within the grid encapsulates the continuous values of u, v, and c, which represent the state of the CA. These values are updated in accordance with Eqs. (1)–(3), embodying the rule principle. The updates are performed conditionally, as indicated by the spaces Ω and Ω_c, shown in Fig. 2. Propagation can occur around cells where the concentration of $c(i, j)$ is greater than zero, in particular within the region defined by Ω_c, which is highlighted in color in Fig. 2. The set of points where propagation is permitted constitutes the evolution enable matrix N.

Furthermore, the neighborhood defined by the Laplacian operator delineates the diffusion processes of u and v across space. As shown in [13], the Laplacian operator is discretized. This makes it possible for the operation, which is based on partial differential equations, to be carried out in the discrete framework of the CA. This discrete form of the Laplacian resembles a Moore neighborhood but with adjusted weights, placing reduced emphasis on the diagonal neighbors of the central cell. This adjustment facilitates a more circular diffusion pattern, a phenomenon commonly observed in natural processes [16]. The Laplacian for the center cell $*(i,j)$, which pertains to the concentrations of either u or v, is expressed in Eq. (5).

$$\nabla^2 *(i,j) = \frac{1}{\Delta x^2} \left(\frac{7}{20} \Big[*(i-1,j) + *(i+1,j) + *(i,j-1) + *(i,j+1) \Big] \right.$$
$$+ \frac{1}{10} \Big[*(i-1,j-1) + *(i+1,j-1) + *(i-1,j+1)$$
$$\left. + *(i+1,j+1) \Big] - \frac{1}{9} *(i,j) \right) \tag{5}$$

Consequently, the extracted kernel that emerges from Eq. 5 is $L1$ and in the same way, the second kernel $L2$ is reproduced to represent the two different CA rules. The two kernels are seen as follows:

$$L1 = \begin{pmatrix} 1/10 & 7/20 & 1/10 \\ 7/20 & -1/9 & 7/20 \\ 1/10 & 7/20 & 1/10 \end{pmatrix}, \quad L2 = \begin{pmatrix} 0 & 1/20 & 0 \\ 1/20 & -1/5 & 1/20 \\ 0 & 1/20 & 0 \end{pmatrix} \tag{6}$$

Two separate rules have been set up to get a dense biomass in the middle and hyphal tips that are spreading out around the edges, which looks a lot like mycelium-based ELMs. More specifically, this evolution pattern is notably similar to the fungus Rhizoctonia Solani [3]. As iterations progress, these rules alternate with the two different kernels $L1$ and $L2$, leading to the emergence of a final evolution pattern characterized by this dual feature.

3 System Design and Architecture

The CA's principles (state, rule and neighborhood) are fully compatible with digital hardware implementations, as they can be effectively represented using memories and logic. These memory units function as matrices which can be modified at each timestep, dictated by the programmed logic. This setup facilitates easy reading from/writing to the memory (state and rule), and efficiently manages local interactions (neighborhood) within each cell of the CA. Moreover, this method is not only highly compatible with existing hardware, but also enhances computational efficiency through significant parallelism [6], where all calculations occur within a narrowly defined time window, specifically the simulation's time step, set by the clock. As a result, the FPGA is a very promising digital platform to implement CAs, incorporating the hardware to effectively support all of the above requirements. The proposed CA was implemented using High-Level-Synthesis (HLS) [10] using the AMD Vitis Software and the Xilinx ZedBoard Zynq 7000 ZC702 Evaluation Board (xc7z020clg484-1).

The algorithm used for the HLS, written in C++ and described in Algorithm 1. The model initiates a loop phase where it's continuously updated at each iteration. Initially, evolution-enabled matrix N is established. Subsequent updates are governed by the *codeselector* variable, a binary vector comprising sequences of 0s and 1s, which determines when the rule update of the CA is altered. The first rule (lines 1–4) facilitates a more circular diffusion by employing a kernel with diminished weights on the diagonal, whereas the second rule utilizes a kernel with equal weights on adjacent and diagonal positions. Equations (1)-(4) and the right Laplacian kernel are used to make changes to the values of the activator u, the inhibitor v, and the main body c. The neighborhood, which is like a circular front, is also multiplied by the RD processes' diffusion term to aid the activator's growth in a radial direction adjacent to the main body. Limiters are then applied to both the activator and inhibitor to prevent excessively large values that could impede the diffusion process. Finally, the states of the activator, inhibitor, and main body are updated.

The memory needed to uphold the state of the CA is divided into 3 two-dimensional matrices containing the activator, the inhibitor, and the main body's values, which utilize the board's BRAM blocks. The rule principle of the CA is implemented in the FPGA, utilizing the board's look-up-tables (LUTs), which are reprogrammable combinational logic blocks containing logic gates capable of handling the complex RD processes. As seen in Fig. 3, the design comprises an *Evolution Module*, which is responsible for computing of the next state of the mycelium's evolution, and a *Top Module*. The *Evolution Module* takes as input the *Evolution Parameters* first initialized in the *Top Module*, and also returns the state of the evolution as *data out*, and the three state tables u, v, and c to it in order to get renewed for the next iteration of the evolution. So, there is a feedback mechanism between the *Evolution Module* and the *Top Module* to take the previously computed state as an input to the next iteration. Finally, it is noted that the values of the utilized parameters are shown in Table 1.

Table 1. Utilized Evolution Parameters.

Parameter	Value	Parameter	Value
θ	2.5	ν	1
ζ	625	ξ	0.49
κ	0.5	a_{max}	20
λ	0.8	i_{max}	30
μ	2.6	$threshold$	0.5
d	30	ρ	0.7

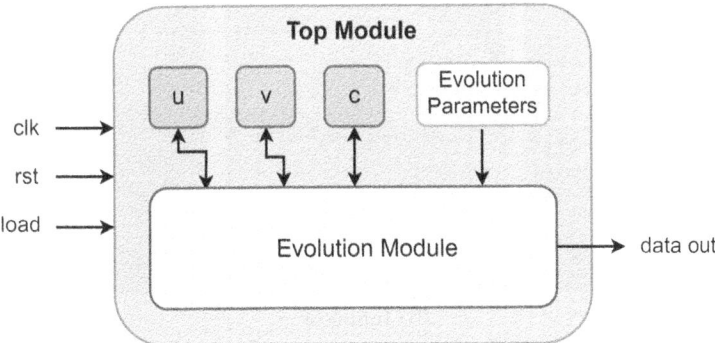

Fig. 3. Qualitative block diagram of the design in the FPGA. (Color figure online)

Algorithm 1. CA RD Model - Iterative Process

1: $N \leftarrow$ evolution enable matrix
2: $n \leftarrow$ substrate nutrient matrix
3: $codeselector = 1$? $L \leftarrow L1 : L \leftarrow L2$
4: $u_{new} \leftarrow u_{old} + N \circ (conv(u_{old}, L)) + n\zeta(\kappa u_{old} + u_{old}^2 - \lambda u_{old} v_{old})$
5: $v_{new} \leftarrow v_{old} + d(conv(v_{old}, L)) + \zeta(\mu u_{old}^3 - v_{old})$
6: $c_{new} \leftarrow c_{old} + \zeta \nu c_{old}(a(u_{old}) - c_{old})(c_{old} - \rho))$
7: $u_{old} \geq threshold$? $a(u_{old}) \leftarrow \xi : a(u_{old}) \leftarrow \xi - \theta(u_{old} - threshold)$
8: $a(u_{old}) < 0$? $c_{new} \leftarrow 1$
9: $u_{new} < 0$? $u_{new} \leftarrow 0$
10: $u_{new} > a_{max}$? $u_{new} \leftarrow a_{max}$
11: $v_{new} > i_{max}$? $v_{new} \leftarrow i_{max}$
12: $v_{old} \leftarrow v_{new}, u_{old} \leftarrow u_{new}, c_{old} \leftarrow c_{new}$

4 Simulation Results

In Figs. 4(d)–(h), the growth of the CA-based mycelium is depicted at various simulation steps, highlighting a dense biomass center marked by both mycelium cells and an activator, with hyphae visible on the stem. The inhibitor's density increases toward the edges and decreases toward the center, illustrating its inverse relationship with both the activator and cell states. Figures 4(a), (b), and (h) show that the patterns of the activator, inhibitor, and the mycelium's main body are closely aligned. Following, Fig. 4(c) displays the substrate, which can affect the mycelium's ability to grow; more conducive areas (indicated by white/bright colors) promote a higher activator concentration and a denser mycelium body.

It is important to note that in order to achieve the most accurate outcomes, a central zone was initialized with minimal concentrations of the activator, the inhibitor, and the main body, representing the initial state of the mycelium, along with all the corresponding RD parameters. It should be also considered that the initialization in the central area is conducted randomly. Considering

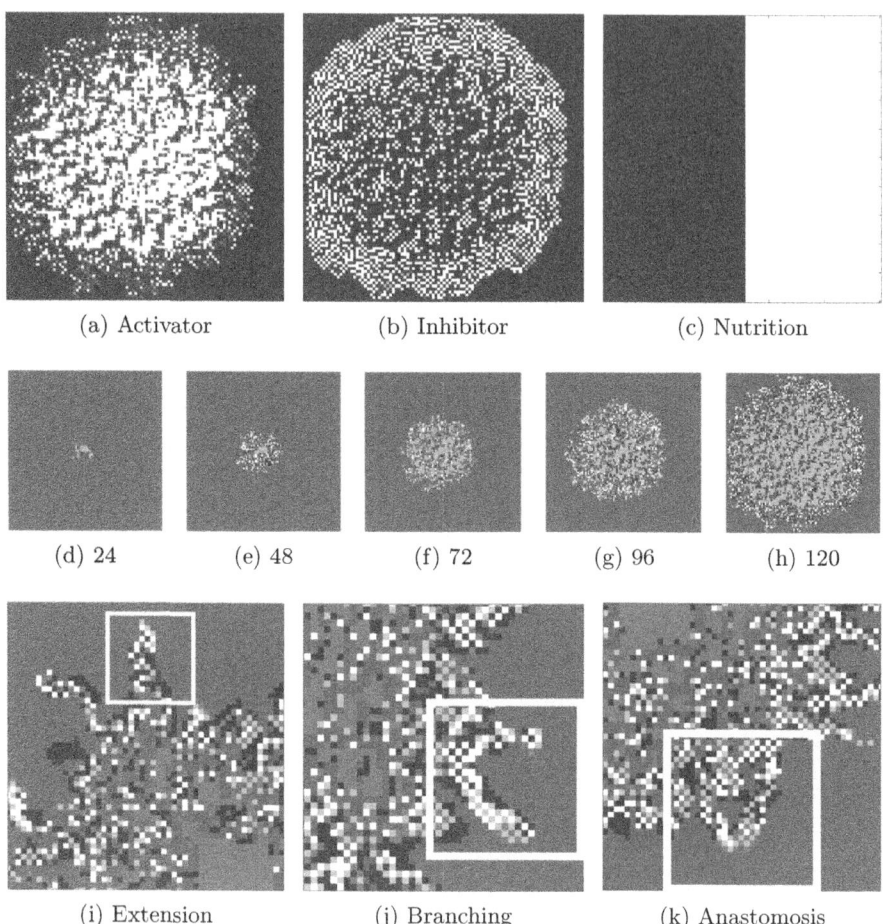

Fig. 4. (a) Activator u, (b) Inhibitor v, and (c) Nutrient concentrations across the grid at 120 simulation's iterations. (d)–(h) Mycelium main body concentration c at different simulation's iterations. In (a)–(h), the lighter in color a cell is, the higher its concentration. (i)–(k) Mycelium's tip mechanisms. Simulations were conducted for 120 iterations at a 100×100 grid. (Color figure online)

the sensitivity of reaction-diffusion systems [22], this initial randomness is propagated across the evolving pattern with each iteration.

Reaction-diffusion models featuring inhibitors and activators offer a compelling framework for simulating the complex behaviors of mycelial growth, including tip extension, branching, and anastomosis, that are readily observable in the mycelium's hyphae. The validation of the result is based on the appearance of these three characteristics (extension, branching, and anastomosis) in every run of the simulation, as it is initialized with different u and v tables. It is noted that the proposed model is a generic reaction-diffusion model that is able,

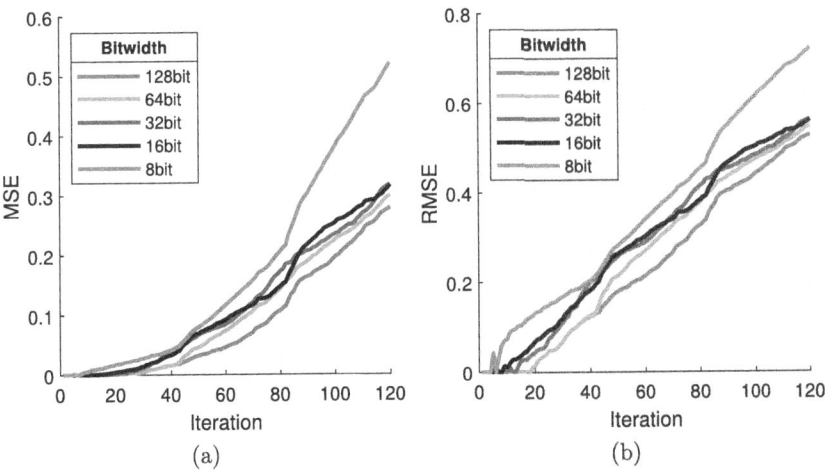

Fig. 5. (a) Mean Squared Error (MSE) and (b) Root Mean Square Error (RMSE) observed during the iterations of mycelium's evolution between software and hardware simulation using different bitwidths. (Color figure online)

with the proper tuning of its parameters, to replicate the evolution of mycelial networks denoted by the aforementioned three characteristics. As referred in Sect. 2, the activator promotes growth and propagation, while the inhibitor regulates and constrains this growth. Tip extension is naturally simulated as the activator diffuses outward, driving the elongation of mycelial hyphae. Branching emerges when local variations in activator and inhibitor concentrations lead to spatial gradients, prompting the formation of new growth points. Anastomosis, the fusion of hyphal strands can be effectively simulated through the interaction of activator fronts, where collisions inhibit further propagation. These mechanisms are observed and showcased in Figs. 4(i)–(k). These processes have been regulated due to mechanisms for nutrient uptake, as indicated by matrix n, which also indicates the density of the activator u.

Other than that, the modeling scale involves discretizing space into square lattice sites, each with a side length of 100 um [9], encompassing a simulated grid area of 1 cm^2. With a small tradeoff in accuracy and adjustment of parameters, even larger mycelium structures can be emulated on the same grid, offering flexibility in modeling various mycelial configurations.

There have been various evolution implementations of the proposed design using different bitwidths for the computations. As one can notice in Fig. 5, using 16 bits, the evolution performs really well with similar MSE and RMSE metrics compared to the higher bitwidth options during the iterations. For 16 bit bitwidth, the hardware resources are seen in Table 2. It has been observed that using 32 bits, the accuracy of the evolution was almost the same as 16 bits, but the utilization of BRAMs and DSPs was doubled, as was the consuming power. For this reason, the 16 bit bitwidth option is considered optimal for the proposed design.

Table 2. Utilization of the FPGA's resources after implementation.

Utilization of the FPGA's resources after implementation			
Resource	Utilization	Total Available	Utilization %
LUT	5787	53200	10.88
LUTRAM	308	17400	1.77
FF	2142	106400	2.01
BRAM	66	140	47.14
DSP	63	220	28.64
IO	23	200	11.50
BUFG	1	32	3.13

Finally, the proposed algorithm has been tested on a CPU (Intel i9 9900k) and also on a GPU (GeForce RTX 2060 Super), as referred to in Table 3. However, FPGA's parallelization capabilities, along with its low power consumption and portability, make it the most efficient candidate. To elaborate more, for the same computation load, FPGA appears to have the lowest energy need between CPU and GPU implementations, while utilizing the smaller area footprint as well. It is noted that the clock of the board is set to 10 ns, providing a running frequency of 100 MHz. Further rationale for the FPGA suitability, including specific advantages and future applications, is discussed in the concluding Sect. 5 of this paper.

Table 3. Efficiency, execution time, and total energy metrics measured on CPU, GPU, and FPGA.

Platform	Power ($Watts$)	Execution Time (ms)	Total Energy (mJ)
CPU	43.1	40	1724
GPU	84.5	4.2	354.9
FPGA	0.37	35.1	12.9

5 Conclusions and Future Work

In this paper, a digital twin of a mycelium-based Engineered Living Material (ELM) within a Field-Programmable Gate Array (FPGA) has been successfully implemented. The model successfully mimics key behaviors of mycelium, such as tip growth, branching, merging, and nutrient absorption, while replicating typical bio-inspired patterns and also utilizing core features of the ELM, which are a dense central biomass and distinctive tips at the mycelium's stem. The proposed implementation showcases a significant speed boost and enhanced energy efficiency compared to basic CPU-powered software execution. This development marks a significant step forward in the simulation and understanding of mycelium-based materials within a controlled digital twin environment and provides an ideal framework for simulating such biological organisms. The current implementation primarily simulates and observes ELM behavior without allowing interactive feedback between biological and digital systems.

In future work, exploring the establishment of a bidirectional communication between the mycelium-based ELM and the FPGA holds great promise. By incorporating analog-to-digital (ADCs) and digital-to-analog converters (DACs) within the FPGA, it could become feasible not only to decode biological signals into digital data for detailed analysis but also to transmit operational commands back to the living material, enabling real-time adjustments based on the ELM's continuous monitoring. The compact form of the FPGA board, along with its showcased efficiency, can make the proposed implementation a really versatile and mobile tool that can easily be deployed for its integration with ELMs. Also, by utilizing the analog ports of the FPGA, the interference with environmental sensors can be exploited to measure metrics like humidity, temperature, and soil quality. So, the digital twin could dynamically reflect and adapt to the conditions affecting the biological counterpart, ensuring a high-fidelity emulation. The model's superior insights can be enhanced by using memristive nanocircuits, which enable non-discrete, continuous computation suitable for analog biological signals. In the end, the FPGA prototype could facilitate the implementation of a specific integrated circuit (ASIC) as an embedded low-power proposal with parallel core processing for mycelium-based ELM emulation.

Acknowledgments. This work has been supported by the framework of the FUNGATERIA project, which has received funding from the European Union's HORIZON-EIC-2021-PATHFINDER CHALLENGES program under grant agreement No. 101071145.

References

1. Adamatzky, A.: Fungal Machines: Sensing and Computing with Fungi, vol. 47. Springer, Heidelberg (2023). https://doi.org/10.1007/978-3-031-38336-6
2. Adamatzky, A., Alonso-Sanz, R., Lawniczak, A.: Automata-2008: Theory and Applications of Cellular Automata. Luniver Press (2008)
3. Boswell, G.P.: Modelling mycelial networks in structured environments. Mycol. Res. **112**(9), 1015–1025 (2008)
4. Bozzini, B., Lacitignola, D., Sgura, I.: Morphological spatial patterns in a reaction diffusion model for metal growth. Math. Biosci. Eng. **7**(2), 237–258 (2010)
5. Chatzinikolaou, T.P., Karamani, R.E., Fyrigos, I.A., Sirakoulis, G.C.: Handling sudoku puzzles with irregular learning cellular automata. Nat. Comput. **23**(1), 41–60 (2024). https://doi.org/10.1007/s11047-024-09975-4
6. Dourvas, N., Tsompanas, M.A., Sirakoulis, G.C., Tsalides, P.: Hardware acceleration of cellular automata Physarum polycephalum model. Parallel Process. Lett. **25**(01), 1540006 (2015)

7. Dourvas, N.I., Sirakoulis, G.C., Adamatzky, A.I.: Parallel accelerated virtual physarum lab based on cellular automata agents. IEEE Access **7**, 98306–98318 (2019)
8. Dourvas, N.I., Sirakoulis, G.C., Tsalides, P.: A GPGPU physarum cellular automaton model. Appl. Math. **10**(6), 2055–2069 (2016)
9. Fricker, M.D., Heaton, L.L., Jones, N.S., Boddy, L.: The mycelium as a network. Fungal Kingdom 335–367 (2017)
10. Gajski, D.D., Dutt, N.D., Wu, A.C., Lin, S.Y.: High-Level Synthesis: Introduction to Chip and System Design. Springer, Heidelberg (2012)
11. Houette, T., Maurer, C., Niewiarowski, R., Gruber, P.: Growth and mechanical characterization of mycelium-based composites towards future bioremediation and food production in the material manufacturing cycle. Biomimetics **7**(3), 103 (2022)
12. Landge, A.N., Jordan, B.M., Diego, X., Müller, P.: Pattern formation mechanisms of self-organizing reaction-diffusion systems. Dev. Biol. **460**(1), 2–11 (2020)
13. O'Reilly, R.C., Beck, J.M.: A family of large-stencil discrete Laplacian approximations in three-dimensions. Int. J. Numer. Methods Eng. 1–16 (2006)
14. Pavlidis, N., Perifanis, V., Chatzinikolaou, T.P., Sirakoulis, G.C., Efraimidis, P.S.: Intelligent client selection for federated learning using cellular automata (2023)
15. Peeters, E., Salueña Martin, J., Vandelook, S.: Growing sustainable materials from filamentous fungi. Biochemist **45**(3), 8–13 (2023)
16. Sirakoulis, G.C., Karafyllidis, I., Thanailakis, A.: A cellular automaton for the propagation of circular fronts and its applications. Eng. Appl. Artif. Intell. **18**(6), 731–744 (2005)
17. Sirakoulis, G., Karafyllidis, I., Mizas, C., Mardiris, V., Thanailakis, A., Tsalides, P.: A cellular automaton model for the study of DNA sequence evolution. Comput. Biol. Med. **33**(5), 439–453 (2003)
18. Sugimura, K., Shimono, K., Uemura, T., Mochizuki, A.: Self-organizing mechanism for development of space-filling neuronal dendrites. PLOS Comput. Biol. **3**(11), 1–12 (2007)
19. Tao, F., Xiao, B., Qi, Q., Cheng, J., Ji, P.: Digital twin modeling. J. Manuf. Syst. **64**, 372–389 (2022)
20. Tsompanas, M.A., Chatzinikolaou, T.P., Sirakoulis, G.C.: Cellular automata application on chemical computing logic circuits. In: Chopard, B., Bandini, S., Dennunzio, A., Arabi Haddad, M. (eds.) ACRI 2022. LNCS, vol. 13402, pp. 3–14. Springer, Cham (2022). https://doi.org/10.1007/978-3-031-14926-9_1
21. Udugama, I.A., Lopez, P.C., Gargalo, C.L., Li, X., Bayer, C., Gernaey, K.V.: Digital twin in biomanufacturing: challenges and opportunities towards its implementation. Syst. Microbiol. Biomanuf. **1**, 257–274 (2021)
22. Van Gorder, R.A.: A theory of pattern formation for reaction-diffusion systems on temporal networks. Proc. Roy. Soc. A: Math. Phys. Eng. Sci. **477**(2247), 20200753 (2021)
23. Zheng, Q., Shen, J., Xu, Y.: Turing instability in the reaction-diffusion network. Phys. Rev. E **102**(6), 062215 (2020)

Author Index

A
Adak, Sumit 10
Adamatzky, Andrew 253, 265
Aravind, A. 109
Ayres, Phil 253, 265

B
Baby, C. J. 147
Baetens, Jan M. 121
Bagnoli, Franco 22, 45, 72, 96, 243
Baia, Michele 45, 72
Bhattacharjee, Kamalika 132, 147
Bilan, Stepan 177
Bosi, Marco 243
Bruno, Odemir M. 121

C
Calidonna, Claudia R. 85
Catelan, Paolo 85
Chatzinikolaou, Theodoros Panagiotis 253, 265
Chatzipaschalis, Ioannis K. 253, 265
Chidichimo, Francesco 85
Chopard, Bastien 219

D
Daly, Aisling J. 121
Dennunzio, Alberto 3
Désérable, Dominique 34
Di Gregorio, Salvatore 85
Dridi, Sara 22

E
El Yacoubi, Samira 22, 189, 203
Elizabeth, M. J. 163

F
Fontaine, Allyx 189, 203
Fujita, Gen 58
Fyrigos, Iosif-Angelos 253, 265

H
Hazari, Raju 163
Hoffmann, Rolf 34

J
John, Anita 109
Jose, Jimmy 109

K
Kamikawa, Naoki 58
Kommineni, Avinash Krishna 163
Kone, Alassane 189, 203

L
Leone, Pierre 219
Lukic, Luka 219
Lupiano, Valeria 85
Lywait, Tarun 132

M
Matteuzzi, Tommaso 45, 72, 243
Modi, Harsh 10
Mokhor, Volodymyr 177
Mouakher, Amira 203

N
Nair, Krishnadas 132

P
Patel, Rahil 10
Pikovsky, Arkady 72

R
Rallis, Konstantinos 265
Rechtman, Raúl 96
Rollier, Michiel 121
Roy, Souvik 10
Rubio, Antonio 265

© The Editor(s) (if applicable) and The Author(s), under exclusive license to Springer Nature Switzerland AG 2024
F. Bagnoli et al. (Eds.): ACRI 2024, LNCS 14978, pp. 277–278, 2024.
https://doi.org/10.1007/978-3-031-71552-5

S

Sağ, Tahir 231
Samburskyi, Volodymyr 177
Seredyński, Franciszek 34
Seredyński, Michal 34
Sirakoulis, Georgios Ch. 253, 265
Srinivasan, Kiran 132
Szaban, Miroslaw 34

T

Tompris, Ioannis 253, 265
Tsompanas, Michail-Antisthenis 253, 265

U

Umeo, Hiroshi 58

SPRINGER NATURE

GPSR Compliance

The European Union's (EU) General Product Safety Regulation (GPSR) is a set of rules that requires consumer products to be safe and our obligations to ensure this.

If you have any concerns about our products, you can contact us on ProductSafety@springernature.com

In case Publisher is established outside the EU, the EU authorized representative is:

Springer Nature Customer Service Center GmbH
Europaplatz 3
69115 Heidelberg, Germany

The manufacturer's authorised representative in the EU is Springer Nature Customer Service Centre GmbH, Europaplatz 3, 69115 Heidelberg, Germany. If you have any concerns regarding our products, please contact ProductSafety@springernature.com

Printed and bound by CPI Group (UK) Ltd, Croydon, CR0 4YY

25/03/2026

02078185-0013